STATISTICS: AN INTRODUCTION

STATISTICS: AN INTRODUCTION

HAROLD J. LARSON

United States Naval Postgraduate School

John Wiley & Sons, Inc.
New York   London   Sydney   Toronto

Library of Congress Cataloging in Publication Data

Larson, Harold J.   1934–
    Statistics: an introduction.
    1. Statistics.   I. Title.
HA29.L2666   519.5   74-13460
ISBN 0-471-51770-4

Printed in the United States of America

10-9 8 7 6 5 4 3 2 1

PREFACE

This book introduces the most frequently applied statistical methods at a noncalculus level. All techniques discussed are illustrated with meaningful examples chosen from a wide variety of subject matter areas, including both the social and the physical sciences. A large number of exercises are presented at the end of each section; each chapter concludes with a summary of the important topics contained in that chapter, and is followed by an additional set of exercises that provide further illustrations of the techniques discussed. Answers to the even numbered exercises are given in the Appendix. Also given there are tables of the binomial, normal, and Poisson probability measures, as well as selected percentiles of the t, χ^2, and F distributions.

The first chapter is mainly devoted the use of the summation notation. It is intended to increase the student's facility with the symbolism necessary in describing manipulations with data. Chapter 2 deals with descriptive statistics, which will always play an important role in any technological society. Basic ideas of tables and graphs are presented, as are the most frequently used measures for summarizing the middle and the variability of a set of data. Simplifications in calculation that result from coding data are also discussed.

The remaining chapters (Chapters 3 to 11) are concerned with topics important to inferential statistics. The concept of a random variable and the operation of expectation are given major emphasis in this presentation (see Chapters 3 and 4). This is done naturally, without requiring the student to wend his way through a morass of counting problems as is typical of other textbooks. Chapter 5 is comments on the three most frequently used probability distributions: the binomial, normal, and Poisson.

Chapter 6 discusses populations and samples. This is where the first problems of inference are encountered. This chapter clearly discusses the basic ideas of sums of random variables and the central limit theorem, leading to the uses of the normal in approximating other probability distributions. In chapter 7, both point and interval estimation of parameters are discussed, and Chapter 8 explores the methodology for testing hypotheses. Chapter 9 presents some basic procedures for comparative experiments, when it is desired to compare two or more populations. Chapter 10 deals with many aspects of least squares and introduces the correlation coefficient as a measure of association. Chapter 11 presents some nonparametric procedures. For students desiring it, the Appendix contains a set-oriented, axiomatic introduction to probability theory; the equally likely case and counting problems are covered here. There is no loss of continuity if this appendix material is not covered. However, for anyone interested in probability theory, it seems logical to read the Appendix before going to Chapter 3.

The material here is ample for a one-semester or two-quarter course, depending partly on whether the probability material in the Appendix is covered. A good one-quarter course could also be constructed by deleting appropriate topics. Natural choices for deletion are the Appendix and part or all of Sections 3.3, 4.3, 5.3, and perhaps Sections 8.3 and 10.3. Whether the material on nonparametric statistic should be covered is for individual instructors to decide.

Anyone writing a textbook such as this one owes a great debt to the many people who first applied and described the various techniques given here. It is not feasible to identify and acknowledge each of them individually. I thank William Archambault, Howard Anton, Donald Berry, Keith Crasswell, Shirley Dowdy, Donald Guthrie, Albert Kingman, Samuel Kodis, Robert Negus, and Goeffrey Watson for comments made on various versions of this manuscript. Tom Burnett provided answers to the even numbered exercises and offered many other helpful comments. I am indebted to Gerhard Brahms, Fred Corey, and Gary Ostedt of Wiley for their encouragement during this project. I also thank Judy Kaitala and Mary Wright for their excellent help in tranforming a handwritten manuscript into a professionally typed product. Tables A.1 and A.2 have been adapted from "Tables of the Binomial Probability Distribution," National Bureau of Standards, Applied Mathematics Series .6, 1949, and from "Tables of Normal Probability Functions," National Bureau of Standards, Applied Mathematics Series .23, 1953, respectively. Tables A.3, A.4, A.5 and A.6 were adapted from "Tables of Common Probability Distributions" by P. W. Zehna and D. R. Barr, Naval Postgraduate School, Technical Report NPS-55ZeBn0091A, September, 1970.

Pacific Grove, California   Harold J. Larson

CONTENTS

CHAPTER 1
NUMBERS AND STATISTICS

The material in this book has evolved over many years and is important in today's modern, technological society. Man's ability to count objects and record their number begins in very ancient times. Among the first uses of numbers and counting were recording the number of people in the provider's household or town, counting the animals in his herd, figuring the yield harvested from his crops, and bargaining with others in the exchange of goods. From these crude beginnings we have progressed very far, but today still use numbers for all of these purposes as well as to describe and predict, for example, the weather, the movements of the stock market, the efficacies of various drugs in the treatment of illnesses, and the performances of athletic teams. Numbers have become such an intimate part of modern life that it is necessary for the health and well-being of each person to have some knowledge (however rudimentary) of numbers and the rules for using them. It is also necessary to have some training in statistics to be able to make sense out of the seemingly endless array of facts and figures with which we are constantly bombarded today. It has often been said that "Figures don't lie, but liars can figure." A knowledge of statistical techniques and reasoning is very useful in avoiding fallacious conclusions or lies that can entrap the unwary when working with data.

A knowledge of high school algebra is sufficient mathematical preparation for the study of this book. Many of the techniques discussed are presented with intuitive motivation but without formal mathematical proof. Where the mathematical proof is simple enough, it is presented in detail. However, this textbook is not intended to establish rigorously every fact presented.

1.1 DESCRIPTIVE AND INFERENTIAL STATISTICS

Two areas of statistics will be presented in this book: *Descriptive statistics* and *inferential statistics*. When the word statistics is used most people think of *descriptive statistics*. This area is concerned with techniques for summarizing and describing a body of numbers. Historically it is by far the older of the two areas. Men have been collecting data for hundreds of years; once the data are collected, decisions must be made on how best to present them to serve the purposes for which they were collected. We are all aware of the reams of facts and data that, recorded through the years, have described the performance of professional baseball teams and individual players. Such data are summarized in many books and newspapers where the original scores, hits made, and walks allowed are presented in a fashion that is hopefully adequate for the reader's use. The tables, graphs, averages, and other summaries used are all examples of descriptive statistics. The United States government publishes a great deal of information about our population, economy, mobility, weather, and many other aspects of our society. The numbers and tables presented are, again, examples of descriptive statistics, simple summarizations of great masses of numbers that have been collected by various agencies. In Chapter 2 we shall study some techniques and quantities that are generally useful in describing data or statistics.

In this textbook we shall put much more emphasis on *inferential statistics*, techniques designed for making inferences from data collected. In general, these techniques are appropriate when the data available represents a sample (or portion) selected from some larger body of data called a population. Based on what is found in the sample, we want to draw inferences (make statements or come to conclusions) that are hopefully true for the whole population, even though the whole population has not been examined. This then is a problem of inference. Generally only a relatively small portion of the population has been examined, but the question demanding an answer refers to the whole population. Many different statistical techniques have been evolved to cope with this type of situation.

A familiar example of inference is given by the opinion polls that publish results in an election year. A candidate for office wants to know the proportion of voters that will vote for him, prior to the actual election. A polling organization contacts a sample of the eligible voters and asks each voter whether he will vote for the candidate. Based on the proportion *in the sample* that will vote for him, statements are made about the proportion of the whole population that will vote for him. Standard statistical methods, some of which are presented in this book, are available for inferring the value of the population proportion, given the value of the sample proportion. Since such an inference is not based on an examination of the whole population, the value guessed for the population proportion may or may not be correct. In any given situation we would prefer to use the method for making the inference that has the best chance of being correct in some sense. Thus, different possible procedures that might be used to make the inference are compared, essentially on the basis of the chances that the inference is correct.

Here is another example of inference. Suppose there are two different methods of teaching the same material (e.g., the old classical approach and the "new math" approach to third grade arithmetic). Which is better? Common sense dictates that the way to answer this question is to try both methods, by using one method with one group of students and the other method with a second group of students. Then, at the end of the period of instruction, the effectiveness of the two methods can be compared and a decision reached about the students used in the two groups. But a bigger question still remains. If method A was clearly better than B, was the difference really due to differences in the two methods of teaching or was it due to differences in the two groups of students used (or due to the differences in the instructors used)? If two new groups of students were selected, and again the two methods used, would similar differences be observed? Again, in this type of situation, questions of inference occur. The groups of students used can be thought of as samples from the population of all possible students. We want to answer questions about the whole population (which method is better for which students) but we have available only the results from a sample of the population.

1.2 NOTATION AND THE SUMMATION OPERATOR

Since we assume that the reader has studied high school algebra, no specific review of basic algebraic concepts will be given here. We shall, however, discuss the mathematical notation for *summation,* since many statistical manipulations are difficult to express without this convention. As has been briefly discussed already, descriptive statistics is concerned with ways of summarizing or describing quantities of data. Many descriptors of sets of data can be easily defined, once we are able to easily denote various types of sums.

Assume that the heights, in inches, of five students in a statistics course are given by 70, 74, 66, 68, and 71. In order to have a notation that is useful in defining manipulations with data, we shall want to adopt a system of symbols to represent the numbers. Thus, we here have five numbers and could represent them by x_1, x_2, x_3, x_4, and x_5, respectively. For the five heights given above, then, $x_1 = 70$, $x_2 = 74$, $x_3 = 66$, $x_4 = 68$, and $x_5 = 71$. Note that the *subscript* on x identifies the order in which the heights were collected (or written down; $x_3 = 66$ means that the third height was 66). We could equally well have used y_1, y_2, y_3, y_4, y_5 or u_1, u_2, u_3, u_4, u_5 to represent these five numbers; the letter adopted is of no special importance. What is important is that the subscript takes on values corresponding to the number of observations made. Thus, x_1, x_2, x_3, x_4, x_5 could be used to represent any set of five numbers. To represent a general set of n numbers, where n could equal 1, 2, 3, or any higher integer, we will use x_1, x_2, $\ldots$, x_n.

The *capital* Greek letter *sigma* (Σ) is used to indicate *summation* of a set of numbers; it is called a *summation operator* because it is the shorthand representation for the operation of summation. To represent the sum or total of five numbers, x_1, x_2, x_3, x_4, x_5, we write $\sum_{i=1}^{5} x_i$. Thus

$$\sum_{i=1}^{5} x_i = x_1 + x_2 + x_3 + x_4 + x_5$$
$$= 70 + 74 + 66 + 68 + 71 = 349$$

for the above heights. The subscript i is called the *index of summation*. Writing $i = 1$ below Σ means that the first term in the sum is given by x_i with i set equal to 1, that is, by x_1. The 5 written above Σ means that the last term in the sum is given by x_i with i set equal to 5, that is, by x_5. The total sum is made up by allowing i to take on all the integer values between the number below Σ and the number above Σ, inclusive. It then is quite easy to represent certain partial sums of a set of data (i.e., totals of only some of the numbers). For example,

$$\sum_{i=1}^{3} x_i = x_1 + x_2 + x_3 = 210$$

$$\sum_{i=2}^{4} x_i = x_2 + x_3 + x_4 = 208$$

for the above heights. Given a general set of n numbers, the grand total then is represented by

$$\sum_{i=1}^{n} x_i = x_1 + x_2 + \ldots + x_n$$

Whenever the range of the index of summation is not given, Σx_i means the total of the full collection of numbers available; the size of that collection will always be clear from the context. Thus, given n numbers, $x_1, x_2, \ldots, x_n$, their grand total will be denoted simply Σx_i.

The summation operator can also be used to represent sums of functions of the numbers in a set of data. For example, given x_1, x_2, x_3, x_4, x_5 the sum of the squares of the x_i's is given by

$$\sum_{i=1}^{5} x_i^2$$

Again, the quantity beside Σ is to be added, the different values being given by i ranging from the number below Σ (here it is 1) to the number above Σ (here 5); in this case the quantity is specified as the square of the value of x_i, x_i^2. For the five heights mentioned earlier,

$$\sum_{i=1}^{5} x_i^2 = (70)^2 + (74)^2 + (66)^2 + (68)^2 + (71)^2 = 24{,}397$$

Note the difference between $\sum_{i=1}^{5} x_i^2 \; (= 24{,}397)$ and $\left(\sum_{i=1}^{5} x_i\right)^2$ $(=121{,}801)$; in the first case the x_i's are to be squared and then added together, while in the second the x_i's are to be added together and the sum then squared. These two quantities will generally not be equal.

Σ operates only on the quantity directly to its right. Thus, if we want to add 1 to each x_i and then total these quantities, we write $\sum_{i=1}^{n} (x_i + 1)$; the parentheses are used to make it plain that Σ is operating on each $x_i + 1$, that is, on the 1 that has been added as well as on the x_i's themselves. The quantity

$$\sum_{i=1}^{n} x_i + 1$$

represents the total of all the x_i's and then 1 is added to that total (in this case Σ is not operating on the 1). For the five heights presented earlier we have $\sum_{i=1}^{5} (x_i + 2) = 359$ and $\sum_{i=1}^{5} x_i + 2 = 351$.

Suppose c is any given constant (e.g., 1, 9, or 3.7). Then when we write $\sum_{i=1}^{n} c$, we mean Σ is operating on the constant c and that c is added to itself n times (since the value of c is the same for each possible i). Thus

$$\sum_{i=1}^{n} c = c + c + \ldots + c = nc$$

This identity will occur several times in our study.

Again, if c is any given constant, and we have a set of n numbers, $x_1, x_2, \ldots, x_n$, $\sum_{i=1}^{n} cx_i$ says we should multiply each x_i by the same constant c and then add all these products together. Note, then, that

$$\sum_{i=1}^{n} cx_i = cx_1 + cx_2 + \ldots + cx_n$$

$$= c(x_1 + x_2 + \ldots + x_n)$$

$$= c \sum_{i=1}^{n} x_i$$

That is, any constant can be factored outside the summation; the same quantity is given by adding all the x_i's together and multiplying their total by c.

We are all used to the idea that addition of numbers is commutative, that is, that the same total is arrived at, even if the numbers are added together in a different order. Thus, given $x_1, x_2, \ldots, x_n$ and $y_1, y_2, \ldots, y_n$,

$$\Sigma(x_i + y_i) = (x_1 + y_1) + (x_2 + y_2) + \ldots + (x_n + y_n)$$

$$= (x_1 + x_2 + \ldots + x_n) + (y_1 + y_2 + \ldots + y_n)$$

$$= \Sigma x_i + \Sigma y_i$$

another result that is at times useful in manipulating summations.

A quantity frequently used in both descriptive and inferential statistics is the *average* or *mean value* of a set of numbers. (We refer to the *arithmetic* average here; see Section 2.2 for additional ways in which the word average is used.) Given a set of n numbers $x_1, x_2, \ldots, x_n$, their (arithmetic) average will be denoted by $\bar{x}$ (read x-bar); it is defined to be

$$\bar{x} = \frac{1}{n} \Sigma x_i$$

the total of the numbers divided by n. As we shall see in Section 2.2, $\overline{x}$ measures in a certain sense the middle of a set of numbers; it can be thought to be typical of the individual numbers in the set of data. For the five heights mentioned earlier, $n = 5$ and $\Sigma x_i = 349$; thus for these data

$$\overline{x} = \frac{1}{n} \Sigma x_i = \frac{1}{5} (349) = 69.8$$

Another quantity that will prove useful both in descriptive and inferential statistics is called the *variance* of a set of numbers. This measures the variability of the numbers. It is denoted by s^2 and is defined to equal

$$s^2 = \frac{\Sigma(x_i - \overline{x})^2}{n - 1}$$

Notice that if all the numbers are equal in value (no variability) then the common value of each number is $x_i = \overline{x}$ and *every* difference $x_i - \overline{x} = 0$, giving $s^2 = 0$. The bigger the differences from one number to another, the bigger the differences $x_i - \overline{x}$ will be and the larger s^2 will become. For the five heights given earlier

$$s^2 = \frac{\Sigma(x_i - \overline{x})^2}{n - 1}$$

$$= [(70\text{--}69.8)^2 + (74\text{--}69.8)^2 + (66\text{--}69.8)^2$$

$$+ (68\text{--}69.8) + (71\text{--}69.8)^2]/4$$

$$= \frac{[36.8]}{4} = 9.2$$

Computing the value of s^2 from this formula is tedious at best, since it requires computing the difference between each x_i and $\overline{x}$, squaring those differences, totaling them, and then dividing the total by $n - 1$. If n is very large, this computation requires a great deal of effort.

Luckily, a shorter computational formula may be used which algebraically gives the same result. We all recall from algebra that

$$(x - y)^2 = x^2 - 2xy + y^2$$

where x and y are any two numbers. Then

$$(x_i - \overline{x})^2 = x_i^2 - 2x_i\overline{x} + \overline{x}^2$$

and we have

$$\Sigma(x_i - \overline{x})^2 = \Sigma(x_i^2 - 2x_i\overline{x} + \overline{x}^2)$$

$$= \Sigma x_i^2 + \Sigma(-2x_i\overline{x}) + \Sigma\overline{x}^2$$

$$= \Sigma x_i^2 - 2\overline{x}\Sigma x_i + n\overline{x}^2$$

$$= \Sigma x_i^2 - 2n\overline{x}^2 + n\overline{x}^2$$

$$= \Sigma x_i^2 - n\overline{x}^2$$

since $n\overline{x} = \Sigma x_i$. Thus, an equivalent formula for s^2 is

$$s^2 = \frac{[\Sigma(x_i - \overline{x})^2]}{n-1} = \frac{[\Sigma x_i^2 - n\overline{x}^2]}{n-1}$$

This latter formula requires much less computational effort than the original, especially if n is very large. For the five heights mentioned earlier we find

$$\Sigma x_i^2 = (70)^2 + (74)^2 + (66)^2 + (68)^2 + (71)^2$$

$$= 24{,}397$$

$$n\overline{x}^2 = 5(69.8)^2 = 24{,}360.2$$

so

$$\frac{[\Sigma x_i^2 - n\overline{x}^2]}{4} = \frac{[24{,}397 - 24{,}360.2]}{4}$$

$$= \frac{36.8}{4} = 9.2$$

which is exactly the result we had earlier. The variance s^2 and other measures of variability are discussed in Section 2.2.

Exercise 1.2

Use the following numbers for questions 1 to 4: 2, 5, 1, 7, 6, 4, 3.

1. Evaluate (a) $\sum_{i=1}^{7} x_i$ (b) $\sum_{i=2}^{5} x_i$ (c) $\sum_{i=1}^{7} (x_i - 3)$ (d) $\sum_{i=1}^{7} x_i - 3$

    (e) $\sum_{i=1}^{7} 2x_i$ (f) $2\sum_{i=1}^{7} x_i$.

2. Evaluate (a) $\sum_{i=1}^{7} x_i^2$ (b) $\left(\sum_{i=1}^{7} x_i\right)^2$.

3. Evaluate (a) $\sum_{i=1}^{3} \sqrt{x_i}$ (b) $\sqrt{\sum_{i=1}^{3} x_i}$.

4. Evaluate $\sum\limits_{i=1}^{7} x_i - \sum\limits_{i=3}^{6} x_i$.

5. The number of tornado deaths in the United States from 1965 to 1968, inclusive, were 299, 105, 116, and 131, respectively.
   (a) Compute the total number of tornado deaths in the United States for these four years.
   (b) Compute the average number of tornado deaths per year for these four years.

6. The amounts of gasoline sold in the United States from 1965 to 1968, inclusive, were 73.8, 76.8, 78.7, and 83.5, respectively, in millions of gallons.
   (a) Compute the total gasoline sales in the United States in these years.
   (b) Compute the average sales per year.

7. Given $x_1, x_2, \ldots, x_n$, where an even number, how could you represent the total of every second number.
   (a) Starting with x_1?
   (b) Starting with x_2?

8. What is necessary for $\sum\limits_{i=1}^{n} x_i^2$ and $\left(\sum\limits_{i=1}^{n} x_i\right)^2 / n$ to be equal?

9. Show that $\sum\limits_{i=1}^{n} (x_i - c) = \sum\limits_{i=1}^{n} x_i - nc$.

10. Show that $\sum\limits_{i=1}^{n} (x_i - c)^2 = \sum\limits_{i=1}^{n} x_i^2 - 2c \sum\limits_{i=1}^{n} x_i + nc^2$.

11. The depreciation suffered by a 1970 automobile in its first four years was \$900, \$500, \$400, and \$300, respectively, by "blue book" prices.
    (a) What was the total depreciation in price over the four years?
    (b) What was the average depreciation per year?

12. In 1969, there were six major disasters in the world caused by floods and tidal waves. The numbers of deaths reported for these six disasters were 91, 218, 41, 500, 189, and 250, respectively.
    (a) Compute the total number of deaths reported during the year, caused by these disasters.
    (b) Compute the average number of deaths reported per disaster.

13. In 1971, the six baseball teams in the American League, Eastern Division, won 101, 91, 85, 82, 63, and 60 games.
    (a) Compute the total number of games won in this year by these teams.
    (b) Compute the average number of games won, per team.

14. An apartment house contains six apartments, of which two rent for \$160 apiece, per month, two rent for \$185 apiece, and the remaining

two rent for $200 apiece.
(a)   What is the total monthly rental income for the owner of the building, assuming all apartments are rented?
(b)   What is the average monthly rental for the six apartments?

15. As of June 15, 1971, the salary of the Governor of California was 49.1, the salary of the Attorney General was 42.5, and the salaries of the Lieutenant Governor, the Secretary of State, the Comptroller, the Treasurer, and the Superintendent of Public Instruction were each 35, all in thousands of dollars.
(a)   What is the total annual amount paid in salaries for the services of these individuals?
(b)   Additionally, California has a 40-man Senate and an 80-man Assembly. Members of each receive a salary of $19,200. What is the total annual salary paid to members of the legislature?

16. Four tires of the same type and manufacture were driven on the same car until the tread depth was 20 millimeters. The numbers of miles traveled, per tire, was 18.4, 22.1, 22.7, and 20.5, in thousands of miles.
(a)   What was the average distance traveled, per tire, to cause the given amount of wear?
(b)   Compute the variance, s^2, of the distances traveled.

17. At the start of the 1971 school year, a college freshman was required to buy four books, whose prices were $8.95, $12.95, $10.95, and $13.25, respectively.
(a)   What was his total expenditure for books?
(b)   What was the average price that he paid per book?
(c)   Compute the variance, s^2, of the prices of these books.

18. An elevator states that its maximum capacity is six people, 990 pounds. What is the implied average weight per person that the elevator is expected to carry?

CHAPTER 2
DESCRIPTIVE STATISTICS

Whenever one is required to work with a large body of data, the sheer mass of numbers, as collected, immediately presents nothing but a sense of confusion to a reader. Consider, for example, the following weights of 60 male college students, presented as they were collected: 185, 188, 241, 160, 180, 181, 202, 190, 203, 190, 215, 203, 210, 210, 189, 136, 208, 204, 225, 209, 205, 196, 204, 177, 188, 203, 221, 186, 207, 219, 185, 212, 245, 195, 179, 194, 194, 193, 201, 158, 206, 160, 131, 214, 197, 177, 201, 190, 187, 169, 171, 164, 190, 197, 192, 179, 203, 176, 188, 193. Such a presentation of raw data has essentially no meaning for any reader. Without doing a considerable amount of work himself the reader is unable to gather any salient facts from the presentation. Many tools have been derived to present data in meaningful ways. We shall discuss several of them in this chapter. These techniques are useful in describing data, no matter what the purpose may have been for the collection of the data. As we shall see, many of these techniques are also useful in problems in statistical inference, to be discussed in subsequent chapters. Section 2.1 presents commonly used *tabular* and *graphical methods* of data presentation. Section 2.2 will then discuss *summary measures* (quantities computed from the data) that are useful descriptors in many cases and that are also useful in problems of inference.

2.1 TABLES AND GRAPHS

Many different tabular summarizations of sets of data are in common usage. We shall discuss several different ways in which tables may be constructed. The first of these is to organize the data into an *ordered array,* a presentation of all the data in order of magnitude of the numbers themselves, rather than in the order in which they were collected. The 60 weights mentioned earlier are presented this way in Table 2.1, from smallest to largest, when read down the columns.

Such a ranking as in Table 2.1 immediately brings some order to the presentation of the data. All of the data is presented, so no information (beyond the order of collection) has been lost in such a presentation. It immediately enables a reader to perceive several aspects of the data, such as the smallest number, the largest number, the value of the "middle" number in the total spread covered, the *range* covered by the data (by taking the difference between the largest and smallest), and so on.

In fact, an ordered array immediately allows the reader to evaluate any *percentile* from the set of data. Percentiles are defined below.

Definition 2.1. Assume a data set contains n numbers. Let $t_{k/(n+1)}$ be the kth number in the ranked array (counting from the smallest). Then $t_{k/(n+1)}$ is the $[100k/(n + 1)]$th *percentile* of the data set, $k = 1,2,3, \ldots , n$.

**Table 2.1 Ordered Array of Weights
of Sixty Male College Students**

131	188	203
136	188	203
158	189	203
160	190	204
160	190	204
164	190	205
169	190	206
171	192	207
176	193	208
177	193	209
177	194	210
179	194	210
179	195	212
180	196	214
181	197	215
185	197	219
185	201	221
186	201	225
187	202	241
188	203	245

At first glance it may seem strange to divide k by $n+1$ instead of n to determine percentiles. If n is large, say 20 or more, there is very little difference between k/n and $k/(n+1)$ [the difference is no more than $1/n + 1$]. More important, this definition is consistent with ideas we shall study in statistical inference. Chapter 11 discusses the uses of percentiles in inference and gives the reasons for this different divisor.

Let us discuss some specific cases. If we have only 2 numbers $(n=2)$, then the smaller number is called the $100(1/3) = 33$rd percentile and the larger number is the $100(2/3) = 67$th percentile. The labels for these percentiles, then, with $n=2$ correspond to taking the whole (100%) and cutting 100% in two places, to give 3 equal parts; the boundaries between the parts are specified by the 33rd and 67th percentiles. Similarly, if $n=3$, the smallest number is the $100(1/4) = 25$th percentile, the middle number is the $100(2/4) = 50$th percentile, and the largest number is the $100(3/4) = 75$th percentile. Again the whole (100%) has been cut into equal pieces, this time 4 pieces by $n=3$ cuts. The places at which the cuts occur are the labels for the percentiles. It is easy to see that this phenomenon continues, no matter what the value of n, the size of the data set. The whole is cut into equal percentage pieces by defining the percentiles in this way. Looked at in this light, the definition seems more natural.

For the ordered array of weights given in Table 2.1 note that the smallest weight is 131, the 21st weight (counting from the smallest) is 188 and the largest weight is 245. Then since the total number of weights available in this set is $n=60$, and $1/61 = .016$, $21/61 = .344$, $60/61 = .984$, we have $t_{.016} = 131$, $t_{.344} = 188$, $t_{.984} = 245$. Percentiles are used frequently in educational and psychological research, especially in reporting test scores. They are used to summarize various aspects of the set of data, as we shall see in the next section.

The array presented in Table 2.1 is still rather bulky and requires some effort to determine percentiles or other aspects of possible interest. A condensed tabular array is frequently useful, in which the scale of measurement (weights in this case) is divided into *classes* and the number of weights per class is then reported, instead of giving all of the individual values themselves. This type of tabular array is presented in Table 2.2, for the 60 weights already discussed. Note that instead of presenting all the original weights, such an array only summarizes the original data. The scale of measurement has been subdivided into intervals, called classes, and the column labled *frequency* records the number of weights that fell in the corresponding class. Thus, for example, nine of the original weights were between 155 and 178 pounds inclusive; it is not possible from this information to know the individual values of the nine weights in the class, so a frequency table loses some information that was contained in the original data. We also see from the table that 19 of the weights were between 203 and 226 pounds, inclusive; the individual values of these 19 weights are not given. This loss of identity of individual values is necessary, of course, if the full collection of numbers is not presented.

The column labeled *cumulative frequency* gives the running total of frequencies for the corresponding class, and all classes that preceded it. Thus, for example, 11 of the weights were no larger than 178 (the upper end point of the class 155–178) and 58 were no larger than 226. Note that all 60 of the weights are no larger than the upper end point of the highest class; this must always be true. These cumulative frequencies, divided by $n+1$ (61 for this case) gives the percentile ranks of the upper end points of the classes. The column labeled midpoint gives the value in the middle of the class, referred to as the *class mark*. This value is of some interest in graphing such tables and in computing summary measures, as we shall see.

Several considerations must be made when constructing a frequency table such as Table 2.2. First, the number of classes used is arbitrary and must be neither too large nor too small. If the number of classes used is too large, then necessarily many of them will have zero frequencies (or very small frequencies) and a tabular presentation of all the data in a ranked array may be equally meaningful (consider using classes of width 1 pound for the 60 weights). On the other hand, if too few classes are used, then the way in which the data is distributed may be seriously masked. A single class, extending from 131 to 245 pounds, with frequency 60, is not a very meaningful summary of the 60 weights, assuming that we want to give the reader more information than just the values of the

Table 2.2   Weights of Sixty Male College Students

Class (End Points Included)	Midpoint x_j'	Frequency f_j	Cumulative Frequency
131–154	142.5	2	2
155–178	166.5	9	11
179–202	190.5	28	39
203–226	214.5	19	58
227–250	238.5	2	60

smallest and largest weights. Thus, some art and much judgment are required in deciding on the number of classes to be used in constructing a frequency table.

Some guidance is available in making this choice. A rule that generally gives a reasonable number of classes is

$$k = 1 + 2.2 \log n$$

where n is the number of data values to be summarized and k is the number of classes to be used in the frequency table. Note that with $n = 60$ weights, this rule suggests the use of

$$k = 1 + 2.2 \log 60 = 4.91 \sim 5$$

classes; in our summary of the 60 college student weights we did in fact use the value suggested by this rule. It is to be stressed that this rule, or any other, is arbitrary and need not be used. It suggests an approximate value for k, the number of classes.

The classes used in Table 2.1 are all of equal width. This, too, is an arbitrary decision. We could have made the first class 40 units long, and each of the remaining 4 classes 20 units long, had we desired. Unless there is a compelling reason for using unequal class widths, it is generally to be recommended that equal class widths be employed. Frequency tables with equal class widths are less easily misinterpreted than are those with variable widths.

Many people prefer *graphical* presentations of data, rather than tabular presentations. A graphical presentation has the advantage that several attributes of the distribution of the data can be quickly comprehended. Figure 2.1.1 presents a *histogram* of the 60 weights we have been discussing. To give a fair picture of the distribution and to be consistent with some considerations that we shall examine in the following chapters, the histogram is constructed according to the following rules:

1. The bars are *centered* over the classes (thus the midpoint of the bar occurs at the class mark).
2. The *area* of a given bar is *proportional* to the *frequency* in the class.

Since the area of a rectangle is given by the product of the lengths of its base and height, rule 2 above simplifies to saying that the height of the bar should be proportional to (and thus can be equal to) the class frequency, *if the class widths are all equal*. The numbers above the bars are optional and give the frequencies of occurrence of the classes. To illustrate the visual distortion that can occur if rule 2 is not followed, Figure 2.1.2 presents a histogram of the

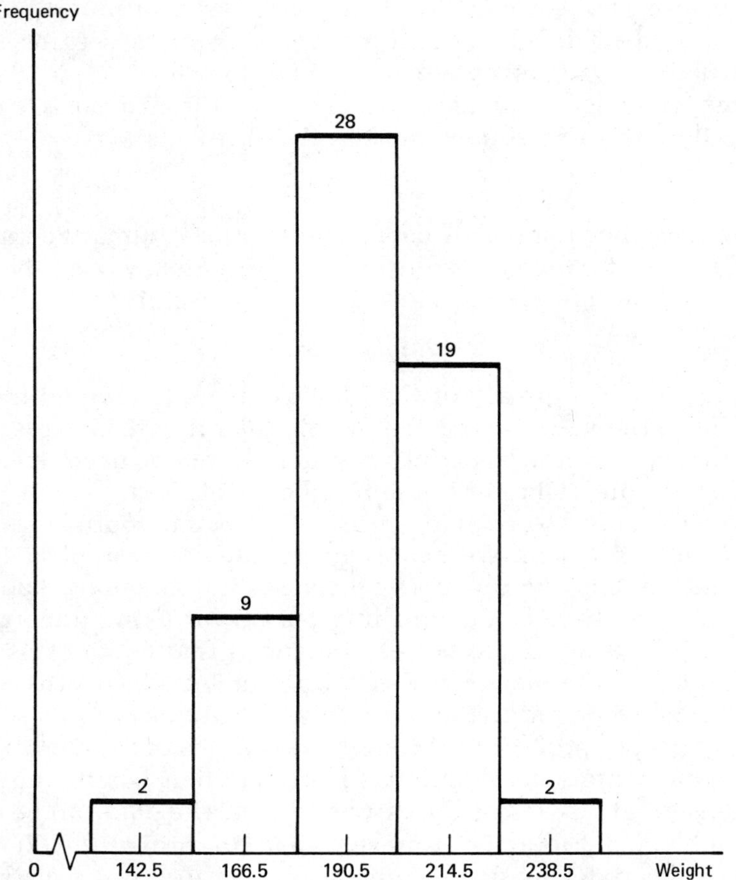

Figure 2.1.1 Distribution of weights of 60 male college students (see Table 2.2).

same weight data, except the first two classes have been combined and the height of the first bar is still taken to be the frequency of the combined classes. In doing this, the areas of all bars are proportional to the class frequencies, except the first, whose area is proportional to twice the frequency, since its base is twice as long as the others. A person glancing at Figure 2.1.2 is likely to get the mistaken impression that there were more weights in the interval from 131 to 178 pounds than there were in the interval from 203 to 250 pounds, which is decidedly not the case.

The cumulative frequencies themselves can also be presented graphically, which then makes the approximate determination of percentiles quite straightforward. Figure 2.1.3 presents a *line graph*

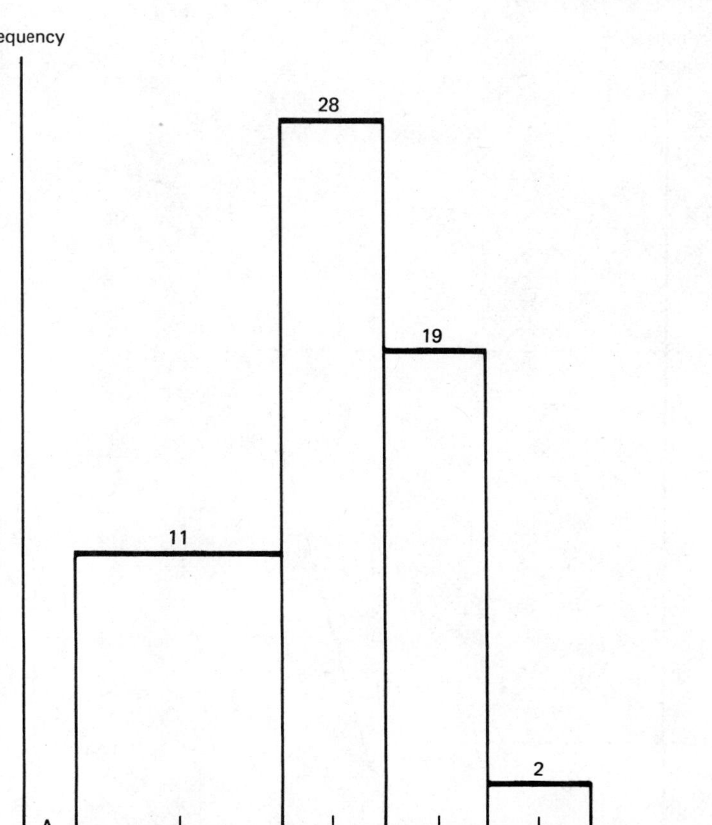

Figure 2.1.2 Deceptive plotting of weight data from Table 2.2.

of the cumulative frequencies from Table 2.2. The cumulative frequency is read from the vertical axis, with the horizontal axis used for the weights. In the figure we see that the cumulative frequency versus the upper class end points is plotted, rather than versus the class marks (or midpoints). This is done because the cumulative frequency of two for the first class says these two weights are no larger than 154, the highest extreme of the class. Similarly the second cumulative frequency of 11 says that all 11 weights were no larger than 178, the upper limit of the class. Thus the plotting is done using the upper class limits, and the individual points are connected by straight lines.

Percentiles can be approximated easily from a plot such as in Figure 2.1.3. It will be recalled that the $(100k)/(n + 1)$th percentile, $t_{k/(n+1)}$, is defined to be the kth value in the ranked array. Since the

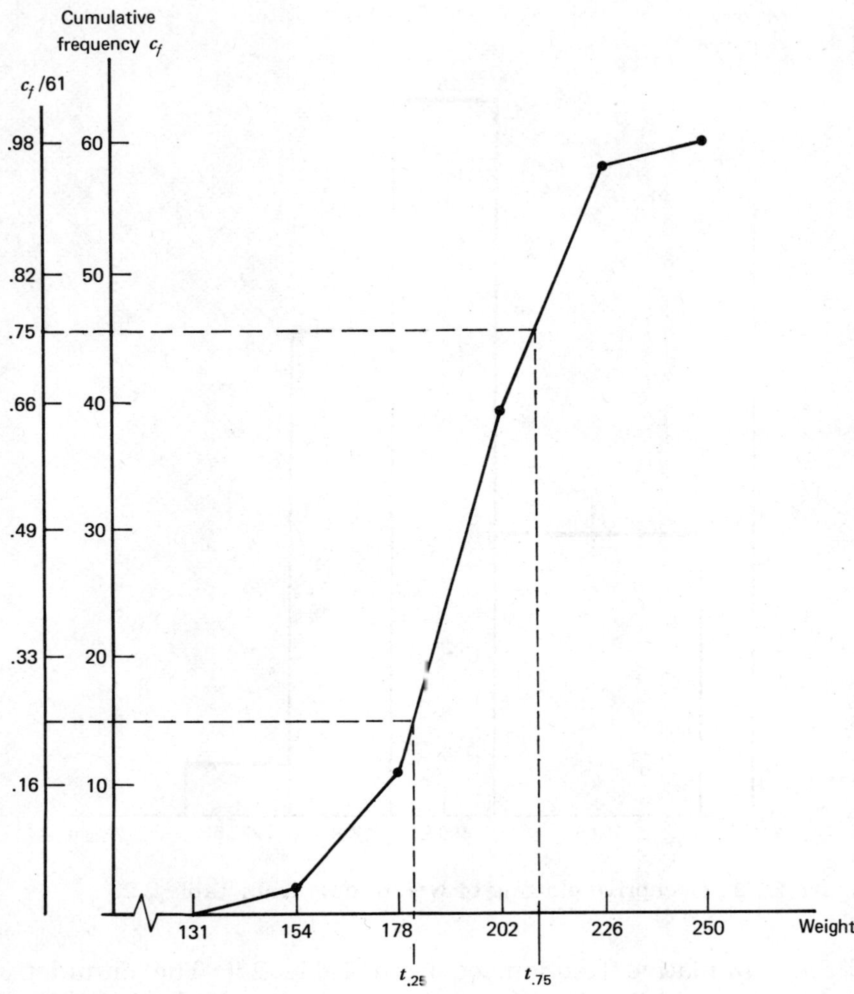

Figure 2.1.3   Cumulative frequencies of weights and approximate evaluation of percentiles.

frequency table presentation, and thus Figure 2.1.3, has summarized the data, and not presented the values for all the individual weights, it is not possible to exactly compute percentiles of the original data from the summarization. However, we can see from the figure that 11 of the weights were no larger than 178 pounds. Thus we can assume that the 11th weight in the ranked array (counting from the smallest) was (approximately) 178; since $11/61 = .18$, then $t_{.18} \doteq 178$ ($\doteq$ means approximate equality). Similarly, 39 of the weights were no larger than 202 pounds, $39/61 = .64$, so $t_{.64} \doteq$ 202. We can approximate any desired percentile by using the alternate vertical scale in the figure; this scale presents the cumulative frequencies c_f divided by 61 $(n+1)$. To read the $100p$th percentile, we need merely find the weight corresponding to $c_f/61 = p$. For example, to find the 75th percentile, we start at $c_f/61 = .75$, proceed horizontally to the line graph and then vertically down to find

$$t_{.75} \doteq 210$$

Similarly, to find the 25th percentile we want the weight corresponding to $c_f/61 = .25$, giving

$$t_{.25} \doteq 181$$

Both these approximations are illustrated in Figure 2.1.3.

In computing percentiles graphically in this way, we are using the upper class limits to approximate their corresponding percentiles. Then since we can get any percentiles in between these by using the straight lines connecting the dots, we are using linear interpolation to determine all the others. The interested reader can easily compare these approximate percentiles with the exact ones, by referring back to the ordered array of all the weights given in Table 2.1.

Exercise 2.1

1. The following table gives the distribution of scores of 120 students on a statistics exam:

Score (Upper End Point Included)	Frequency
60–65	6
65–70	13
70–75	17
75–80	25
80–85	23
85–90	16
90–95	12
95–100	8

(a) Compute the cumulative frequencies.
(b) Draw a histogram describing this data.
(c) Draw the cumulative line graph and use it to find the 25th, 50th, and 75th percentiles for this data.

(d)   If you were in this class and had a score of 91, estimate your percentile standing in the class.

2.   The game of roulette uses a spinning wheel, marked into 38 slotted spaces, numbered 00, 0, 1, 2, ... , 36. The 00 and the 0 are green; half the odd numbers (1, 3, 5, ... , 35) are black, the other half are red; half the even numbers (2, 4, 6, ... , 36) are red; the other half are black. The croupier spins the wheel and drops the ball onto it; in a few seconds the ball drops into one of the slotted numbers, which then is the winner. The following table summarizes the number of dollars bet by the players on 100 plays of the game.

Dollars Bet (Upper End Point Included)	Frequency
No more than 100	47
100–200	22
200–300	17
300–400	9
400–500	5

(a)   Draw the histogram for this data. This distribution is said to be *skewed* to the right, since it has a long "tail" extending to the right.

(b)   Plot the line graph of the cumulative relative frequencies.

(c)   Use the plot in (b) to find the 50th percentile. Half the time (approximately) this total amount was bet on a single play.

3.   The following table summarizes the amount of money the casino made on the same 100 plays of roulette described in problem 2. The amount the casino makes is defined to be the amount lost by the players less the amount the casino must pay out for winnings. Thus a negative number means the casino lost money on that play and a positive number means it won money on that play.

Casino's Gain (Upper End Point Included)	Frequency
−100–0	65
0–100	20
100–200	9
200–300	4
300–400	2

(a)   Plot the histogram for this distribution. Note it is also skewed to the right.

(b)   Plot the cumulative frequency line graph and use it to estimate the 50th percentile for this distribution.

(c)   Note that this 50th percentile is negative, which means that half the time (and actually more than half) the casino paid out more money than it received. Why might the game still be attractive to the casino?

4. The following table summarizes the cumulative frequencies of annual rainfall in Monterey, California, over a 25-year period.

Amount in Inches (Upper End Point Included)	Cumulative Frequency
10–12	3
12–14	7
14–16	12
16–18	20
18–20	23
20–30	25

Use this data to find the frequencies in the classes and draw the corresponding histogram. (Note the classes are not of equal length.)

5. Between January 1, 1965 and February 9, 1971, inclusive, a period of 2231 days, a total of 163 earthquakes were recorded somewhere in the world, each of magnitude 4 or more on the Richter scale. If we know the calendar date at which each quake occurred (local time), it is then possible to compute the number of days between successive quakes. (Since local time is used, an error of one day could occur in such differences, relative to actual time.) The 163 quakes then will give a total of 162 differences, for days between quakes. The following table summarizes the number of days between quakes, of magnitude 4 or more, between January 1, 1965 and February 9, 1971.

Number of Days Between Quakes	Frequency
0–4	50
5–9	31
10–14	26
15–19	17
20–24	10
25–29	8
30–34	6
35–39	6
40 or more	8[a]

[a] These eight values were 40, 43, 44, 49, 58, 60, 81, and 109 days.

Draw both the histogram (with the last class wider than the others) and the cumulative line graph for this data.

6. Compute (and compare) the 10th, 50th, and 90th percentiles of the 60 college student weights using (a) the ranked array in Table 2.1 and (b) Figure 2.1.3.

7. Find the 50th percentile for the earthquake data given in problem 5. Half the time this number of days, or less, passed between earthquakes of magnitude 4.

8. The numbers of sales made by a department store for 42 consecutive days were, respectively, 432, 1236, 847, 812, 819, 1315, 908, 481, 1157, 830, 746, 833, 1282, 896, 512, 1194, 892, 825, 795, 1266, 927, 490, 1302, 865, 850, 847, 1294, 935, 452, 1216, 872, 794, 852, 1302, 865, 467, 1248, 880, 808, 765, 1219, 901.
   (a) Make a histogram for this data.
   (b) Find the 25th, 50th, and 75th percentiles for this distribution of sales.

9. Did you notice the definite pattern in the data presented in problem 8? The 42 days covered are Sunday to Saturday for each of six weeks; the store is open from 12 P.M. to 5 P.M. on Sundays, from 9:30 A.M. to 9:30 P.M. on Mondays and Fridays, and from 9:30 A.M. to 5:30 P.M. the other four days of the week. Thus this data might more reasonably be separated into three parts, corresponding to the three different lengths of days that the store uses.
   (a) Draw a histogram for the Tuesday to Thursday plus Saturday data (a total of 24 days), and determine the 50th percentile for the number of sales on these days.
   (b) Combine the data for Mondays and Fridays (a total of 12 days) and find the 50th percentile for the number of sales on these days.
   (c) Find the 50th percentile for the number of Sunday sales (six days).
   (d) The above 50th percentiles are quite variable because of the differing number of hours that the store is open. Divide each 50th percentile by the number of hours the store is open that day. Do you find these numbers closer to being equal?

10. The weights of the 43 Miss America contest winners, through 1971, were, in reverse chronological order, 121, 110, 125, 135, 116, 115, 124, 124, 115, 118, 116, 120, 114, 130, 120, 116, 124, 132, 118, 143, 119, 106, 140, 130, 123, 135, 125, 130, 118, 120, 120, 126, 128, 120, 114, 120, 112, 115, 118, 138, 137, 140, 108. Make a histogram for this data and find the 25th and 75th percentiles of these weights.

2.2 SUMMARY MEASURES

In the last section we discussed some tabular and graphical methods for summarizing sets of data. These procedures, while not reproducing the full set of data, do keep the "flavor" of the data and enable a reader to determine many aspects of the full collection of numbers. In this section we shall discuss more drastic summarizations than either tables or histograms; these are more drastic because they are single numbers whose value is computed from the original set of data. The value of the number summarizes a specific aspect of the set of data; thus it is called a summary measure (measuring that aspect of the data set). Such measures can be constructed for any desired property of the data set. We shall discuss the two that are most commonly employed: measures of location (or the middle) of a data set and measures of variability of a data set.

Undoubtedly, we have all heard the word *average* used many times. In baseball, hitters are described by their batting average and pitchers are described by their earned run average. Climates in different areas are described and compared in terms of their average rainfall, as well as their average high and low temperatures. The earning potential for a person studying a certain field, or adopting a given profession, is frequently described by an average annual salary. A student's progress in college is described by his grade point average.

In each of these cases the average used is meant to summarize a body of data and, in a sense, to condense it all down to a single number: the value quoted for the average. This average value is a typical value, a number that measures the *middle* of the full set of data that has been described and gives an idea of the location of the data on a numerical scale. As we shall see there are several different measures of the middle of a set of data, or ways in which an average can be computed from the given numbers.

Undoubtedly the most commonly used average is the *arithmetic mean* of a set of data. Given the numbers $x_1, x_2, \ldots, x_n$, we shall denote their arithmetic mean by $\bar{x}$; it is defined as the total of the numbers divided by n, the number of data points given. Thus

$$\bar{x} = \frac{1}{n} \sum_{i=1}^{n} x_i$$

This quantity was briefly discussed in Section 1.2. To compute the arithmetic mean of a data set, then, we need only the total of the numbers, Σx_i and n. The ratio of these two quantities is $\bar{x}$. For the weights of 60 male college students discussed in Section 2.1, $\Sigma x_i =$ 11,576, $n = 60$, and thus $\bar{x} = (11,576)/60 = 192.9$. If the total amount of rain that fell in your own hometown over an $n = 6$ year span was 194.5 inches, then the average annual rainfall for your town during this period was $\bar{x} = (194.5)/6 = 32.4$. If an engineering society polls

10 of its members and finds the total of their annual salaries to be $179,200, then the average annual salary for these 10 people is ($179,200)/10 = $17,920. In this book, if we refer to the *average* of a set of numbers, with no other qualification, it is understood that the average is $\overline{x}$, the arithmetic mean; $\overline{x}$ will also be called simply the *mean* of the data, without the adjective arithmetic. As we have seen, the evaluation of the mean $\overline{x}$ requires knowledge of the total of all the numbers in the data set. Given a frequency table summarization of a data set, then, we cannot exactly evaluate the total of all the numbers and, thus, cannot exactly evaluate the mean $\overline{x}$ of the data. We can, however, evaluate the mean of the frequency table, which is itself a good approximation to $\overline{x}$. We shall use $\overline{x}'$ to denote the mean of the frequency table. It is defined as follows. Let us refer specifically to the frequency table of 60 weights discussed earlier. The classes, class marks x_j', and frequencies f_j are reproduced as table 2.3.

Also given in Table 2.3 are the products of the class marks times the frequencies, $f_j x_j'$. Granted that nine of the weights are in the interval from 155 to 178, we can approximate the total of these weights by $9(166.5) = 1498.5$, the product of the frequency with the midpoint of the interval. Similarly the total of the 28 weights in the interval 179 to 202 is approximated by the product of the frequency times the class midpoint, $28(190.5) = 5334.0$. If we sum these products over all classes, the total then approximates the actual total of the full data set. The mean of the frequency table is the ratio

$$\overline{x}' = \frac{(11,670.0)}{60} = 194.5$$

Notice that it does not differ markedly from the actual mean ($\overline{x} = $

Table 2.3

Class (End Points Included)	Class Mark x_j' (Midpoint)	Frequency f_j	Class Total $f_j x_j'$
131–154	142.5	2	285.0
155–178	166.5	9	1498.5
179–202	190.5	28	5334.0
203–226	214.5	19	4075.5
227–250	238.5	2	477.0
Total		60	11,670.0

192.9) of the original data. In more general terms the mean of the frequency table is

$$\bar{x}' = \frac{(\Sigma f_j x_j')}{(\Sigma f_j)} = \frac{(\Sigma f_j x_j')}{n}$$

since the sum of the frequencies equals n, the number of data points in the set. This quantity gives a good approximation to $\bar{x}$, the mean of the original unsummarized data.

Graphically, $\bar{x}'$ locates the point at which the histogram of the frequency table would balance on a knife edge. See Figure 2.2.1 for a graphic presentation of this idea. Thus, $\bar{x}'$ is said to locate the center of gravity of the histogram. The mean of the original unsummarized data, $\bar{x}$, locates the center of gravity of the original numbers; granted that the actual total of the numbers in each class used in the frequency table is not necessarily $f_j x_j'$, $\bar{x}$ and $\bar{x}'$ are not necessarily equal.

The *median, m,* of a set of data gives a different measure of location (the middle) of the data. It is by definition the 50th percentile, $t_{.5}$, so

$$m = t_{.5}$$

Recall that the $(100k)/(n+1)$th percentile is given by the kth value in the ordered array. Thus with $k/(n+1) = .5$, we want $k = (n+1)/2$; that is, to determine the median m we find the $(n+1)/2$th number in the ordered array. If n is odd this leads us to the (unique) middle number in the ranked array and, if n is even, the median is the midpoint between the two middle numbers in the ranked array. In either case, (about) half the numbers are less than m and half are larger than m; thus m measures the middle of the data set in a straightforward way.

For the original ranked array of 60 male college students' weights, given in Table 2.1, $n = 60$; therefore the median is given by the $(n+1)/2 = 61/2 = 30.5$th number; as mentioned above this is interpreted to be the value halfway between the 30th and 31st numbers, the two middle numbers. We find these two numbers to be 193 and 194, respectively; thus the median for this data is

$$m = \frac{(193 + 194)}{2} = 193.5$$

Given a set of data summarized into a frequency table, it is not possible to exactly evaluate m, the median of the original set of data. It is, however, straightforward to determine the value m' that cuts the histogram into two equal areas. This quantity, m', is the

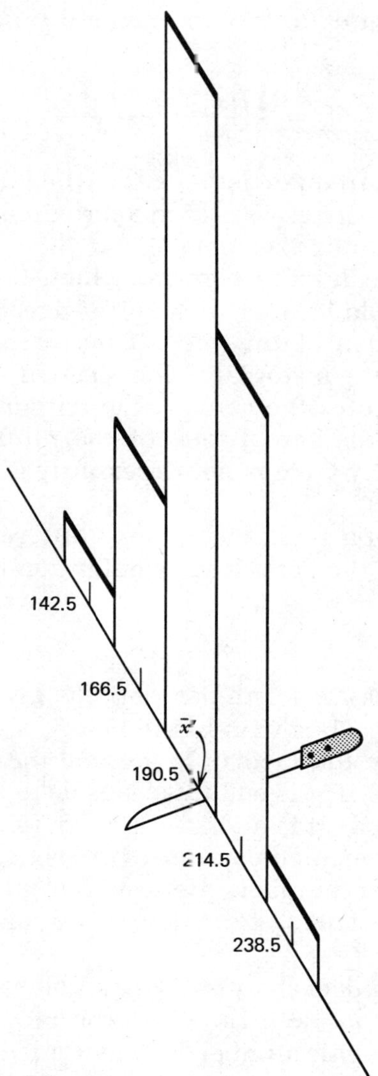

Figure 2.2.1 Balance point for histogram of weights.

median of the table and will give a good approximation to m, the median of the original data. We saw in Section 2.1 that percentiles of frequency tables can be read from the line graph of the cumulative frequencies, by finding the value corresponding to the desired value of $k/(n+1)$. Setting $k/(n+1) = .5$, in Figure 2.1.3, we find $m' = 194$, which is very close to the original median of the full data set.

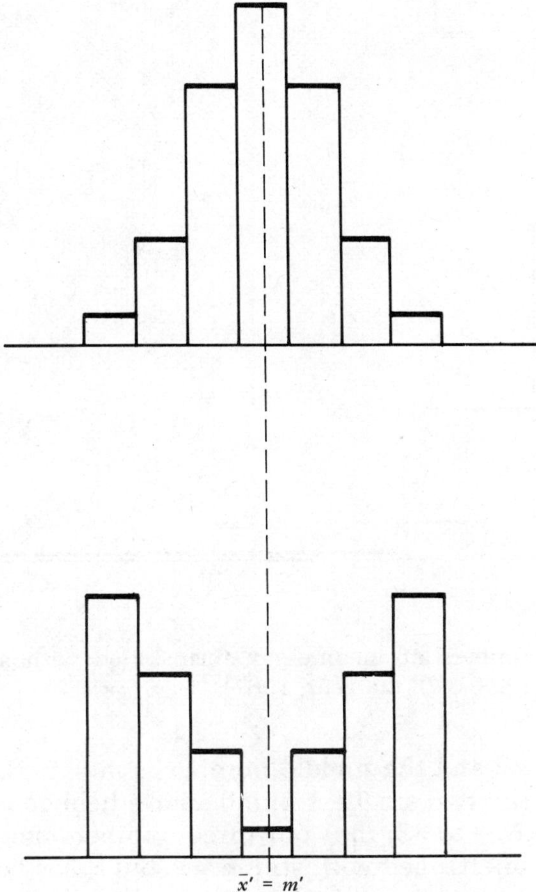

Figure 2.2.2 Mean and median are equal for symmetric histograms.

Granted we have two measures of location, $\bar{x}$ and m, we might inquire when the two are equal. Remembering the geometric interpretation of the two, it is not difficult to see when they are equal. The mean $\bar{x}$ measures the center of gravity and m is the middle-ranked number; thus they will be equal if the center of gravity falls right in the middle of the data set. This will happen if and only if the numbers are symmetrically distributed about the middle. Similarly, $\bar{x}'$ and m' for a frequency table will coincide only if the histogram is symmetric (Figure 2.2.2). In both these histograms $\bar{x}' = m'$ because of the symmetry.

What happens, then, if the data set or the histogram is not symmetric? Consider the three numbers 1, 2, 3; the mean of these

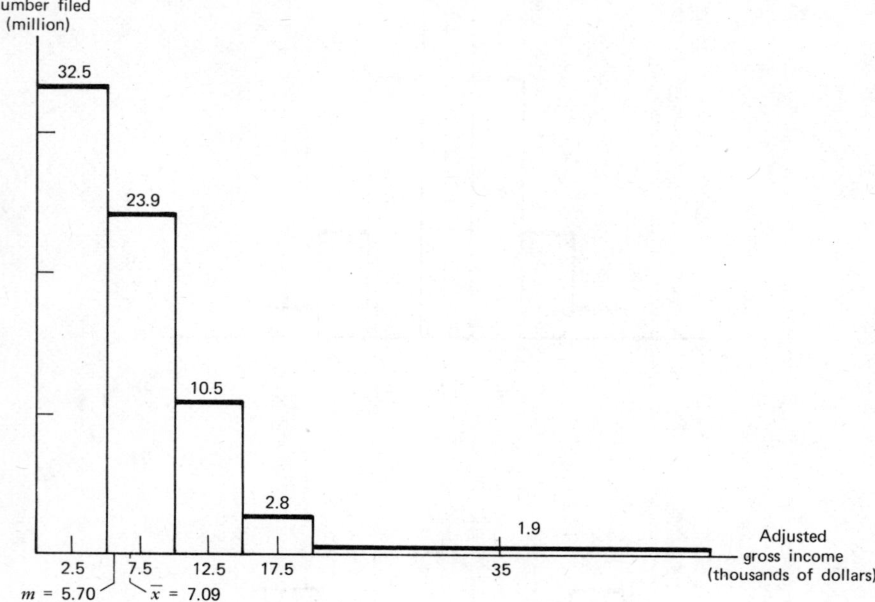

Figure 2.2.3 Numbers of income tax forms filed with adjusted gross income less than \$50,000 (tax year, 1967).

numbers is $\bar{x} = 2$ and the middle number is $m = 2$, that is, $\bar{x} = m$. Now, suppose the two smallest numbers are held constant but the largest is increased to 33; thus our three numbers now are 1, 2, 33. The median is unchanged with still $m = 2$, but $\bar{x}$ has become 36/3 = 12, a shift of 10 from its original value of 2. Thus, with this change of a single extreme value, m remained fixed, but the value for $\bar{x}$ was shifted considerably in the same direction as the number that was changed. This can be shown always to be true. The value of $\bar{x}$ will be shifted in the same way as the data. If the original distribution was symmetric, then $\bar{x} = m$; if we keep the values less than m constant and increase one or more of the values larger than m, then $\bar{x}$ will be increased and m will not. If we hold all values greater than m constant and decrease one or more of the values smaller than m, then $\bar{x}$ will decrease and will be smaller than m. Thus, it is said that $\bar{x}$ is shifted in the direction of the longer tail. This wording is best illustrated with a histogram such as Figure 2.2.3, where we see a histogram of the adjusted gross incomes for tax year 1967, taken from people who filed with an adjusted gross income of less than \$50,000. Notice that this figure is far from symmetric, with the longer tail trailing off to the right; the mean, $\bar{x}$, is also "pulled" to a value larger than the median m. This phenomenon (being skewed to the right) is quite common in studies devoted to measuring incomes; in cases of such extremely skewed data, the median is generally used as a measure of the middle, since it is in a sense "more typical" of the data than is the mean $\bar{x}$. If the distribution is not perfectly symmetric, $\bar{x}$ and m will not be equal; $\bar{x}$ is more affected by extreme values and will be displaced from m in the direction of the longer tail, as in Figure 2.2.3.

Both the mean, $\bar{x}$, and the median, m, are called measures of location for a set of data; they measure the "middle" of the data in two different senses and, thus, either $\bar{x}$ or m could be taken to be typical values of the full data set. Thus they are called measures of location. A different aspect of a set of data is its variability. Regardless of where the data is located or of its typical value, is the set of data relatively variable or constant? Do the individual values in the data set vary a lot about the typical value or do they vary relatively little? Measures of variability are used to measure how variable the data set is and, thus, to answer the question just posed.

Consider the following three sets of five numbers:

Data set 1: 10, 10, 10, 10, 10
Data set 2: 5, 10, 10, 10, 15
Data set 3: 0, 1, 10, 19, 20

each has its mean and median equal to 10; from this point of view, then, they are summarized in the same way. But it is obvious that the three sets of data are not identical. One reason they are not identical, even though their means and medians are equal, is that they are *not* equally variable about their middle values. The numbers in data set 1 are all equal to 10; there is no variability among the five numbers. Data set 2 has three values equal to 10 and two values that differ from 10 by five; thus it is certainly more variable than data set 1. Relatively speaking, data set 3 is the most variable; it has only one value equal to its mean of 10, and the other four all differ from 10 by more than five. We shall now discuss methods of measuring the variability of the values in a set of data.

If we think of the values, $x_1, x_2, \dots , x_n$, in the data set as locating points on a line, then the distance between two of them, say x_1 and x_2, is given by the difference $x_1 - x_2$. The distance is measured negatively if x_1 is to the left of x_2 ($x_1 < x_2$) and is measured positively if x_1 is to the right of x^2 ($x_1 > x_2$). The mean, $\bar{x}$, then will locate a point somewhere toward the middle of all the points $x_1, x_2, \dots, x_n$. One quantity that might seem to give a good measure of variability is the sum of the distances $x_i - \bar{x}$, for all the individual values in the data set. This particular measure is always zero, for any set of data, since

$$\sum_{i=1}^{n} (x_i - \overline{x}) = \sum_{i=1}^{n} x_i - \sum_{i=1}^{n} \overline{x}$$

$$= \sum_{i=1}^{n} x_i - n\overline{x}$$

$$= \sum_{i=1}^{n} x_i - \sum_{i=1}^{n} x_i = 0$$

when we recall that $\overline{x} = \dfrac{1}{n} \sum_{i=1}^{n} x_i$ and thus $n\overline{x} = \sum_{i=1}^{n} x_i$. Since this quantity is *always* zero, it certainly is of no use in describing the variability of a set of data.

The reason that $\sum_{i=1}^{n} (x_i - \overline{x}) = 0$, of course, is that some of the differences $x_i - \overline{x}$ are negative and some are positive; they just exactly take on the right magnitudes to cancel each other off, no matter how variable the data points are. We can get around this cancellation effect of positive versus negative differences by squaring the $(x_i - \overline{x})$'s and basing our measure on $\Sigma(x_i - \overline{x})^2$. Since $\Sigma(x_i - \overline{x})^2$ is the sum of n terms, it will naturally tend to increase with n; as more squares are added we expect a larger total. To get a measure of variability that does not directly increase with n, we could average the squares and divide by n. For reasons that will become plain as we consider problems of statistical inference, we shall instead divide by $n - 1$. If n is large, there is very little difference between dividing by n or $n - 1$. Thus, the *variance* of a set of data is denoted by s^2 and defined by

$$s^2 = \frac{\Sigma(x_i - \overline{x})^2}{n - 1}$$

If the original measurements have units attached to them (pounds, feet, hours), then s^2 is measured in the square of those units. The *standard deviation* is the positive square root of s^2 and thus is measured in the same units as the original numbers; we define then

$$s = \sqrt{s^2}$$

to be the standard deviation of the set of data.

Notice that the smallest possible value for s^2 (and s) is 0; this value occurs only if all the x_i's are equal to each other (and thus equal to $\bar{x}$; thus $x_i - \bar{x} = 0$ for all i). As the differences, $x_i - \bar{x}$, increase, so will the values for s^2 and s. Thus larger values of s (or s^2) mean greater variability.

For data set 1 above we find

$$s^2 = \tfrac{1}{4}[(10-10)^2 + (10-10)^2 + (10-10)^2 + (10-10)^2 + (10-10)^2] = 0$$

and thus $s = \sqrt{0} = 0$ as well.

For data set 2,

$$s^2 = \tfrac{1}{4}[(5-10)^2 + (10-10)^2 + (10-10)^2 + (10-10)^2 + (10-15)^2]$$

$$= \frac{50}{4} = 12.5$$

and $s = \sqrt{12.5} = 3.54$.

For data set 3,

$$s^2 = \tfrac{1}{4}[(0-10)^2 + (1-10)^2 + (10-10)^2 + (19-10)^2 + (20-10)^2]$$

$$= \frac{362}{4} = 90.5$$

and $s = \sqrt{90.5} = 9.51$.

Note that the sizes of these variances (and standard deviations) do reflect the relative variability of these three sets of data. The standard deviation (or variance) of a set of data does not have as simple a geometric interpretation as do the mean and median. The more variable the data and, thus, the "fatter" or wider the histogram, the larger that s will become.

We saw in Chapter 1 that

$$\sum_{i=1}^{n} (x_i - \bar{x})^2 = \sum_{i=1}^{n} x_i^2 - n\bar{x}^2$$

This can also be written

$$\sum_{i=1}^{n} (x_i - \bar{x})^2 = \sum_{i=1}^{n} x_i^2 - \frac{1}{n}\left(\sum_{i=1}^{n} x_i\right)^2$$

Either of these equivalent forms can be used in computing s^2 (or s) and when n is fairly large, these involve less computational effort than does the original form. You may verify that these formulas give the same values for s^2 for the three data sets mentioned earlier.

Given a set of data that has been summarized by a frequency table, then the variance s^2 of the original data cannot be computed exactly. It will be well approximated, though, by s'^2, the variance of the frequency table. Assume that we are given a frequency table with k classes, x_1', x_2', ... , x_k' are the class marks, $f_1, f_2, \ldots, f_k$ are the class frequencies, and $\overline{x}'$ is the mean of the table. Then, the difference between each individual data value in class j and the mean $\overline{x}$ is approximated by $x_j' - \overline{x}'$, the difference between the class mark and the table mean. The squares of these differences are approximated by $(x_j' - \overline{x}')^2$ and, granted f_j values in this class, the total of the squared differences for this class is approximately $f_j\,(x_j' - \overline{x}')^2$. Adding these quantities over all the classes then gives an approximation to the sum of the squares of the differences between the individual data values, x_i, and their mean, $\overline{x}$. Thus, we define the variance of the frequency table to be

$$s'^2 = \frac{\displaystyle\sum_{j=1}^{k} f_j\,(x_j' - \overline{x}')^2}{n - 1}$$

Computational formulas for the numerator of this quantity are given by

$$\sum_{j=1}^{k} f_j\,(x_j' - \overline{x}')^2 = \sum f_j\, x_j'^2 - n\overline{x}'^2$$

$$= \sum f_j\, x_j'^2 - \left(\sum f_j\, x_j'\right)^2 / n$$

The standard deviation of the frequency table is $s' = \sqrt{s'^2}$.

Table 2.4 summarizes the number of days between adjustments in the rate of exchange for the United States dollar and the Brazilian cruzeiro, for the period August 27, 1968 to March 19, 1971. In this period the rate was adjusted a total of 21 times, always in favor of the dollar. Let us use this data to illustrate the computation of the variance and the standard deviation for a frequency table. We find

Table 2.4 Number of Days between Adjustments in the Exchange Rate of the Dollar and the Cruzeiro

Days between Changes	Class Mark x_j'	Frequency f_j
14–27	20.5	2
28–41	34.5	6
42–55	48.5	9
56–60	58.0	4

$$\overline{x}' = \frac{\Sigma f_j x_j'}{21}$$

$$= \frac{1}{21}\,[2(20.5) + 6(34.5) + 9(48.5) + 4(58)]$$

$$= 43.64$$

The variance of this frequency table is

$$s'^2 = \frac{[2(20.5)^2 + 6(34.5)^2 + 9(48.5)^2 + 4(58)^2 - 21(43.64)^2]}{20}$$

$$= 130.7$$

and the standard deviation is $s' = \sqrt{130.7} = 11.43$ days.

Let us examine two additional measures that are sometimes used to describe the variability of a set of data. The first of these is the *range* of the data, which is simply the difference between the largest and the smallest numbers in the set of data. For data set 1 given above the range is 0, for data set 2 it is 10, and for data set 3 it is 20. Note that it also correctly indicates the relative amount of variability in these three data sets, since the values of the range are increasing from data set 1 to set 2 to set 3. For a tabular summary of frequency counts, the range is defined to be the difference between the upper limit of the highest class and the lower limit of the lowest class. Thus, for Table 2.4 the range is $60 - 14 = 46$ days. Note that the range, like s', has the units of the original data.

The range is much simpler to calculate than is s, since it involves the difference of only two values in the data. But, since it does depend only on the two extremes of the data, it is a much less sensitive measure of variability than is s. If we assume the two extreme values are fixed, the range will not change in value no matter what may be the values for the other $n - 2$ data points, whereas s does depend on the particular values they may assume, as well as on the values for the two extremes. This is one reason the standard deviation is in much more common use, especially in problems of inference, than is the range.

A third measure of variability, sometimes used in educational and psychological research, is the *interquartile range*. The interquartile range is defined to be the difference between the 75th percentile and the 25th percentile for a distribution. Thus, since 75% of the data is no larger than the 75th percentile, and 25% is no larger than the 25th percentile, the interquartile range brackets 50% of the data; the 50% that is bracketed is centrally located since it is centered at the median.

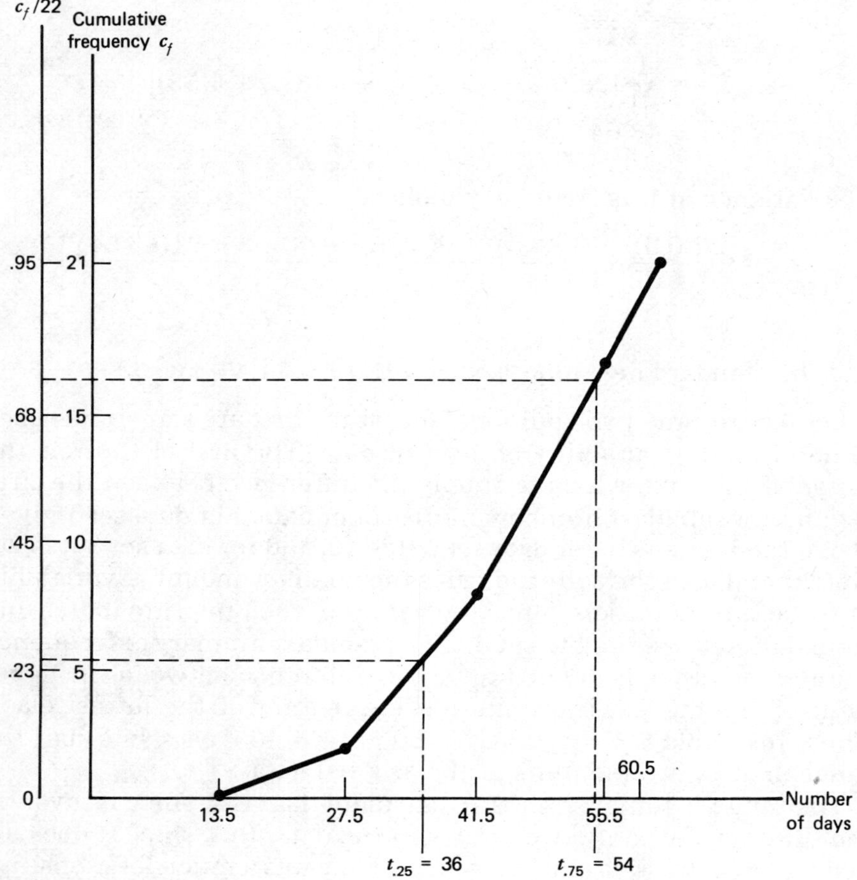

Figure 2.2.4 Days between shifts in exchange rate, United States dollar and Brazilian cruzeiro.

Figure 2.2.4 presents the cumulative line graph for the days between the dollar-cruzeiro changes, summarized in Table 2.4. Note that the 75th percentile is about 54 days, the 25th is about 36 days and, thus, the interquartile range for this set of data is $54 - 36 = 18$ days. Thus, 50% of the time the exchange rate shifted within nine days of the median time for changes, over this period of time.

The interquartile range again is computed as the difference between two numbers but, unlike the range, the position of the two numbers is affected by more than just the two extremes; thus the interquartile range can take on many different values, even if the largest and smallest values are fixed, and it is also a more sensitive measure of variability than the range. However, it does not prove as useful in problems of inference as does the standard deviation. Of the three measures of variability discussed, the standard deviation s (or equivalently the variance s^2) is by far the most frequently used and the most important for our purposes.

Exercise 2.2

1. The following table presents the fraction of the total population of the United States in 1930 that had ages in the indicated ranges:

Age	Fraction
Under 14	.273
14–19	.113
20–24	.089
25–44	.295
45–64	.175
65 and over	.055

Realizing that the fraction of the population is the same as relative frequency (f_i/n, where n is the size of the population), compute the average age of the individuals in the United States population in 1930. What was the (approximate) median age? (For the first class, use 7 as a class mark and for the last use 72 as the class mark.)

2. In 1965, the fraction of the total United States population in the same age brackets was as follows:

Age	Fraction
Under 14	.289
14–19	.106
20–24	.070
25–44	.242
45–64	.200
65 and over	.093

Compute the mean and median ages of the United States population in 1965 and compare these values with those from 1930 in question 1. (Use the same class marks as in question 1.) Can you think of any plausible reasons why these should be different?

3. The following table gives the same display of fractions of the United States population in the various age brackets in 1900.

Age	Fraction
Under 14	.322
14–19	.120
20–24	.097
25–44	.282
45–64	.138
65 and over	.041

Again compute the mean and median ages for the United States population in 1900 (using the same class marks) and compare these values with those derived in questions 1 and 2.

4. We have defined

$$s'^2 = \frac{1}{n-1} \sum_{j=1}^{k} f_j \, (x_j' - \overline{x}')^2$$

for data summarized in frequency tables. Compute the values for $(n - 1)s'^2/n$ for the distributions given in questions 1, 2, and 3 (since n is quite large, $(n - 1)/n$ is approximately 1 and these numbers can be taken as the variances of the numbers of people in the given age brackets in these specific years).

5. In a period of two weeks, in competition, a university broad jumper made 10 jumps. The distances he jumped on these 10 occasions were 268, 250, 292, 270, 272, 256, 266, 244, 284, and 296 inches, respectively. Compute the mean, variance, and standard deviation for this set of data.

6. Verify the computational formulas for $\Sigma f_j(x_j' - \overline{x}')^2$ given in the text.

7. Compute the mean, median, standard deviation, and interquartile range of the table of statistics exam scores given in problem 1 of Exercise 2.1.

8. Compute the mean and median of the amount of money bet at roulette, per spin, described in problem 2 of Exercise 2.1.

9. Compute the standard deviation and interquartile range for the data mentioned in problem 8 above.

10. Compute the mean and median gain to the casino for the data summarized in problem 3, Exercise 2.1.

11. Let $\overline{x}$ be the mean of a set of data and let $\overline{x}'$ be the mean of the frequency table summarizing the same set of data; assume that equal width classes have been used in the frequency table. Show that

$$|\overline{x} - \overline{x}'| \le \frac{w}{2}$$

where w is the width of the classes used. (*Hint.* At one extreme all values in the same class may equal the upper class boundary, and at the other extreme all values may equal the lower class boundary.)

2.3   CODING OF DATA

We have discussed several measures of the middle of a set of data and of the variability of a set of data. If the computations of these measures must be done by hand, without the use of a desk calculator or a computer, then frequently the labor necessary can be shortened considerably by *coding* the data before making the computations. Mathematically, the coding of data refers to a linear transformation of the numbers that shifts the origin or changes the scale of measurement. Practically, coding the data means that we will add or subtract the same constant from each number, as well as multiply each by the same number (which can be 1).

Assume we have n numbers $x_1, x_2, \ldots, x_n$, and let a and b be any two constants; the only restriction we insist on is $b \neq 0$. Then for each x_i we can define a y_i by

$$y_i = a + bx_i, \qquad i = 1, 2, \ldots, n$$

Note that the *same* a and b are used to define each value of y_i. We find then, by using the rules of summation discussed in Chapter 1, that

$$\sum_{i=1}^{n} y_i = \sum_{i=1}^{n} (a + bx_i)$$

$$= \sum_{i=1}^{n} a + \sum_{i=1}^{n} bx_i$$

$$= na + b \sum_{i=1}^{n} x_i$$

Note that

$$\bar{y} = \frac{1}{n} \sum_{i=1}^{n} y_i = \frac{1}{n} \left(na + b \sum_{i=1}^{n} x_i \right)$$

$$= a + b \left(\frac{1}{n} \sum_{i=1}^{n} x_i \right)$$

$$= a + b\bar{x}$$

that is, $\bar{y}$ is related to $\bar{x}$ in exactly the same way that each y_i is related to its corresponding x_i. From this equation we can solve for $\bar{x}$ to find $\bar{x} = (\bar{y} - a)/b$. Thus, if we can choose a and b so that $\bar{y}$ is easy to compute, we can then use its value (together with a and b) to find the value of $\bar{x}$. Let us illustrate this reasoning and procedure with the following weights of 10 girls in kindergarten: 44, 58, 42, 48, 46, 52, 44, 54, 56, and 50. Notice that 50 is a number in the middle of these numbers, and that if we were to subtract 50 from each, we would have much smaller numbers to work with (both positive and negative). Furthermore, every weight (and thus every weight less 50) is an even number; if we divide each difference by two, we will still have whole numbers to work with, and they will further be reduced in size. Letting x_i, $i = 1, 2, \ldots, 10$, represent the original data then, what we are proposing is to define

$$y_i = \tfrac{1}{2}(x_i - 50) = -25 + \tfrac{1}{2}x_i$$

Thus using our earlier notation, $a = -25$, $b = 1/2$. The values for y_i, $i = 1, 2, \ldots, 10$, then, are $-3, 4, -4, -1, -2, 1, -3, 2, 3$, and 0, respectively. We easily find $\sum\limits_{i=1}^{10} y_i = -3$, and thus,

$$\bar{y} = \frac{-3}{10} = -.3$$

Then

$$\bar{x} = \frac{1}{b}(\bar{y} - a)$$

$$= \frac{1}{1/2}[\bar{y} - (-25)]$$

$$= 2[\bar{y} + 25]$$

so

$$\bar{x} = 2[-.3 + 25]$$

$$= 2[24.7]$$

$$= 49.4$$

That this is, in fact, the value of $\bar{x}$ can be verified from the original x_i's.

Coding is also very useful in reducing the computations necessary to compute s^2. As before, assume we are given the numbers $x_1, x_2, \ldots, x_n$ and define $y_i = a + bx_i$, $i = 1, 2, \ldots, n$. We have just seen that

$$\overline{y} = a + b\overline{x}$$

note then that

$$y_i - \overline{y} = (a + bx_i) - (a + b\overline{x})$$
$$= b\,(x_i - \overline{x})$$

for each value of i. Thus, the variance of the y_i's, the coded values, is

$$s_y{}^2 = \frac{1}{n-1} \sum_{i=1}^{n} [y_i - \overline{y}]^2$$

$$= \frac{1}{n-1} \sum_{i=1}^{n} [b(x_i - \overline{x})]^2$$

$$= \frac{b^2}{n-1} \sum_{i=1}^{n} (x_i - \overline{x})^2$$

$$= b^2 s_x{}^2$$

Thus, we find $s_x{}^2 = (1/b^2)\, s_y{}^2$, and the variance of the original x_i's is easily determined from the variance of the y_i's. Note that the constant a that was added to each observation does not enter into this computation.

For the weights of the 10 kindergarten girls given earlier, again let

$$y_i = -25 + \tfrac{1}{2}\, x_i$$

leading to the same 10 values of y quoted earlier. By using the computational formula for the variance of the y's, we find

$$s_y{}^2 = \frac{1}{9} \left[\sum_{i=1}^{10} y_i{}^2 - \frac{(\Sigma y_i)^2}{10} \right]$$

$$= \frac{1}{9} \left[(-3)^2 + (4)^2 + \ldots + (0)^2 - \frac{(-3)^2}{10} \right]$$

$$= \frac{1}{9} \left[69 - \frac{9}{10} \right]$$

$$= \frac{1}{9}\, [68.1] = 7.57$$

Then

$$s_x{}^2 = \frac{1}{b^2}\, s_y{}^2$$

$$= \frac{1}{(1/2)^2}\, (7.57)$$

$$= 4(7.57) = 30.28$$

Note that $s_x = (1/|b|)\, s_y$ and, for this numerical example, $s_x = 5.5$. These coded computations are much easier to carry out, as long as a and b are chosen wisely.

Exercise 2.3

1. Using coding, recompute $\bar{x}$ and s^2 for the data given in problem 5 of Exercise 2.2.

2. Use coding to recompute $\bar{x}$ and s^2 for the data given in problem 10 of Exercise 2.1.

3. Over a seven day period, Joe slept an average of 7.5 hours; the variance of the number of hours he slept was 1. What is the mean and variance of the number of minutes he slept, over this period?

4. Coding may also be used on frequency tables to simplify the computation of $\bar{x}'$ and s'^2. Define new class marks by

$$y_j' = a + bx_j'$$

   Show that $\bar{x}' = (1/b)\,(\bar{y}' - a)$ and $s_x' = (1/|b|)\, s_y'$.

5. Use coding to recompute $\bar{x}'$ and s' for the statistics exam scores summarized in problem 1 of Exercise 2.1.

6. Use coding to recompute $\bar{x}'$ and s' for the rainfall data given in problem 4 of Exercise 2.1.

7. Why is the restriction $b \neq 0$ needed in the discussion on coding?

2.4 SUMMARY

Procedures of descriptive statistics are used to describe and summarize sets of data. A ranked array is a listing of the full set of data in order of magnitude. A frequency table summarization presents the numbers of individual data values that fell in various classes that partition the range of values observed. A frequency table summarization allows at least approximate evaluation of many aspects of the original set of data and brings organization into the collected data. Histograms are used as graphical presentations of frequency tables and allow pictorial representations of many aspects of a set of data. Cumulative frequency line graphs are useful summarizations of data sets or frequency tables and provide easy evaluation of percentiles.

Summary measures of a data set are single numbers computed from the set of data as measures of various aspects of the data set. The mean of a set of data locates the center of gravity of the data and is a measure of location. The median is also a measure of location and is equal to the middle number in the ranked array.

Measures of variability of a set of data are given by the variance (or standard deviation), the range and the interquartile range. For our purposes the standard deviation and its square, the variance, are the most important of these measures.

CHAPTER 3
RANDOM VARIABLES

We have briefly looked at methods used in descriptive statistics, procedures that are appropriate in describing or summarizing certain features of large masses of data. The description of bodies of data in meaningful ways is how statistics originated and is probably the only meaning visualized by the typical layman today. Indeed, the word statistics was originally coined some 200 years ago to describe bodies of data used by heads of states to measure various aspects of their countries; the tables of data were called statistics and the people in charge of gathering and summarizing them were called statisticians. Measurements were made of the size of the population (people and various domestic animals), of the amounts and value of various types of crops raised, of the amounts of various goods produced, of the value of commerce internally and externally, and on and on. These quantities were (and still are) of use to the state in describing its current position and in planning future actions. All of these are aspects of *descriptive statistics*.

Descriptive statistics continue to play an important role in the world today. Indeed the complex technological society that we live in could not function without the reams of data that are used to allocate resources by both private industry and various levels of government. In spite of the continuing, growing, important role of descriptive statistics, many other important applications involve *inferential statistics*, methods by which inferences (or generalizations) are made from an examination of a sample or portion; the inferences are based on the sample examined and hopefully are correct for the parts of the whole that have not been examined. Some specific examples of this type of situation are described below.

Medical science abounds with situations requiring inferences. Suppose a specific medication (e.g., a pill) is to be used to cure a headache. The manufacturer of the medication has given it to a large number of people with headaches and has observed the number of cases in which it was effective (headache is gone) as well as the number of cases in which it was not (no effect on the headache). Then, based on these observations, the company must make several inferences about the medication before marketing it and expecting the public to buy it. First, if the same people used in this study have headaches in the future, and the medication is applied again, would the results be the same (in terms of the numbers helped and not helped)? Second and more important, what will the effect be on the general public? Are the results observed in this study representative of the results to be observed if the public is free to buy and use the product? Will the proportion that are helped by the medication in the study be at all similar to the proportion in the whole population that will be helped? Are there undesirable side affects to the medication? There are many such questions that are of interest; they all concern making inferences (deriving statements) that are true for the whole population, based only on the cases observed in the study itself.

As a second example, most students have at one time or another taken an aptitude test, designed to measure their capacity for learning specific types of subject matter or material appropriate for specific trades. The use and interpretation of such a test also involves many problems of inference. Generally, many specific questions are placed in the test, and the test is taken by people with known aptitudes in given areas; their scores are then known for the items included in the test. If a person of unknown aptitude takes the test, his scores will be known and can be compared with those made by the original test group. The "closeness" of the new person's scores to those made by plumbers, mathematicians, and historians, for example, is then used to judge the person's aptitudes for these different types of jobs. In this sort of procedure inferences are made about the aptitudes of each new person who will take the test. Which particular questions should be included in the test? Which particular individuals of known aptitudes are used in the original test group? Which way should closeness of scores be measured? All of these factors and others will have an effect on the accuracy of the resulting inference about a specific individual's aptitude. The basic problem, though, is still one of inference: What can be done to measure correctly the aptitudes of individuals before they have actually tried a specific field (and possibly rejected it)?

These are two specific instances of problems of inference from a very large group of problems. If a physical measurement (such as the speed of sound) has a specific value under given conditions, will it always have the same value under those conditions? If the conditions change, does the speed of sound also change? If a Westinghouse 40-watt light bulb lasted 720 hours in a given socket, will its replacement also last 720 hours? Is a specific insecticide effective against a garden pest? If you play a gambling game for two hours in a Las Vegas casino, will you come out ahead or behind?

We shall begin studying statistical inference in this chapter; this body of material is useful in answering all of the questions mentioned above. The first topic we shall study is the random variable, which is defined in Section 3.1. As the name indicates, random variables are quantities that vary; in a sense, a study of random variables is a study of uncertainty and its implications.

3.1. RANDOM VARIABLES AND THEIR RANGES

We are all acquainted with, and affected by, events that cannot be predicted in advance. For example, we don't know for sure what the weather will be like tomorrow, how many questions we will get correct the next time we take an exam, who will win a game played by two-matched teams, how many votes will be gotten by a candidate in an election, and so on. A close examination of the world around us shows that we are surrounded by uncertainty, events that cannot be predicted in advance. Indeed, the world would be quite dull if everything in it were predictable and stable. We shall use the phrase "determined by chance" to describe phenomena that cannot be predicted in advance.

In this chapter we shall begin studying events that are affected or determined by chance. Even with such uncertainty, there are some elements of regularity that can be exploited in making inferences. Essentially all problems of inference can be couched in terms of random variables; this important concept is defined below.

> **Definition 3.1.** A *random variable* is a numerical quantity whose value is not known in advance. Its value is determined by chance. We shall use capital letters at the end of the alphabet, such as U, V, W, X, and Y, to denote random variables. Some specific random variables are discussed in the following examples.

Example 3.1.1

Assume that a college football team will play 10 intercollegiate games in

1976. If we let U be the number of these games that the team will win, then U is a random variable, a numerical quantity whose value is not known in advance.

Example 3.1.2

One hundred families in the same city will be telephoned between 8 and 9 P.M. next Tuesday night. If we let V be the number of these families that have a television set turned on, somewhere in their residence at the time they are called, then V is also a random variable since the value it will equal is not known in advance.

Example 3.1.3

Suppose we let W be the amount of rain that will be recorded in your hometown during the coming calendar year. Since the value of W is not known in advance, it is a random variable.

Example 3.1.4

Let X be your statistics teacher's age at death. Since the total span of time that any individual will live is not known in advance, then again X is a random variable.

In each of the examples just discussed (and in all others we shall examine) the random variable *will* be equal to some number, but which specific value it will have, we do not know in advance. A time will arrive at which the random variable's value is known. For example, we will know after the season is over how many games the football team won; we will know how many residences had television sets on after we have called all 100 residences; and we will know how much rain fell in any specific locale during a specific year after the period is over. Before the specified time arrives, we do not know the value of the random variable, but after that specified time, we do know which value it finally was equal to. This value it eventually equals is called its *observed value.*

Before the observed value of a random variable is known, we shall in general be able to specify the full *possible* collection of observed values it may have. This collection of possible values the random variable may equal is called the *range* of the random variable, as is given in the following definition.

Definition 3.2   The set of possible values a random variable X may equal is called the range of the random variable; the range of X will be denoted by R_X.

We shall reserve the symbol R to stand for the range of a random variable; the subscript identifies the specific random variable being discussed. We shall now examine the ranges of the random variables, U, V, W, and X, already discussed in the preceding examples.

Example 3.1.5

In Example 3.1.1, U was defined as the number of games that a college football team will win in 1976; the team considered will play 10 games that year. Then the range of U is

$$R_U = \{u: u = 0,1,2,\ldots, 10\}$$

Since R_U is a set, we have used set notation in its definition. The braces "$\{\ \}$", or set builders, are used to begin and end the definition of the set. The definition of the set is given between the two braces. Thus, for R_U we have the statement "$u: u = 0,1,2, \ldots , 10$" between the braces. We read, then, that R_U is the set of elements u such that u equals 0 or 1 or 2 or ... or 10 (the colon "$:$" is read "such that"). This is merely a convenient way of saying that, even though U is a random variable, we know that it must end up equal to one of the integers between 0 and 10, inclusive. In Example 3.1.2, V was defined to be the number of families, among 100 called, who had a television set turned on at a specified time. Again, then, V is a random variable and surely must end up equaling one of the integers between 0 and 100 inclusive; that is, the range of V is

$$R_V = \{v: v = 0,1,2,\ldots, 100\}$$

   The ranges of the two random variables discussed in Example 3.1.5 both consisted of sets of integers or whole numbers; the range of neither random variable included fractions or any values between the whole numbers. As we shall see, random variables of this type are called *discrete* (because their ranges are discrete sets of numbers). The way in which their behavior is described will be discussed in the next section.

Example 3.1.6

In Example 3.1.3, W was defined as the amount of rain that will fall in a certain town in a year. The amount of rain that may be recorded is certainly not restricted to a whole number of inches (or any other unit); that is, it is possible that we might observe 12.64, 15.2, or 31.459 inches of rain in the year. Thus we would not want to use just a set of integers as the range for W. The set of possible values for W must also include values between integers; we will use a set that contains all the numbers in a continuous interval as the range for a random variable like W. Assuming that your hometown has never recorded less than 10 inches nor more than 40 inches of rain in a single year, we should be safe if we use

$$R_W = \{w: 0 \leq w \leq 100 \text{ inches}\}$$

as the range of W; note that R_W does include all the numbers in a continuous interval and, so long as your hometown is not deluged with more than 100 inches of rain next year, the actual value that will be observed is an element of R_W. In Example 3.1.4, X was defined to be your statistics teacher's age at death. Time (and age) also is not restricted to taking on only integer values (whether a year, month, or day is used as a unit); thus to be sure R_X does contain the observed value for X, we would want it to contain all the numbers in a continuous interval. Let us use

$$R_X = \{x: a \leq x \leq 200 \text{ years}\}$$

where a is your teacher's current age; surely R_X does contain the number that will be your statistics teacher's age at death (assuming he or she does not live more than 200 years).

The two random variables discussed in Example 3.1.6 both had ranges that contain all the numbers in a continuous interval. This was necessary because of the nature of the numerical quantity being studied; the numerical quantity itself was not restricted to being equal to integers and, thus, the range of the random variable must contain more than just whole numbers. Random variables like W and X are called *continuous* random variables; their description is accomplished in a slightly different manner than is the description of discrete random variables. We shall study methods appropriate for describing both types of random variables in the following sections.

As mentioned earlier, the value that a random variable eventually equals is called the observed value of the random variable. This observed value will necessarily be some element from the range of the random variable and will thus be denoted by a lower case letter.

Exercise 3.1

1. A baseball player will be at bat four times in a game. Let W be the number of hits he makes.
    (a)   Is W discrete or continuous?
    (b)   What is R_W?

2. In driving from your residence into town you must pass through seven intersections with stoplights. Let V be the number of stoplights that are red when you arrive at the corresponding intersection.
    (a)   Is V discrete or continuous?
    (b)   What is R_V?

3. Let X be the amount of rain to be recorded in San Francisco in December next year.
   (a) Is X discrete or continuous?
   (b) What is R_X?

4. Let Y be the amount of time necessary for you to drive your car from your residence to the supermarket the next time you go.
   (a) Is Y discrete or continuous?
   (b) What is R_Y?

5. A new golf ball has been designed. It will be hit from a tee by a professional golfer. Let U be the distance the ball travels.
   (a) Is U discrete or continuous?
   (b) What is R_U?

6. A clairvoyant will try to guess whether each of nine cards is red or black, without seeing them. Let X be the number he identifies correctly.
   (a) Is X discrete or continuous?
   (b) What is R_X?

7. Each of 75 housewives is asked whether she prefers detergent 1 or detergent 2 for cleaning clothes. Let Y be the number of wives that prefer detergent 1.
   (a) Is Y discrete or continuous?
   (b) What is R_Y?

8. A 12-inch piece of string is cut into two pieces; the cut may occur anywhere along its length. Let U be the length of the shorter piece.
   (a) Is U discrete or continuous?
   (b) What is R_U?

9. In question 6, let W be the number of cards *not* identified correctly.
   (a) Is W discrete or continuous?
   (b) What is R_W?

10. In question 8, let V be the length of the longer piece.
    (a) Is V discrete or continuous?
    (b) What is R_V?

3.2. DISCRETE RANDOM VARIABLES AND PROBABILITY MEASURES

We have seen several examples of discrete random variables and their ranges. Since a random variable is a numerical quantity whose value cannot be predicted in advance, it is not possible to be absolutely certain of the particular number it will equal. It is, however, straightforward to list a set of values (the range) that we assume contains the particular number that the random variable will in fact equal. What more might be done to describe the random variable?

Since the value of the random variable cannot be predicted in advance, there must be some chance that each of the values in its range might be the one to be observed. We will find it helpful to be able to express these chances in some manner. For example, assume that a 50-cent piece has a zero painted on the heads side and a one painted on the tails side; the coin will be flipped once, and we define Y to be the number that is uppermost. Then $R_Y = \{y: y = 0, 1\}$ and we would find it helpful to have a measure of the chance that $Y = 0$ versus the chance that $Y = 1$. This measurement of the chances that various values in the range of the random variable may be the one observed is provided by the *probability measure* for the random variable. This is defined below.

> **Definition 3.3.** The *probability measure* of a random variable X expresses the chances (or probability) that individual elements or sets of elements from R_X will be observed.

Something is "probable" if it is likely to occur; the word probability is used in the measurement of the likelihood of various possible events. If an event has "high" probability, that means it is quite likely to occur; if an event has "low" probability, it is not very likely to occur. Two events with equal probability are equally likely to be observed.

For classical statistical procedures, such as those we shall study, probabilities may be given a relative frequency interpretation. Suppose that the probability of an event occurring is .7; the relative frequency meaning of probability states that over repetitions of the same experiment the event, then, should be observed 70% of the time (and 30% of the time it should not). If the probability of an event is .05, then it should be observed only 5% of the time and 95% of the time it should not. This relative frequency interpretation for probability is by far the most commonly used. If the weather forecaster in your area states that the probability for precipitation tomorrow is .6 or 60%, he is using a relative frequency interpretation for the probability statement. In the past, when your location had "identical" weather conditions with those today, 60% of the tomorrows brought rain and 40% of them did not. Thus the .6 is measuring the relative proportion of the past days like today that were followed by a day with rain. Notice that a probability equal to 1 would correspond to a relative frequency of 100% and a probability equal to 0 would correspond to a relative frequency of 0%. Thus if we assume the probability is 1 that a random variable equals 5, say, then we are assuming that the random variable is certain to equal 5; if we assume the probability is 0 that it equals 5, then the relative frequency of 5 is 0 and we are certain it will not equal 5. Intermediate values for this probability correspond to situations between these two extremes.

If we return for a moment to the 50-cent piece, with a zero painted on one side and a one painted on the other, it would seem reasonable to assume that over repeated flips of the coin, 50% of the time we should observe a head and 50% of the time we should observe a tail. Thus, letting Y equal the number to be observed when the coin is flipped, it would seem reasonable to assume 1/2 as the probability that $Y = 0$ and also to assume 1/2 is the value of the probability that $Y = 1$. Such an assumed assignment of probabilities of occurrence for the values of the random variable Y is a probability measure for Y. Note that we got the values of 1/2 for the probabilities of $Y = 0$ and of $Y = 1$ by assumption; since a 50-cent piece appears symmetric and well-balanced, there does not appear to be any physical reason to expect a preponderance of either heads or tails over repeated flips (repetitions of the experiment). Thus we would expect equal frequencies of occurrence for the two values of Y, leading to relative frequencies of 1/2 for each. We have in fact constructed a model or description of the situation and then that model led us to the values for the two probabilities involved. As is discussed more fully in the Appendix, numerical values for probabilities and probability measures are in general derived from models of physical experiments. The model used then will imply specific numerical values for the probabilities involved.

Having discussed probability itself to some extent, let us return to the idea of a probability measure of a random variable, given in Definition 3.3. The reader will recall that there are two distinct types of ranges for random variables, discrete sets or sets that consist of all real numbers in an interval. The corresponding random variables are called *discrete* or *continuous*, respectively. The details of the computations and usages of probability measures are different for these two types of random variable; thus they will be discussed and illustrated separately. First, let us consider what a probability measure for a discrete random variable must be.

> **Definition 3.4.** Assume X is a discrete random variable with discrete range R_X. The *probability measure* for X is a table or a formula that gives the probability of occurrence of each element x in the range R_X. The probability that X equals a specific value x will be denoted by $P(X = x)$. For any random variable X it is necessary that
>
> (i) $P(X = x) \geq 0$ for each x in R_X.
> (ii) $\sum_{R_X} P(X = x) = 1$. (This means summation over all elements in R_X.)

Let us examine some examples of discrete random variables and their probability measures.

Example 3.2.1

As discussed earlier, assume a 50-cent piece has a zero painted on the head side and a one painted on the tail side. Let Y be the number uppermost if the coin is flipped once. Then, $R_Y = \{y: y = 0,1\}$ and some possible probability measures for Y are

$$\text{(i)} \quad P(Y = 0) = \frac{1}{2} \quad P(Y = 1) = \frac{1}{2}$$

$$\text{(ii)} \quad P(Y = 0) = \frac{1}{4} \quad P(Y = 1) = \frac{3}{4}$$

$$\text{(iii)} \quad P(Y = 0) = 0 \quad P(Y = 1) = 1$$

Assuming the coin is symmetric and does not favor the occurrence of either face, the first measure (i) is the logical one to use. Measures (ii) and (iii) would be appropriate if we thought (or assumed) the coin was biased. Note that for each of these three measures we have $P(Y = 0) \geq 0$, $P(Y = 1) \geq 0$, and $P(Y = 0) + P(Y = 1) = 1$; thus we do satisfy requirements (i) and (ii) of Definition 3.4. [We could in fact use $P(Y = 0) = p$, $P(Y = 1) = 1 - p$, where p is any number between 0 and 1, inclusive, as the probability measure for this random variable Y.]

Example 3.2.2

In Example 3.1.1 we assumed that a college football team will play 10 intercollegiate games in 1976. U was defined as the number of games the team would win and then $R_U = \{u: u = 0,1,2,\ldots,10\}$. A probability measure for U must assign a nonnegative probability to each u belonging to R_U such that the sum of the 11 probabilities used is 1. The table below presents some possible probability measures for U. The entries in the table are the values for $P(U = u)$.

Elements of R_U
$u =$

	0	1	2	3	4	5	6	7	8	9	10
Measure 1	1/11	1/11	1/11	1/11	1/11	1/11	1/11	1/11	1/11	1/11	1/11
Measure 2	0	0	0	0	0	0	0	0	0	0	1
Measure 3	1	0	0	0	0	0	0	0	0	0	0
Measure 4	.001	.010	.044	.117	.205	.246	.205	.117	.044	.010	.001

Measure 1 assigns equal probability to each element of R_U; this assignment is appropriate if it seemed reasonable that this team is as likely to win 0 games as it is to win 1 game or 2 games, and so on, up to 10 games. Measure 2 has assigned one as the probability that $U = 10$ and zero as the probability for all other values u in R_U; this measure would be appropriate if we wanted to assume that this team is certain to win all its games. Measure 3 has assigned one as the probability that $U = 0$ and zero as the probability for all other values of u in R_U; this measure would be appropriate if we wanted to assume that this team is certain to lose all its games. Measure 4 is a different assignment of probabilities than any of the first three; as we shall see, it is appropriate if we wanted to assume that the team has probability 1/2 of winning each of the 10 games and that the team's chance of winning any given game is unaffected by whether it won or lost the games preceding. Notice again, then, that there are many different possible probability measures for U; which one is appropriate depends on the assumptions we feel are justified.

The detailed, complete justification for the reason that one probability measure should be more appropriate than another for a given random variable involves judgment and a thorough knowledge of the experiment concerned and of probability theory. The emphasis in this book is on understanding what a probability measure is and what it can be used for. The Appendix presents a discussion of probability theory and its applications to some simple problems; this material is useful in justifying when a particular probability measure should be used.

Probability functions for discrete random variables can also be presented graphically by histograms. To do so, a bar is centered above each element from the range of the random variable; each bar is one unit in width, and its height is equal to the value of the probability function for that element. That is, the height then is equal to the probability that the given element will be the observed value for the random variable. In plotting this way, the *area* of the bar centered at a given element is also equal to the probability that the element will be the observed value of the random variable since

$$\text{Area} = \text{width} \times \text{height}$$

$$= 1 \times \text{probability}$$

$$= \text{probability}$$

Figure 3.2.1 graphs the three probability measures discussed for Y in Example 3.2.1, and Figure 3.2.2 graphs the four probability measures for U discussed in Example 3.2.2. This graphical presentation of a probability function provides a quick comparison of the relative chances of the different values in the range of the random variable to be the observed value; the higher the bar (and greater the area) for any element, the larger the probability that it will be the observed value. If we want to compute the probability that the observed value of a discrete random variable will lie in a specific interval, this is given by the sum of the probabilities of occurrence of the whole numbers in the interval; thus, graphically, it is given by the sum of the areas of the bars whose centers lie in the interval.

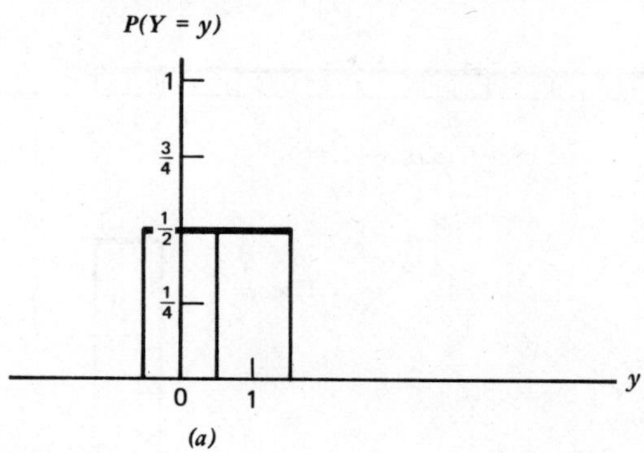

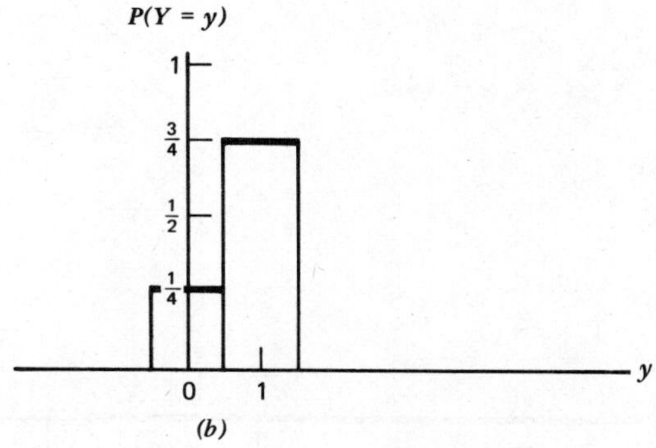

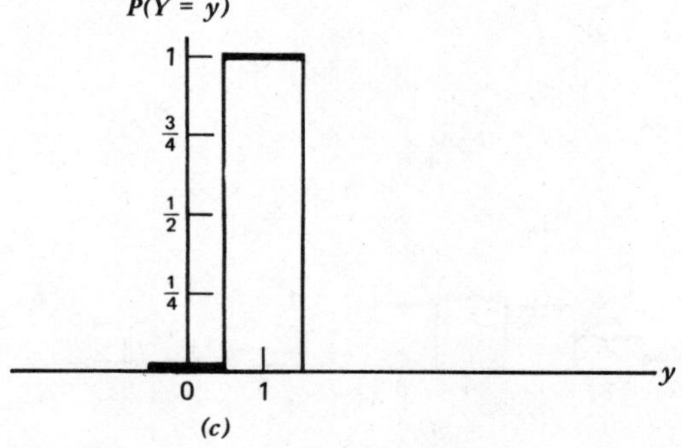

Figure 3.2.1 Probability measures for Y, Example 3.2.1. (*a*) Probability measure (i). (*b*) Probability measure (ii). (*c*) Probability measure (iii).

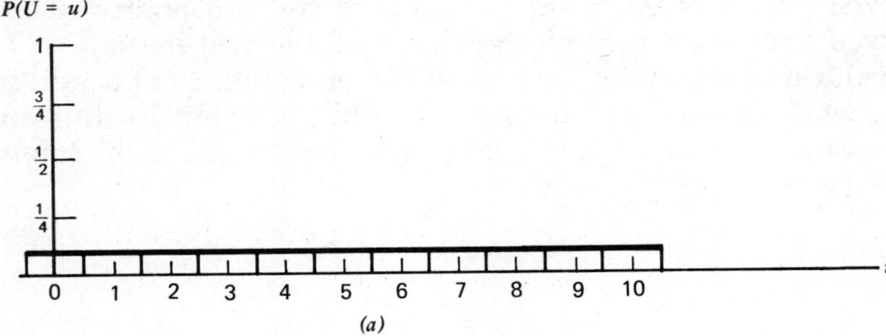

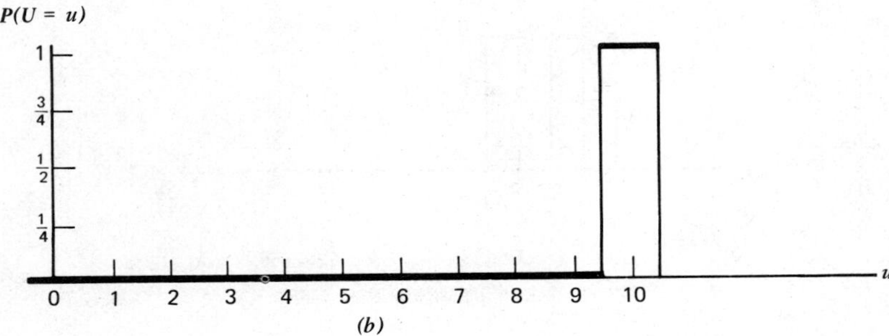

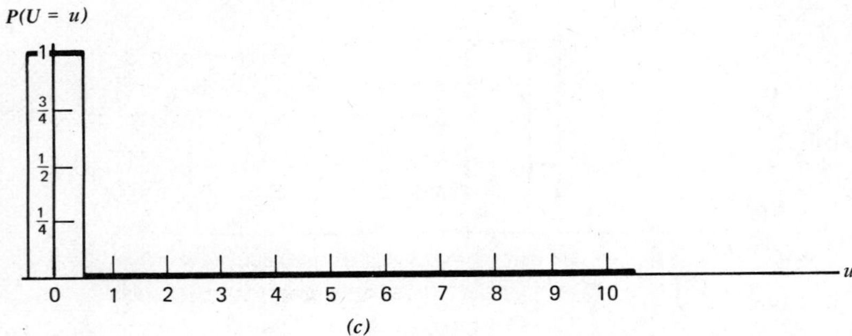

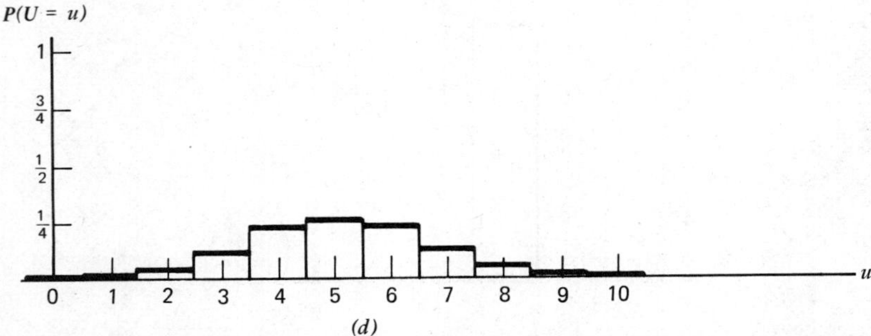

Figure 3.2.2 Probability measures for U, Example 3.2.2. (a) Probability measure 1. (b) Probability measure 2. (c) Probability measure 3. (d) Probability measure 4.

For example, consider the random variable U, the number of football games won in 1976 by the college team. The probability that U lies between 4 and 6, inclusive, is

$$P(4 \le U \le 6) = P(U = 4) + P(U = 5) + P(U = 6)$$

the sum of the areas of the three bars at $u = 4, 5, 6$. The numerical magnitude of this probability, of course, depends on which probability measure is used. By using measure 1, we see that $P(4 \le U \le 6) = 3/11$; with measures 2 or 3, $P(4 \le U \le 6) = 0$, while with measure 4, $P(4 \le U \le 6) = .656$. In each of these cases, the value of the probability is given by the sum of the areas of the bars at 4, 5, and 6.

Exercise 3.2

1. A die is a six-sided cube (dice is the plural of die). The six faces are distinguished by having on them one spot or two spots or any number up to six spots. Suppose a die is rolled once, and we let V be the number of spots uppermost when the die stops.
    (a) What is R_V?
    (b) If the die were weighted so that the face with one spot always ended uppermost, what would you use as a probability measure for V?
    (c) If the die were weighted so that the face with four spots always ended uppermost, what would you use as a probability measure for V?
    (d) If you felt each of the six faces were equally likely to be uppermost, what would you use as a probability measure for V?

2. (a) For each of the three measures described in question 1, what is the probability that V is no larger than 3?
    (b) For each of the three measures described in question 1, what is the probability that V equals a number between 2 and 5, inclusive?

3. A family is going to drive a camper from New York to California and return. Let T be the number of flat tires they will have on the trip. The probability measure for T is given below:

Observed Value t	0	1	2	3
Probability	.8	.1	.07	.03

    (a) What is the probability they have at least one flat tire on the trip?
    (b) What is the probability they have at most one flat tire on the trip?

4. Let V be the number of television sets sold per day by a small store. The probability measure for V is as follows:

Observed Value v	0	1	2 or more
Probability	.7	.25	.05

(a) What is the probability that this store will sell at least one set on a given day?

(b) What proportion of business days will they sell at most one set?

5. The number of traffic accidents per week on a 10-mile stretch of highway is a random variable W. The probability measure for W is

Observed Value w	0	1	2	3	4 or more
Probability	.497	.348	.122	.028	.005

(a) What is the probability of at least two accidents on this road in a given week?

(b) What is the probability of three or fewer accidents on this road in a given week?

(c) What proportion of all weeks will have no traffic accidents?

6. A new novel is being set in type. Let U be the number of misprints that occur per page. The probability measure for U is

Observed Value u	0	1	2 or more
Probability	.905	.091	.004

(a) What is the probability of at least one misprint on any specified page?

(b) What proportion of the pages would you expect to have no misprints?

7. As in question 1, suppose a six-sided die is rolled one time, except now two sides have one spot, two sides have two spots, and the remaining two sides have three spots. Let Y be the number of spots uppermost when the die stops rolling.
   (a) What is R_Y?
   (b) What would you suggest as a probability measure for Y?

8. Now assume that the die mentioned in question 7 has three faces with one spot, two faces with three spots and one face with two spots. Let X be the number of spots uppermost when it stops rolling.
   (a) What is R_X?
   (b) What would you use as a probability measure for X?

9. A dime has a one painted on the heads side and a two painted on the tails side; a quarter is painted the same way. Both coins are tossed onto a table. Let V be the sum of the two numbers on the faces turned up. The range for V, then, is $R_V = \{v: v = 2,3,4\}$. Two possible probability measures for V are

Observed Value v	2	3	4
Probability measure 1	1/3	1/3	1/3
Probability measure 2	1/4	1/2	1/4

Which of these two do you feel is the better measure and why?

3.3. CONTINUOUS RANDOM VARIABLES AND PROBABILITY MEASURES

We have seen the way in which probability measures are defined for discrete random variables. In this section we shall see how probability measures are defined for continuous random variables.

It will be recalled that a continuous random variable X is one whose range R_X contains all the numbers in an interval; the observed value for X is not restricted to being an integer or whole number. Any value between whole numbers could also occur for the observed value for X. For such continuous random variables there are so many different possible observed values that we are forced to assign zero as the probability of each individual point in R_X. If we did not do this, there are so many points that the sum of the individual probabilities would have to exceed 1, which would lead to contradictions with the relative frequency interpretation for probability. What, then, does a probability measure consist of for continuous random variables?

Instead of assigning probabilities to individual points in R_X, a probability measure for a continuous random variable must assign probabilities to subintervals of R_X, that is, to all possible intervals contained in R_X. We then may make statements and calculations that particular intervals will contain the observed value of X, rather than that particular points will equal the observed value of X. This situation is quite analogous to the idea of length in geometry. For example, if we picture a line one inch long, then the length (measure of that line) is 1, using an inch as a unit, a line two inches long has length (measure) 2, and a line 1/4 inch long has length (measure) 1/4. Any one of these lines is surely composed of a large number of individual points; but it will be recalled from geometry that each individual point has length 0. Then, a line one inch long has length 1, but the length of 1 is not the sum of the lengths of the individual points making it up. The measure "length" must take on the value 0 for each individual point or contradictions will arise. The same situation holds for the measure "probability" for a continuous random variable X, the measure for individual points must equal 0 or contradictions will be encountered. Instead of assigning probabilities to individual points in R_X, as is done if X is discrete, we must assign probabilities to intervals contained in R_X. The fact that all individual points, in geometry, have length 0 does not imply intervals have length 0; similarly a probability of zero for individual points in an interval does not imply the probability the interval contains the observed value must be 0. The definition of a probability measure for a continuous random variable is given below.

Definition 3.5. Assume X is a continuous random variable with range R_X. The probability measure for X is a rule which gives the probability that any possible interval will contain the observed value for X. These probability assignments are expressed as the area between the x-axis and a curve $f(x)$, called the density for X. For any continuous random variable X, it is necessary that

(i) $f(x) \geq 0$ for all x in R_X and $f(x) = 0$
      for all x not in R_X.

(ii)   The total area under $f(x)$, over R_X, is 1.

The two requirements quoted above for the density are necessary if we are to be consistent with the relative frequency interpretation for probabilities. We shall write $P(a < X < b)$ for the probability that the observed value of X will lie in the interval between a and b. The following examples illustrate the computations of probabilities for continuous random variables.

Example 3.3.1

Assume that an electric clock will stop sometime between 12 noon and 12:01 P.M. The clock has a sweep second hand; let V be the position of the second hand when the clock stops. Then

$$R_V = \{v: 0 < v \leq 12\}$$

since the second hand could be anywhere to the right of the 12 all the way around past 1, 2, 3, and so on up to 12 again on the clock face, and the exact place it stops could be any point around the circumference. A probability density for V is given in Figure 3.3.1. Notice that $f(v)$, the density for v, is positive for all v belonging to R_V and that the area under $f(v)$ is $12 \times (1/12) = 1$; thus we do satisfy the two requirements given in Definition 3.5. The probability that the second hand is between the 1 and the 2 when the clock stops is

$$P(1 < V < 2) = \text{area under } f(v), \text{ between 1 and 2}$$

$$= 1 \times \frac{1}{12} = \frac{1}{12}$$

The probability the second hand is between any other two consecutive numbers is also 1/12. The probability that it lies between the 4 and the 7 is

$$P(4 < V < 7) = \text{area under } f(v), \text{ between 4 and 7}$$

$$= 3 \times \frac{1}{12} = \frac{1}{4}$$

The probability that it lies between the 1 and the 11 is

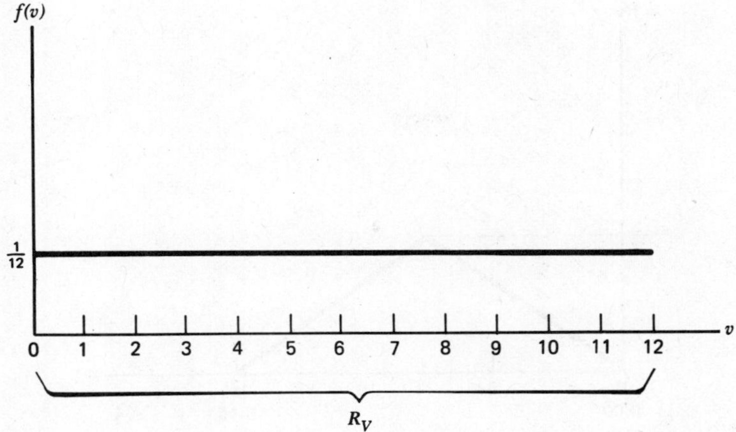

Figure 3.3.1   Density for V, Example 3.3.1.

$$P(1 < V < 11) = \text{area under } f(v), \text{ between 1 and 11}$$

$$= 10 \times \frac{1}{12} = \frac{5}{6}$$

Note also that the probability that the second hand is pointing exactly at the 2 is zero, since the area under $f(v)$, over the single point 2, is zero. The probability that the second hand is in a short interval centered at 2 (almost equal to 2) is not zero and would be the length of the interval times (1/12); as the length of the interval shrinks to zero so does the probability.

The shape of the density assumed for V in the above example was arbitrary; it is not necessary that the density be equal to the same constant for all values in R_V. Just as with discrete random variables the values of probabilities will depend on the assumed shape of the probability measure (the density function for continuous random variables).

Example 3.3.2

Assume that you drive to work five mornings a week and that you always take the same route, which passes through one intersection with a stoplight. The stoplight is red one minute, then green for two minutes, red for one, and so on, with the red cycle beginning on the hour every hour. You always arrive at the stoplight between 8:00 A.M. and 8:15 A.M. Let W be the time you arrive at the stoplight on your next trip, measured in minutes past 8:00 A.M. Then

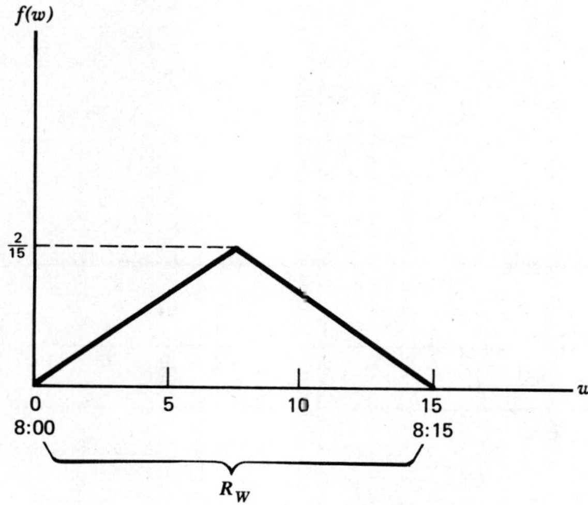

Figure 3.3.2 Density for W, Example 3.3.2.

$$R_W = \{w: 0 \leq w \leq 15\}$$

Let us assume the density for W pictured in Figure 3.3.2. Notice that $f(w) \geq 0$ for every w in R_W and that the total area under $f(w)$ equals

$$\frac{1}{2} \times 15 \times \frac{2}{15} = 1$$

so that this is a legitimate probability measure. The probability that you arrive at the light between 8:00 and 8:05 A.M., then, is

$$P(0 < W < 5) = \text{area under } f(w), \text{ between 0 and 5}$$

$$= \frac{1}{2}(5)\frac{4}{45} = \frac{2}{9}$$

since the height of $f(w)$ at 5 is 4/45 [= (2/3)(2/15)]. (See Figure 3.3.3). Because of the symmetry of the density we also have

$$P(10 < W < 15) = 2/9$$

and then the probability that you arrive at the light between 8:05 and 8:10 is

$$P(5 < W < 10) = 1 - 2/9 - 2/9 = 5/9$$

The light is red between 8:00 and 8:01, between 8:03 and 8:04, between 8:06 and 8:07, between 8:09 and 8:10, and between 8:12 and 8:13. If you arrive at the light in any of these intervals, you will have to stop. Thus, the probability you will have to stop at the light is

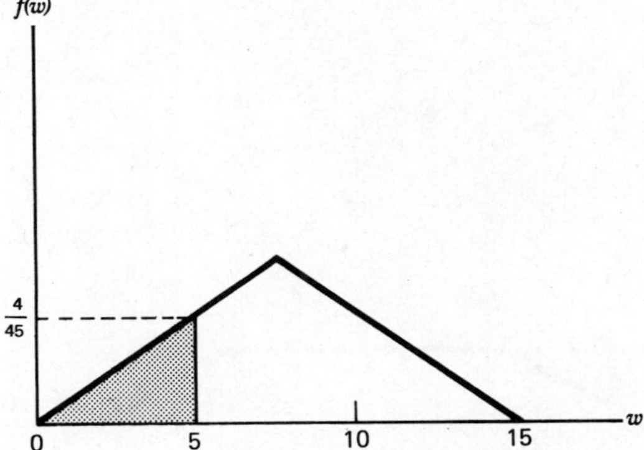

Figure 3.3.3 $P(0<W<5)$, Example 3.3.2.

$$P(0 < W < 1) + P(3 < W < 4) + P(6 < W < 7)$$

$$+ P(9 < W < 10) + P(12 < W < 13) = \frac{74}{225} = .329$$

These areas are shown in Figure 3.3.4.

When computing the probability that a continuous random variable lies in an interval, it does not matter whether the end points of the interval are included or not, since the probability for any particular value occurring is zero. For example, the probability that the time you arrive at the stoplight is between 8:05 and 8:10 A.M. is unchanged, regardless of whether the two instants, 8:05 and 8:10 A.M. are included in the interval. For discrete random variables, it does matter whether the end points of the interval are included, if those end points have a positive probability of occurring.

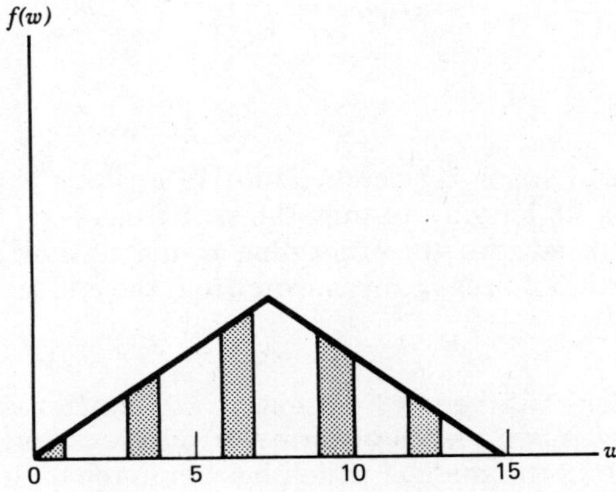

Figure 3.3.4 P(light is red), Example 3.3.2.

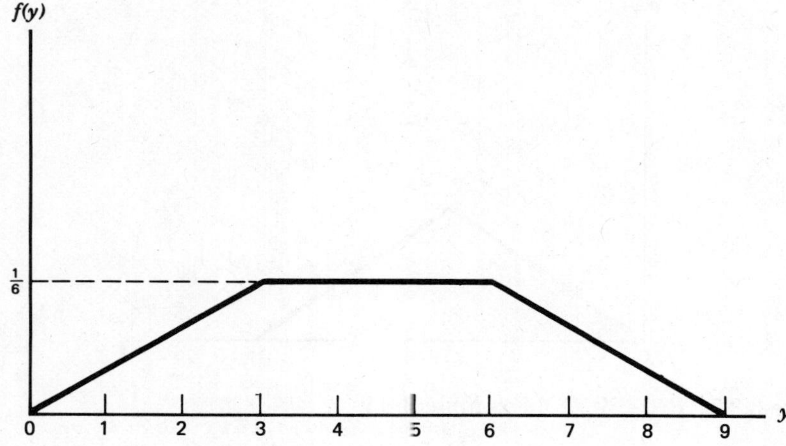

Figure 3.3.5 Density for Y, Example 3.3.3.

As we have seen, the evaluation of probabilities for continuous random variables depends on being able to evaluate areas underneath density functions. For complete generality, of course, density functions are not restricted to being rectangular or triangular, as in the preceding two examples. We assumed those particular shapes for the density to simplify the computation of areas. Calculus techniques are appropriate for evaluating areas beneath more arbitrarily shaped curves but, since it is not assumed that the reader has had preparation in calculus, we shall restrict our attention here to familiar shapes for which areas are easily calculated. Let us close our discussion of continuous random variables with one further example.

Example 3.3.3

A nine-inch piece of string is stretched until it breaks; the break can occur at any point along its length. Assume the string has been laid out flat and that one end is labeled zero, the other nine (using an inch as a unit). Let Y be the point at which it breaks, measuring from the end labeled zero. Then

$$R_Y = \{y: 0 < y < 9\}$$

We assume the density given in Figure 3.3.5 for Y. After the string breaks we will be left with two pieces of string of different lengths (unless the break occurs right in the middle, which has zero probability of occurring). Let us compute the probability that the longer piece is at least twice as long as the shorter, using the assumed density. We have to discover first where

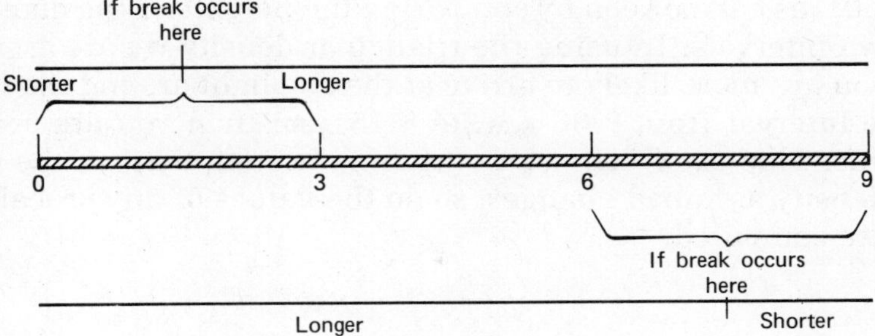

Figure 3.3.6 Location of break so that the longer piece is at least twice as long as the shorter.

the break must occur if the longer piece is to be at least twice the length of the shorter. By referring to Figure 3.3.6, we can see that if the string breaks between 0 and 3, the shorter piece is on the left and is no more than half the length of the longer; similarly if the break occurs between 6 and 9, then the longer piece is on the left and is at least twice as long as the shorter; thus

P (longer is at least twice as long as shorter)

$$= P(0 < Y < 3) + P(6 < Y < 9)$$

$$= \text{area under } f(y) \text{ between 0 and 3}$$

$$+ \text{ area under } f(y) \text{ between 6 and 9}$$

$$= \frac{1}{4} + \frac{1}{4} = \frac{1}{2}$$

For continuous random variables, the height of the density function controls the relative amount of probability for various intervals. In Example 3.3.1 the density was flat, parallel to the horizontal axis. Notice that with such a density the area above an interval of fixed length (say 1 unit) is the same no matter where the interval is located within the range of the random variable. Thus in using this density, the probability that an interval of fixed length contains the observed value depends only on the length of the interval and not on where the interval is located. Intervals of equal length are assigned equal probability.

For other densities, such as the triangular one in Example 3.3.2, all intervals of the same length are not assigned the same probability. The one-minute interval between 8:07 and 8:08 has much higher probability of occurring than does the interval between 8:00

and 8:01, as can be seen by comparing the heights of the density for the two intervals. In using the triangular density we are assuming that you are more likely to arrive at the stoplight around the middle of the interval from 8:00 A.M. to 8:15 A.M. than you are at either extreme. This is, of course, an arbitrary assumption; as the particular density assumed changes, so do the values of any probabilities that are computed.

Exercise 3.3

1. Assume X is a random variable with range

$$R_X = \{x: 0 < x < 3\}$$

and that its density is as pictured in Figure E.1.
What must the value of a be?

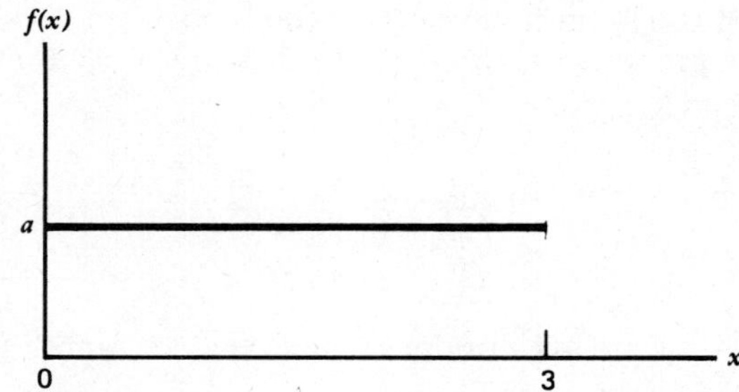

Figure E.1

2. Suppose Y is a random variable with range

$$R_Y = \{y: 0 < y < 4\}$$

and that its density is as shown in Figure E.2.
What must the value of b be?

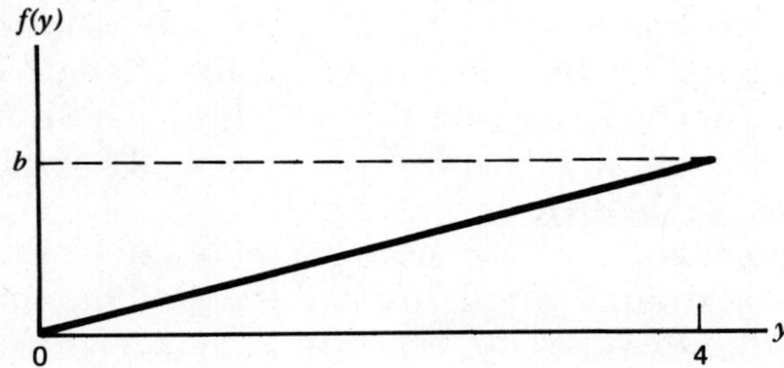

Figure E.2

3. For the random variable V defined in Example 3.3.1, using the density given there, evaluate
   (a) $P(2 \le V \le 4)$
   (b) $P(2 < V < 4)$
   (c) $P(V \le 6)$

4. Instead of the density used for V in Example 3.3.1, assume $f(v)$ is as pictured in Figure E.4.

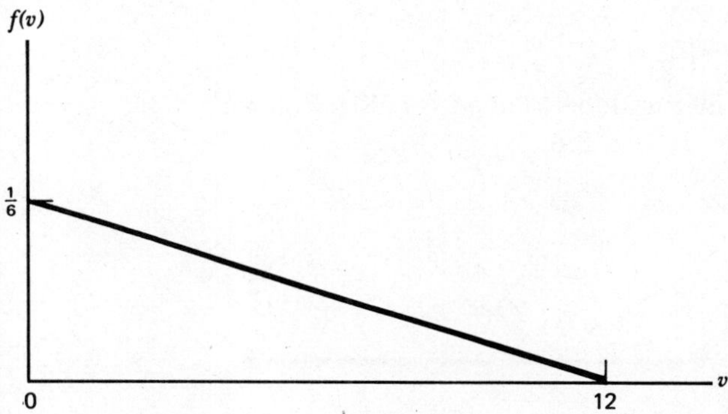

Figure E.4

Using this density evaluate the three probabilities computed in the example.

5. Using the random variable W and the density given in Example 3.3.2, compute
   (a) $P(7 < W < 8)$
   (b) $P(14 < W < 15)$
   (c) $P(W > 15)$

6. Instead of the density used for W in Example 3.3.2, assume $f(w)$ is as shown in Figure E.6.

Figure E.6

Use this density to evaluate
(a)　$P(0 < W < 5)$
(b)　$P(5 < W < 10)$
(c)　$P(\text{You have to stop at the light})$

7. Using the random variable Y and the density given in Example 3.3.3, compute
(a)　$P(0 < Y < 3)$
(b)　$P(3 < Y < 6)$
(c)　$P(6 < Y < 9)$
(d)　$P(2 < Y < 5)$

8. Instead of the density used for Y in Example 3.3.3, assume $f(y)$ is as shown in Figure E.8.

Figure E.8

Use this density to compute the probability that the larger piece is at least twice as long as the shorter.

9. Referring to Example 3.3.3, compute the probability that the longer piece is at least three times as long as the shorter using
(a)　The density given in Question 8.
(b)　The density given in the example.

10. What must be true of the density for Y in Example 3.3.3 if we want the probability to equal 1 that the longer piece is at least twice as long as the shorter?

3.4　SUMMARY

In this chapter we have introduced the important concept of a *random variable*, a numerical quantity whose value cannot be predicted in advance. Even though the particular value a random variable will equal can not be predicted in advance, it is assumed that the complete set of possible values it might equal can be specified; this set is called the *range* of the random variable.

The behavior of a random variable is described by its *probability measure*. The probability measure for a discrete random variable gives the probabilities of occurrence of individual values in the range of the random variable. The probability measure for a continuous random variable gives the probabilities that intervals in the range of the random variable will contain the observed value. Probability measures give numerical values for probability statements; the particular numerical values for probabilities depend on the probability measure used.

Exercise 3.4

1. A precinct has 1000 registered voters. Let U be the number of these that are Democrats.
   (a) Is U discrete or continuous?
   (b) What is R_U?

2. X is a continuous random variable. It is known that

   $$P(1 \leq X \leq 3) = \frac{1}{3} \qquad P(2 < X \leq 3) = \frac{1}{4}$$

   Evaluate $P(1 \leq X \leq 2)$.

3. A nurseryman plants 115 seeds in a flat. Let V be the number that germinate.
   (a) Is V discrete or continuous?
   (b) What is R_V?

4. A California farmer has 40 acres of land planted with Cabernet Sauvignon grapes. Let W be the weight of his crop from this land next year. Is W discrete or continuous?

5. A student takes four courses in one quarter. Let Y be the number of A's he gets that quarter and assume the probability measure for Y is

Value for Y	0	1	2	3	4
Probability	$1/a$	$1/a$	$1/a$	$1/a$	$1/a$

   (a) What is the value of a?
   (b) What is the probability he gets at least two A's?

6. The number of hours a television set will operate until it needs repair is a random variable T. Assume T has the density shown in Figure E.9.

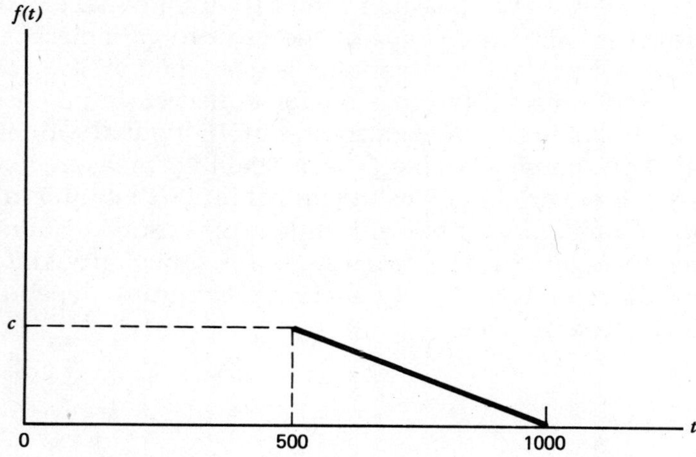

Figure E.9

    (a) What is the value of c?
    (b) What is the probability it operates at least 600 hours before repair?
    (c) What is the probability that the time it needs repair lies between 700 and 800 hours?

7. The number of trips a family makes to the beach per month is a random variable B with probability measure as presented below.

Value for B	0	1	2	3
Probability	1/10	5/10	3/10	1/10

    (a) What is the probability this family goes to the beach in a given month (at least once)?
    (b) What is the probability they go to the beach an odd number of times?

8. The number of movies a person attends per month is a random variable M with probability measure specified below.

Value for M	0	1	2	3	4
Probability	1/b	2/b	3/b	2/b	1/b

    (a) What is the value for b?
    (b) What is the probability this person attends at least two movies in a month?
    (c) What is the probability that this person attends at most two movies in a month?

9. A discrete random variable X has range $R_X = \{1, 2, 3, 4\}$. It is known that $P(1 \le X \le 2) = 3/4$, $P(2 \le X \le 3) = 7/24$
$P(3 \le X \le 4) = 1/4$. Give two different values for $P(X = 1)$.

10. Assume that a continuous random variable U has the density pictured in Figure E.10. Evaluate
(a) $P(0 \le U \le 1)$
(b) $P(1 \le U \le 2)$
(c) $P(2\ 1/2 \le U \le 3\ 1/2)$

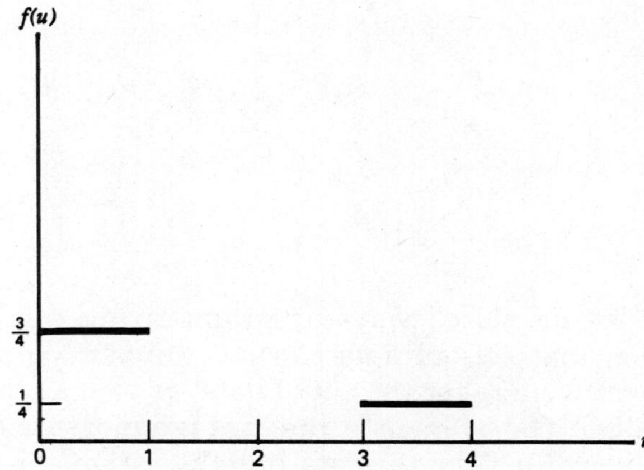

Figure E.10

CHAPTER 4
EXPECTED VALUES

In Chapter 2 we discussed ways of summarizing sets of data; it will be recalled that sets of data can be summarized by computing the mean or average for the set of data, as well as the variance for the set of data. The average of the data is a measure of the middle or a "typical" value from the data, and the variance is a measure of how variable the numbers are in the set of data. Similar quantities are of interest for random variables. Since the particular value that a random variable will equal is not known in advance, it is useful to have an idea of the value we may expect the random variable to equal, as well as an idea of how variable the actual value may be about that expected value. These topics are discussed in the following sections. Since it is not assumed that the reader has had preparation in calculus, we shall necessarily concentrate most of our effort on discrete random variables.

4.1 MEASURES OF LOCATION

Let us start by examining a particular random variable and see what we can say about the value we would expect it to equal. A die is a six-sided cube; the six faces all have different numbers of spots on them, from one spot to six spots. Suppose such a die is rolled one time, and we let X be the number of spots uppermost when the die stops; then the range of X is

$$R_X = \{x: x = 1,2,3, \ldots 6\}$$

Let us also suppose that the die appears symmetric and well balanced so that we would not expect any single face to appear more frequently than any other. Then a reasonable probability measure for X would be

$$P(X = x) = 1/6, \qquad x = 1, 2, \ldots, 6$$

That is, we are assuming the six different possible values for X are equally likely to occur. Let us ask then what value might we expect X to equal on a given roll?

This question may be answered by thinking of some large number of rolls we might make, say n, and then we ask what might be the average value of X over these rolls. Granted we do roll the die n times, we would expect, for example, to find $X = 1$ about 1/6 of the time and to find $X = 2$ about 1/6 of the time. In short, we would expect to observe each of the six faces about $n/6$ times in the n rolls. The average or mean value of X over the n rolls, then, would be the expected total value for X divided by n, the number of rolls made. Since we expect to find $X = 1$ $n/6$ times, $X = 2$ $n/6$ times, etc., the expected total value for X is

$$\frac{n}{6} \cdot 1 + \frac{n}{6} \cdot 2 + \frac{n}{6} \cdot 3 + \frac{n}{6} \cdot 4 + \frac{n}{6} \cdot 5 + \frac{n}{6} \cdot 6 =$$

$$n\left(\frac{1}{6} + \frac{2}{6} + \frac{3}{6} + \frac{4}{6} + \frac{5}{6} + \frac{6}{6}\right) = \frac{21n}{6}$$

using the above probability measure for X. The average value we would expect for X, then, is

$$\frac{21n}{6} \bigg/ n = \frac{21}{6} = 3.5$$

This is called the *expected value* for X. Note that this "expected value" was derived by multiplying the different values X might equal (each value from R_X) by the expected number of times it would occur in the n rolls ($n/6$ for each with this measure), adding these quantities together to get the expected total for X and then dividing by n, the number of observations to get the expected or average value for X. The expected number of times a given value, x, would occur was taken to be $nP(X = x)$, since $P(X = x)$ equals the relative proportion of the time that we should find $X = x$. Thus, for this case, the expected value is

$$\sum_{R_X} x n P(X = x)/n = \sum_{R_X} x P(X = x)$$

This is, in fact, the general definition for the expected value of a discrete random variable, as given below.

Definition 4.1   Assume X is a discrete random variable with range R_X and probability measure $P(X = x)$ for x in R_X. The *expected value* for X (written $E[X]$ or μ_X) is defined to be

$$E[X] = \sum_{R_X} x P(X = x)$$

This expected value for X is also called the *mean* or *average* value for X. Notice that $E[X]$ does not have to be an element of R_X, as in the case of the die discussed above; $E[X]$ will, however, always be between the smallest and largest values in R_X. The value for $E[X]$, of course, depends on the assumed probability measure for X; as the values of $P(X = x)$ change, so will $E[X]$. It is important to realize that $E[X]$ is a constant, given the probability measure. It describes one aspect of the probability measure for X.

Example 4.1.1

Let us return to the college football team that will play 10 intercollegiate games in 1976. U was defined as the number of games it will win. Four different probability measures for U were discussed in Example 3.2.2. Measures numbered 2 and 4 are reproduced below.

Elements of R_U

$u =$

	0	1	2	3	4	5	6	7	8	9	10
Measure 2	0	0	0	0	0	0	0	0	0	0	1
Measure 4	.001	.010	.044	.117	.205	.246	.205	.117	.044	.010	.001

Using measure 2, we compute $E[U]$ as the sum of the products of the values u times $P(U = u)$; this gives

$$E[U] = 0(0) + 1(0) + 2(0) + \ldots + 9(0) + 10(1) = 10$$

Using measure 2 the expected number of games the team will win is 10. This is not surprising since this measure assumes $P(U = 10) = 1$ and that all other values have probability zero of occurring. If we use measure 4, the expected value for U is

$$E[U] = 0(.001) + 1(.010) + 2(.044) + \ldots + 9(.010) + 10(.001)$$
$$= 5$$

a different value than we get using measure 2. It was pointed out in Example 3.2.2 that measure 4 is appropriate if we wanted to assume that the team had probability 1/2 of winning each game; under this assumption the expected value of U is 1/2 times the number of games played. This is a particular example of what is called a *binomial* random variable that we shall study in the next chapter.

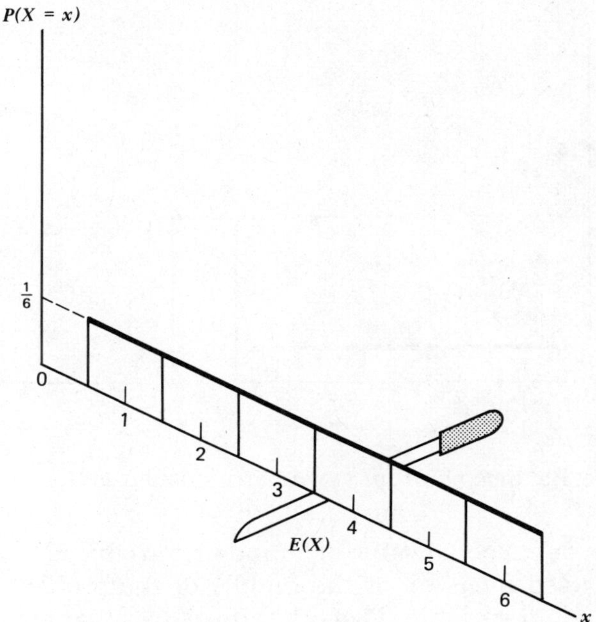

Figure 4.1.1 $E[X]$ measures the place at which the histogram balances.

The expected value of a random variable is a measure of the "middle" of the distribution of the random variable X. In fact, if we think of plotting the histogram of the probability measure of X, the mean or expected value of X measures the place at which the histogram would balance, if placed on a knife edge (see Figure 4.1.1). So long as the histogram is symmetric, as it is for the die with $P(X = x) = 1/6$ for each $x \epsilon R_X$, the point of balance or $E[X]$ will be right at the middle of the range (3.5 for the probability measure used for X). However, if the histogram of the probability measure is not symmetric, the point at which it balances ($E[X]$) will not be in the middle of R_X. For example, assume for the die that

$$P(X = 1) = P(X = 2) = P(X = 3) = P(X = 4) = P(X = 5) = 1/10$$

$$\text{and } P(X = 6) = 1/2. \text{ Then}$$

$$E[X] = 1 \, (1/10) + 2(1/10) + 3(1/10) + 4(1/10) + 5(1/10) + 6(1/2)$$
$$= 4.5$$

This histogram and its balance point are drawn in Figure 4.1.2. For the random variable U above, and probability measure 2, $E[U] = 10$, one of the extremes of R_U.

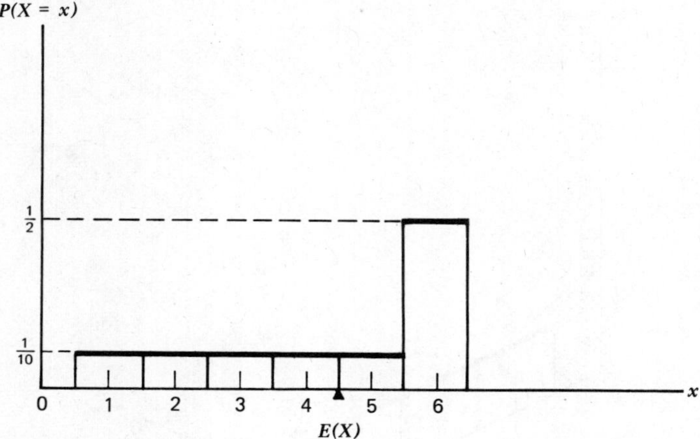

Figure 4.1.2   Balance point for asymmetric histogram.

The mean or expected value of a random variable is by far the most frequently used measure of the middle of the probability distribution of a random variable. Two other measures that are occasionally used are the *median* and the *mode*. Both the median and the mode were discussed earlier as descriptive statistics; it will be recalled that the median m_X was the middle number in the ranked set of data. Thus, half the numbers were smaller and half were larger than m_X. The mode was the most frequently occurring value. Their definitions are the same for random variables, translated into terms involving the probability measure, as given below.

> **Definition 4.2.**   The *median* for a discrete random variable X is defined to be the value m_X that splits the histogram into two equal areas. The *mode* is the value with the highest probability of occurrence.

Recalling that areas correspond to probabilities, the median then has the property of being in the middle of the distribution in a probability sense; the random variable would have probability .5 of being no larger than m_X and also probability .5 of being no smaller than m_X.

Example 4.1.2

Refer to Figure 4.1.1, the distribution of the number of spots uppermost on a die, assuming all faces are equally likely. The median for this random variable is $m_X = 3.5$, since a line through that point cuts the histogram into two areas each of 1/2; thus for that case $m_X = E[X]$. A little reflection will reveal that m_X and $E[X]$ will always be equal if the histogram is symmetric, since with symmetry the point at which the histogram will balance coincides with the point that cuts it into two equal areas. Now let us refer to the distribution for X given in Figure 4.1.2. In that case the histogram is not symmetric and $E[X] = 4.5$, as we saw earlier; it is also easy to see that the area to the left of 5.5 is 1/2 as is the area to the right of 5.5. Thus for this case $m_X = 5.5$ and m_X and $E[X]$ are unequal. Whenever the histogram is not symmetric, we will find that m_X and $E[X]$ are not equal.

Example 4.1.3

Let us use the same two probability measures for X to discuss the mode. For the distribution pictured in Figure 4.1.1 the probability is 1/6 for each value in R_X; thus there is no single number that has the highest probability of occurrence. The mode is not well defined in a case like this, which is one of the reasons it is very seldom used. For the probability measure pictured in Figure 4.1.2 the probability that $X = 6$ is larger than the probability that X equals any other value in R_X. Thus in that case the mode of X is 6. Notice that the mode is not necessarily equal to either m_X or $E[X]$, in cases in which it is well defined. Thus it is a different measure of the middle of the probability distribution than either m_X or $E[X]$.

Frequently it is of interest to compute the expected value of a function of a random variable, rather than just the expected value of the random variable itself. Before giving the general definition, we show in the following example a specific case where we are interested in the expected value of a function of a random variable.

Example 4.1.4

Suppose we gamble on the single die discussed above. Specifically, the die will be rolled one time and if an even number (2,4,6) occurs we win \$1; if an odd number (1,3,5) occurs we lose \$1. What would we expect to win (or lose) if we play this game? This question involves the expected value of a function of a random variable. With X as the number of spots uppermost when the die is rolled, as before, define

$$h(X) = \ 1 \quad \text{if} \quad X = 2,4,6$$
$$= -1 \quad \text{if} \quad X = 1,3,5$$

$h(X)$ then gives the amount we will win if we play this game and we want to find the expected value of $h(X)$. Reasoning as before, we will win $h(1)$ dollars with probability 1/6, $h(2)$ dollars with probability 1/6, etc., out to $h(6)$ dollars with probability 1/6. Thus the amount we would expect to win is given by the product of the values of $h(X)$ times the probabilities of occurrence of the corresponding values for X; that is,

$$E[h(X)] = h(1)P(X = 1) + h(2)P(X = 2) + \ldots + h(6)P(X = 6)$$

$$= (-1)(1/6) + (1)(1/6) + \ldots + (1)(1/6) = 0$$

We would expect to win \$0 if we played this game, that is, we would expect to come out even. A game is called *fair* if the expected amount won (or lost) is zero; thus this would be an example of a fair game.

Below we define the expected value of a function of a discrete variable.

> **Definition 4.3.**   Assume X is a discrete random variable with range R_X and probability measure $P(X = x)$. Let $h(X)$ be a function of X. Then the expected value of $h(X)$ is $E[h(X)] = \sum_{R_X} h(x)P(X = x)$.

This definition will prove useful in defining measures of variability in the next section, as well as in computing other quantities.

Example 4.1.5

Life insurance policies and the gain that the insurance company expects to make provide good examples of expected values of functions of random variables. There are, of course, many kinds of life insurance policies available. Perhaps the simplest (and cheapest) type is called term insurance. Under this type of policy, the insurance company will pay the beneficiaries of the insured person a fixed amount of money (the policy value) if the insured person dies within a specified term of years; the insurance company collects a fixed fee at the beginning of the term (the policy premium). If the insured person does not die during the specified term, it keeps the premium and pays the beneficiaries nothing; if the insured person does die during the specified term, the company keeps the premium and pays the beneficiaries the policy value. Let us examine a simple example of term insurance. Recent records of American lifetimes show that the proportion of Americans that died during their 20th year (between ages 19 and 20) was .00174. The proportion of Americans that had reached age 19 and did not die until sometime after their 20th birthday was

$$1 - .00174 = .99826$$

Assume that a life insurance company will sell a \$1000 one-year term life insurance policy to a person on his 19th birthday for \$5. Thus, if the person dies during the next year the company will pay his beneficiaries \$1000 (and the company is behind by \$995); if the person does not die by his 20th birthday, the company keeps the \$5 and pays nothing. Let Y be the number of years this person will live beyond his 19th birthday. Then the historical records give

$$P(0 < Y \leq 1) = .00174$$

$$P(Y > 1) \quad = .99826$$

The gain the insurance company will make on this policy is a function of Y. In fact, if we define

$$h(Y) = \text{insurance company's gain}$$

then

$$h(Y) = -995 \quad \text{if} \quad 0 < Y \le 1$$

$$h(Y) = 5 \quad \text{if} \quad Y > 1$$

Then the expected gain for the insurance company is

$$E[h(Y)] = (-995)P(0 < Y \le 1) + (5)P(Y > 1)$$

$$= (-995)(.00174) + (5)(.99826)$$

$$= \$3.26$$

The insurance company can expect to make \$3.26 for each such policy it sells. Out of this expected gain, it must pay all costs of selling and administering this policy.

The following theorem gives some results about expected values that are frequently useful in computations.

Theorem 4.1

Assume X is a discrete random variable with probability measure $P(X = x)$. $h(X)$ is a function of X.
(a)   If $h(X) = c$ (a constant for all values of X), then $E[h(X)] = c$, that is, $E[c] = c$.
(b)   If $h(X) = g(X) + k(X)$, then $E[h(X)] = E[g(X)] + E[k(X)]$.
(c)   If $h(X) = aX + b$, where a and b are constants, then $E[h(X)] = aE[X] + b$.

Each of these results follows in straightforward fashion from the definition of $E[h(X)]$. First, if we assume $h(X)$ equals a constant c for all values of X, then

$$E[h(X)] = \sum_{R_X} h(x)P(X = x)$$

$$= \sum_{R_X} cP(X = x)$$

$$= c \sum_{R_X} P(X = x)$$

$$= c$$

since $\sum_{R_X} P(X = x) = 1$. This establishes result (a) above.

If we assume $h(X)$ can be written as the sum of two functions, $g(X)$ and $k(X)$, that is, $h(X) = g(X) + k(X)$ for all X, then

$$
\begin{aligned}
E[h(X)] &= \sum_{R_X} h(x)P(X = x) \\
&= \sum_{R_X} [g(x) + k(x)]P(X = x) \\
&= \sum_{R_X} g(x)P(X = x) + \sum_{R_X} k(x)P(X = x) \\
&= E[g(X)] + E[k(X)]
\end{aligned}
$$

which establishes result (b) above. You are asked in question 11 of Exercise 4.1 to establish result (c) in the same manner. All of these results are also true for continuous random variables; however, to establish them for that case requires calculus.

The following example illustrates a usage of result (c).

Example 4.1.6

A newsboy sells papers to passersby on the same city corner every day. Each day he buys 100 papers, at 8¢ apiece, and then sells them on the corner at 15¢ apiece. Any papers left over are a loss to him. Let X be the number he sells on a given day; we assume $E[X] = 90$. Let Y be the amount of gain he makes on the given day. He paid $(.08)(100) = \$8$ for the papers and will receive 15¢ for each one he sells. Thus, his gain Y is

$$
Y = -8 + .15X
$$

using a dollar as a unit, where X is the number of papers he sells. For example, if he sells $X = 50$ papers (and thus must throw 50 away) he paid $8 and made $(.15)(50) = \$7.50$ for a gain of $-.50$ (a loss of 50¢). If he sells $X = 100$ papers, he makes $(.15)(100) = \$15.00$ for a gain of $7.00. His expected gain, using result (c), is

$$
E[Y] = E[-8 + .15X] = -8 + .15E[X]
$$

$$
= -8 + .15(90) = \$5.50
$$

Exercise 4.1

1. Compute the expected value for T, whose probability measure is given in question 3 of Exercise 3.2.

2. Compute $E[V]$ for each of the two probability measures given in question 9, Exercise 3.2.

3. A baseball player will be at bat four times in the same game. Let H be the number of hits he will make; the probability measure for H is

$h=$	0	1	2	3	4
Probability	.13	.35	.35	.15	.02

Compute $E[H]$, the expected number of hits he will make in the game.

4. Each of eight rats is placed individually in a straight corridor; at the end of the corridor each rat must turn left or right. Let U be the number that turn right and assume the following probability measure for U.

$u=$	0	1	2	3	4	5	6	7	8
Probability	.10	.27	.31	.21	.09	.02	0	0	0

Compute $E[U]$, the expected value for U.

5. Instead of the probability measure given for U in problem 4, assume the following measure and compute $E[U]$.

$u=$	0	1	2	3	4	5	6	7	8
Probability	0	.03	.11	.22	.28	.22	.11	.03	0

6. Each of four people with the same skin condition is given the same medicine. Let W be the number that are cured. Assuming the following probability measure for W, compute $E[W]$, the expected number that will be cured.

$w=$	0	1	2	3	4
Probability	0	.03	.15	.41	.41

7. An operator of a small sweetshop buys his goods from a large bakery. He buys three cheese pastries every day and, because of spoilage, must throw away any that are not sold the same day. Let X be the number of pastries he will sell on any given day and assume the probability measure for X is

$x=$	1	2	3
Probability	$\frac{1}{8}$	$\frac{1}{2}$	$\frac{3}{8}$

He pays 10¢ for each pastry and sells them for 20¢ apiece. Compute his expected daily gain from stocking this item.

8. A roulette wheel has 38 spots, of which 18 are red, 18 are black, and 2 are green. Assume you bet \$1 on the occurrence of a red number; thus, if a red number occurs, you end up \$1 ahead and if a red number

does not occur, you end up $1 behind. Assume the probability that a red number occurs is 18/38, and that a red number does not occur is 20/38. Define V to be the amount you win in playing this game and compute $E[V]$.

9. Assume the probability that an American will die in his 21st year is .00179. If an insurance company sells an American a $1000 one-year term life insurance policy on his 20th birthday for $5.25, what is the expected gain for the company?

10. A church holds a raffle to benefit their building fund. A $6000 automobile is the prize to be awarded the lucky ticket holder. A total of 12,000 tickets are sold at $1 apiece; you buy three tickets. Assuming that the probability you win the car is 3/12,000, what is your expected gain?

11. Show that result (c) in Theorem 4.1 is true.

12. The expected snowfall per year in a small Iowa town is 3 feet. Assuming each foot of snow equals 3 inches of rain, how much annual rainfall is expected from snow in this town?

4.2   MEASURES OF VARIABILITY

Random variables are quantities that vary, whose values are not known in advance. In the last section we considered the idea of an expected or average value for a random variable; it gives a measure of the middle of the distribution of the random variable; in a sense it is a typical value for the random variable.

To describe or summarize the probability distribution of a random variable more is required than just the expected value. Random variables with equal expected values can still have quite diverse probability distributions. Having studied some methods of descriptive statistics perhaps the natural second aspect of a probability distribution that we might try to describe is its variability. How variable are the possible observed values about $E[X] = \mu_X$, the mean of the distribution?

The most commonly used measure of the variability of a random variable is its standard deviation. This quantity is quite analogous to the standard deviation of a set of data that we studied in Section 2.3. The standard deviation and the variance of a random variable are defined below.

Definition 4.4.   Assume X is a random variable.
(a) The *variance* of X is $\sigma_X^2 = E[(X - \mu_X)^2]$.
(b) The *standard deviation* of X is $\sigma_X = \sqrt{\sigma_X^2}$.

Notice that the variance of X is again an expected or average value; rather than being the expected value of X itself, notice it is the expected value of a function of X, namely $(X - \mu_X)^2$. The difference $X - \mu_X$ measures the "distance" between X and its mean μ_X; since μ_X is a measure of the middle of the distribution of X we know that X will generally be smaller than μ_X some of the time and that it will be larger than μ_X some of the time. When $X < \mu_X$, then $X - \mu_X$ is negative, and when $X > \mu_X$, then $X - \mu_X$ is positive. By squaring $X - \mu_X$ we get rid of these differences in signs and $(X - \mu_X)^2$ is always nonnegative. Thus, $\sigma_X{}^2 = E[(X - \mu_X)^2]$ measures the average squared distance between the random variable and its mean; the larger $\sigma_X{}^2$ gets, the larger this average squared distance; and the smaller $\sigma_X{}^2$ gets, the smaller this average squared distance and the less variable X is about μ_X. Just as discussed in Section 2.3, $\sigma_X{}^2$ will be measured in the square of the units of X. To get a measure of variability that is in the same units as X itself, we take the positive square root of $\sigma_X{}^2$; this gives the standard deviation σ_X. It can be shown that σ_X is a natural scale factor for any random variable for measuring how much probability X has of equaling values in intervals centered at μ_X.

Example 4.2.1

Let us compute the variance and the standard deviation for the random variable X that has a probability 1/6 of equaling each of the integers 1, 2, 3, 4, 5, 6 that was discussed in the last section. We found there that $\mu_X = E[X] = 21/6$; thus, from the definition,

$$\sigma_X{}^2 = E[(X - \mu_X)^2]$$

$$= \left(1 - \frac{21}{6}\right)^2 \left(\frac{1}{6}\right) + \left(2 - \frac{21}{6}\right)^2 \left(\frac{1}{6}\right) + \ldots + \left(6 - \frac{21}{6}\right)^2 \left(\frac{1}{6}\right)$$

$$= \frac{630}{216} = 2.92$$

and

$$\sigma_X = \sqrt{2.92} = 1.71$$

As in the preceding example, if we compute $\sigma_X{}^2$ from the definition we must subtract μ_X from each element of R_X, square this difference, and then multiply each squared difference by its probability, and add them up. This can be quite tedious but, luckily, a computational formula is available that requires much less labor; this result is contained in the following theorem.

Theorem 4.2

For any random variable X

$$\sigma_X{}^2 = E[(X - \mu_X)^2]$$
$$= E[X^2] - \mu_X{}^2$$

This result can easily be seen to be true using Theorem 4.1 from the previous section. The function $(X - \mu_X)^2$ is identical with $X^2 - 2X\mu_X + \mu_X{}^2$; thus

$$E[(X - \mu_X)^2] = E[X^2 - 2X\mu_X + \mu_X{}^2]$$
$$= E[X^2] + E[-2X\mu_X] + E[\mu_X{}^2]$$

As we know, μ_X is a constant for a given probability measure; thus it factors outside the expectation and we have

$$E[X^2] + E[-2X\mu_X] + E[\mu_X{}^2] = E[X^2] - 2\mu_X E[X] + \mu_X{}^2$$
$$= E[X^2] - 2\mu_X{}^2 + \mu_X{}^2$$
$$= E[X^2] - \mu_X{}^2$$

since $E[X] = \mu_X$; this establishes the result. In general, it is easier to compute $E[X^2]$ and $\mu_X{}^2$ and then to form their difference, than it is to compute $E[(X - \mu_X)^2]$ directly.

Let us apply this computational formula to the random variable X discussed in Example 4.2.1 above and see that we do get the same numerical value for $\sigma_X{}^2$. Thus, for the probability measure given there,

$$E[X^2] = 1^2 \left(\frac{1}{6}\right) + 2^2 \left(\frac{1}{6}\right) + \ldots + 6^2 \left(\frac{1}{6}\right)$$

$$= \frac{91}{6}$$

then

$$E[X^2] - \mu_X{}^2 = \frac{91}{6} - \left(\frac{21}{6}\right)^2$$

$$= \frac{630}{216}$$

the same result we computed earlier as, of course, it must be.

Example 4.2.2

In Example 4.1.1 we considered two different probability measures for U, the number of intercollegiate football games a team will win in 1976, out of 10 games played. These two probability measures are presented again below.

u

	0	1	2	3	4	5	6	7	8	9	10
Measure 2	0	0	0	0	0	0	0	0	0	0	1
Measure 4	.001	.010	.044	.117	.205	.246	.205	.117	.044	.010	.001

Using measure 2 we found earlier that $\mu_U = 10$. We need to find $E[U^2]$ as well to compute $\sigma_U{}^2$; thus

$$E[U^2] = 0^2 \, (0) + 1^2 \, (0) + 2^2 \, (0) + \ldots + 9^2 \, (0) + 10^2 \, (1) = 100$$

and we have

$$\sigma_U{}^2 = 100 - 10^2 = 0$$

The fact that $\sigma_U{}^2 = 0$ with this probability measure says that U never varies about μ_U, that it is always equal to μ_U. The only time that a random variable has zero variance is when it is equal to a specific value with probability 1; that specific value is necessarily its expected value. Using measure 4, we found earlier that $\mu_U = 5$; again, to evaluate $\sigma_U{}^2$ with this measure the only additional quantity we need to evaluate is $E[U^2]$; we find

$$E[U^2] = 0^2 \, (.001) + 1^2 \, (.010) + 2^2 \, (.044) + \ldots + 9^2 \, (.010) + 10^2 \, (.001)$$

$$= 27.5$$

and

$$\sigma_U{}^2 = E[U^2] - \mu_U{}^2 = 27.5 - 25 = 2.5$$

The standard deviation is $\sigma_U = \sqrt{2.5} = 1.58$. With this measure the variance and the standard deviation are not zero, because there is more than one value in R_U with a positive probability.

Other measures of variability, in addition to the standard deviation and variance, can also be defined. We shall examine one further measure of the variability of discrete random variables. It will be recalled that $X - \mu_X$ measures the distance between the random variable and its mean value, where the sign of the difference is maintained. By squaring $(X - \mu_X)$, we can get rid of differences in sign and then average these quantities to get the variance, which is a squared measure. Taking the square root of this average gives us the standard deviation, measured in the same units as the original X. Thus, to get rid of the plus and minus signs of $(X - \mu_X)$ we squared, then averaged, and took the square root of this average to get back to the original units. The absolute deviation is another measure of variability that uses the same units as the original random variable. Instead of squaring $X - \mu_X$ to avoid cancellation of negative and positive differences, it averages the absolute values (magnitudes ignoring signs) of these differences. It is defined below.

> **Definition 4.5.** Assume X is a random variable. The *absolute deviation* of X is
>
> $$a_X = E[|X - \mu_X|]$$

Example 4.2.3

Let us compute the absolute deviations for the random variables X and U considered in the two examples above. For the random variable X in Example 4.2.1 we have

$$a_X = |1 - 3.5| \left(\frac{1}{6}\right) + |2 - 3.5| \left(\frac{1}{6}\right) + \ldots + |6 - 3.5| \left(\frac{1}{6}\right)$$

$$= \frac{9}{6}$$

For this random variable we found $\sigma_X = 1.71 > a_X$. By using measure 2 for the random variable U of Example 4.2.2, we find

$$a_U = |0 - 10| \, (0) + |1 - 10| \, (0) + \ldots + |9 - 10| \, (0) + |10 - 10| \, (1)$$

$$= 0$$

By using the same measure for U, we found earlier that $\sigma_U = 0 = a_U$. Using measure 4 for this same random variable, we have

$$a_U = |0 - 5| \, (.001) + |1 - 5| \, (.010) + \ldots + |9 - 5| \, (.010) + |10 - 5| \, (.001)$$

$$= 1.23$$

With this same measure we had

$$\sigma_U = 1.58 > a_U$$

Notice that in each of the three cases computed above we found the absolute deviation no larger than the standard deviation. This can be shown always to be true. For any random variable X it is necessarily the case that $a_X \leq \sigma_X$. The absolute deviation measures variability in the same terms as does σ_X; the larger that a_X gets, the more variable X is about μ_X; $a_X = 0$ if and only if $P(X = \mu_X) = 1$. Because the absolute deviation involves the absolute value signs, it is more unwieldy than σ_X from an analytic point of view, and σ_X is the more commonly used measure of variability.

We saw in the last section that the expected value of a linear function of a random variable is the same linear function of its expected value, that is, that $E[aX + b] = aE[X] + b$. The following theorem gives the relationship of the variance (and standard deviation) of a linear function to these same quantities for the original random variable.

Theorem 4.3

Assume X is a random variable and let $Y = aX + b$, where a and b are any constants. Then

$$\sigma_Y^2 = a^2\, \sigma_X^2$$
$$\sigma_Y = \sigma_X |a|$$

We saw in the last section that

$$\mu_Y = a\mu_X + b$$

if $Y = aX + b$. Then, by definition,

$$\sigma_Y^2 = E[(Y - \mu_Y)^2]$$
$$= E[(aX + b - a\mu_X - b)^2]$$
$$= E[(aX - a\mu_X)^2]$$
$$= a^2 E[(X - \mu_X)^2]$$
$$= a^2 \sigma_X^2$$

Taking square roots of both sides of this equation gives the relationship between the standard deviations of Y and X.

Example 4.2.4

Consider the random variable X discussed in Example 4.1.6, the number of newspapers a newsboy sells each day. The gain he makes per day is

$$Y = .15X - 8 = aX + b$$

Assuming $\mu_X = 90$ we found $\mu_Y = \$5.50$. If we assume the variance of X is

$$\sigma_X^2 = 9$$

then the variance of his gain is

$$\sigma_Y^2 = a^2 \sigma_X^2 = (.15)^2\, (9) = .2025$$

and the standard deviation of his gain is

$$\sigma_Y = |a|\, \sigma_X = (.15)\, (3) = .45$$

Exercise 4.2

1. Compute the variance and standard deviation for H, whose probability measure is given in problem 3, Exercise 4.1.

2. Compute the variance and standard deviation for U, whose probability measure is given in problem 4, Exercise 4.1.

3. Assume Y has range $R_Y = \{-h, h\}$, where h is a positive constant, and $P(Y = -h) = P(Y = h) = 1/2$. Compute σ_Y^2, σ_Y, and a_Y.

4. Assume W has range $R_W = \{-h, 0, h\}$ and probability measure

$w =$	$-h$	0	h
Probability	p	$1 - 2p$	p

   where h is a positive constant and $0 \le p \le 1/2$.
   (a) Compute a_W and σ_W.
   (b) For what values of p are a_W and σ_W equal?

5. Compute the variance of X, whose probability measure is given in problem 7, Exercise 4.1. Compute the variance of the gain.

6. Compute σ_V for the random variable defined in problem 8, Exercise 4.1.

7. Compute the standard deviation of the insurance company's gain for the policy described in problem 9, Exercise 4.1.

8. Assume the random variable X has range $R_X = \{-1, 1\}$. Can you find a probability measure for X such that $\sigma_X = 1$? Can you find a probability measure such that $\sigma_X > 1$?

4.3 SUMMARY MEASURES AND CONTINUOUS RANDOM VARIABLES

The quantities just discussed in the preceding sections—for example, the mean, median, standard deviation, and absolute deviation—are all used to summarize aspects of the distribution (or probability measures) for random variables. Thus they are frequently called summary measures. Our discussions of these quantities has been up to now concerned only with discrete random variables. For discrete random variables these summary measures are straightforward to evaluate, although possibly tedious, by using methods of algebra. The same quantities are useful in describing or summarizing the distributions of continuous random variables. Their definitions for continuous random variables generally involve the use of calculus, which is not assumed as a prerequisite for this book, so

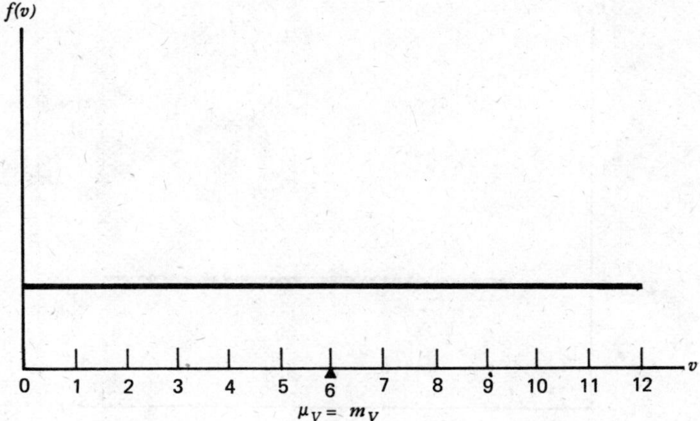

Figure 4.3.1   Mean and median for V, Example 4.3.1.

that our discussions of these quantities for continuous random variables will necessarily be brief.

It will be recalled that probabilities for continuous random variables are given by areas underneath curves, called density functions. Thus, the probability measure for a continuous random variable is specified by its density function. The *expected value* of a continuous random variable locates the middle of its distribution in a center-of-gravity sense, as it also does for a discrete random variable; thus it locates the point at which the density function would balance, if placed on a knife-edge. The *median*, on the other hand, locates the point that splits the area under the density into two equal parts (both of value 1/2). If the density is symmetric, of course, then the point at which the density balances and the point that cuts it into two equal areas will coincide and the mean and median are equal.

Example 4.3.1

In Example 3.3.1, V was defined as the position of the sweep second hand on a clock if it stops between 12 noon and 12:01 P.M. The density assumed for V is given in Figure 4.3.1. The mean μ_V for this random variable is the point at which the density will balance and the median m_V is the point that cuts the density into two equal areas. Because this density is symmetric, these two quantities are both equal to 6, as shown. Thus, on the average we would expect the second hand to be pointing at the 6, the midpoint around the circumference.

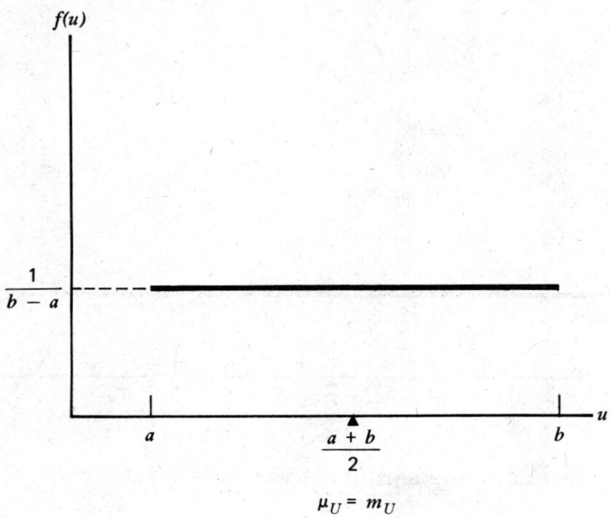

Figure 4.3.2 The mean and median for a general uniform random variable U.

The random variable V discussed above is a special case of a *uniform random variable*. This name is used because the density function is uniform, or constant, over its range. A general uniform random variable U has range

$$R_U = \{u : a \le u \le b\}$$

that is, it can equal any value between two arbitrary constants a and b (for V above, $a = 0, b = 12$). The density function for U is constant over the interval R_U and, thus, must have height $1/(b - a)$, so that the total area under the density will be 1. This constant density is symmetric, no matter what the values for a and b; thus the mean, μ_U, and the median, m_U, will be equal; their common value is the midpoint of the interval from a to b, that is,

$$\mu_U = m_J = \frac{a + b}{2}$$

Figure 4.3.2 pictures a general uniform density, as well as its mean and median.

Example 4.3.2

Let T be the time required for you to drive from your residence to your job in the morning. Assume

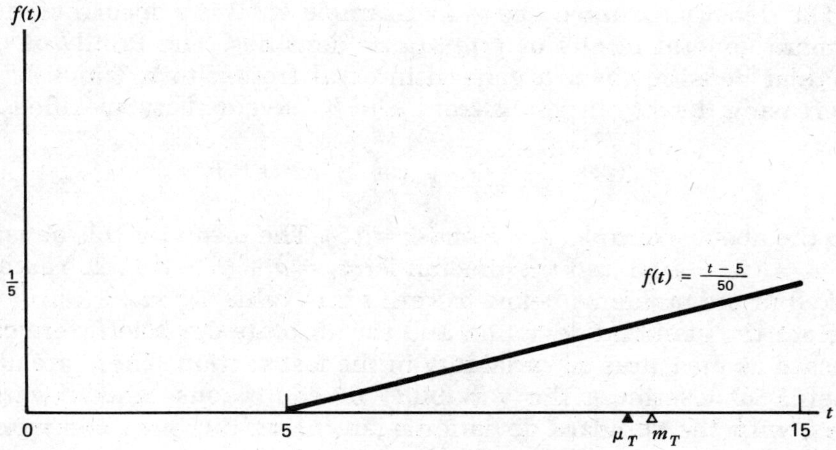

Figure 4.3.3 Density, mean, and median for T, Example 4.3.2.

$$R_T = \{t: 5 \le t \le 15\}$$

(thus it always takes you at least 5 minutes and never more than 15 minutes to get there) and assume the density function for T is as shown in Figure 4.3.3. (This is one type of *triangular* density.) Notice that in using a density of this shape we are assuming that it is more likely for you to take between 14 and 15 minutes to get there, than it is for you to take between 5 and 6 minutes to get there, because the area between 14 and 15 is bigger than the area between 5 and 6. Assuming this density, the median time, m_T, for you to drive to work is given by the value that splits the density into two equal areas. Thus, in particular, the area to the left of m_T must be 1/2. The geometric figure to the left of m_T is necessarily a triangle, with this density. The area of a triangle is equal to 1/2 times the product of the length of the base and the height. Thus, since the length of the base is $m_T - 5$ and the height (at m_T) is $(m_T - 5)/50$, the value of m_T is defined by

$$\frac{(m_T - 5)^2}{100} = \frac{1}{2}$$

the solutions to this equation are

$$m_T = 5 \pm \sqrt{50}$$

We obviously must have $m_T > 5$; thus the value of m_T is $5 + \sqrt{50} = 12.07$, and your median driving time using this density is 12.07 minutes. Half of the time you will get there in 12.07 minutes or less and half of the time you will not. The place at which this triangle will balance can be shown to be $11\frac{2}{3}$ (using calculus techniques); thus, $\mu_T = 11\frac{2}{3}$ and your average time required to get there is then $11\frac{2}{3}$ minutes using this density. Since the density is not symmetric, $\mu_T \ne m_T$, and neither μ_T nor m_T is equal to the midpoint of R_T.

The density assumed above in Example 4.3.2 is a special case of a more general family of *triangular* densities. The family of triangular densities over a general interval from a to b, which have increasing density in going from a to b, have density specified by

$$f(t) = \frac{2(t-a)}{(b-a)^2} \qquad \text{for} \qquad a \le t \le b$$

(in the above example, $a = 5$ and $b = 15$). The mean for this density is $\mu_T = (2b + a)/3$ and the median is $m_T = a + (b-a)/\sqrt{2}$. You are asked in the problems below to verify this value for m_T.

Both the standard deviation and the absolute deviation were discussed as measures of variability in the last section. These are also useful in describing the variability of continuous random variables, with the standard deviation again being the more commonly used of the two measures. Their evaluation for specific densities requires calculus; for the uniform random variable U with density

$$f(u) = \frac{1}{b-a}$$

it can be shown that $\sigma_U = (b-a)/2\sqrt{3}$ and that the absolute deviation is $a_U = (b-a)/4$. For the triangular random variable T with range $R_T = \{t : a \le t \le b\}$ and density

$$f(t) = \frac{2(t-a)}{(b-a)^2}$$

it can be shown (using calculus) that $\sigma_T = (b-a)/3\sqrt{2}$; the absolute deviation, a_T, for this triangular density is a cumbersome function of a and b and will not be given.

Example 4.3.3

The random variable V in Example 4.3.1 was uniform on the interval from $a = 0$ to $b = 12$; thus its standard deviation is

$$\sigma_V = \frac{b-a}{2\sqrt{3}} = \frac{12}{2\sqrt{3}} = 2\sqrt{3}$$

and its absolute deviation is

$$a_V = \frac{b-a}{4} = \frac{12}{4} = 3$$

The random variable T in Example 4.3.2 was triangular with $a = 5$, $b = 15$; thus its standard deviation is

$$\sigma_T = \frac{b-a}{3\sqrt{2}} = \frac{10}{3\sqrt{2}} = \frac{5\sqrt{2}}{3}$$

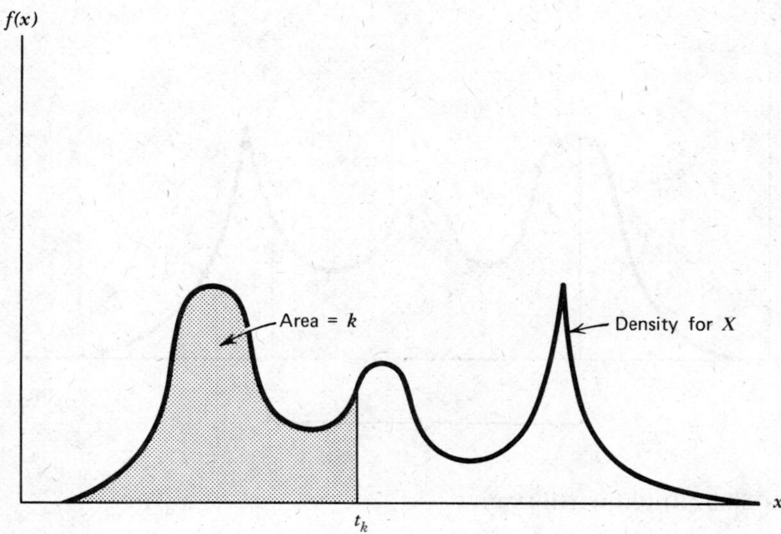

Figure 4.3.4   (100k)th percentile for a random variable X.

   The probability distributions for continuous random variables may also be described by the percentiles of the distribution. These quantities were introduced in Chapter 2 in terms of summarizing sets of data. The percentiles for a continuous random variable are defined below.

   Definition 4.6.   Assume X is a continuous random variable with density $f(x)$. The value t_k is defined to be the (100k)th *percentile* for X if the area under $f(x)$ to the left of t_k is k, where $0 < k < 1$; that is, the (100k)th percentile has the property that

$$P(X \leq t_k) = k$$

Figure 4.3.4 gives a graphical representation of t_k for a general density function. As t_k shifts, so does the area to its left. The way in which the area changes as t_k moves depends on the particular density function assumed for X. We shall see below what the value for t_k is for our uniform and triangular families of density functions.

   With our notation the subscript for t_k gives the area to the left of t_k. Thus, for example, the area to the left of $t_{.40}$ is .4 and the area to the left of $t_{.25}$ is .25. The commonly used values for 100k are the integers between 0 and 100 (which is the reason for the name). It will be recalled that the median m_X, which was introduced as a measure of the middle of the distribution, splits the area under the density into two equal parts; in particular, then, since the area to

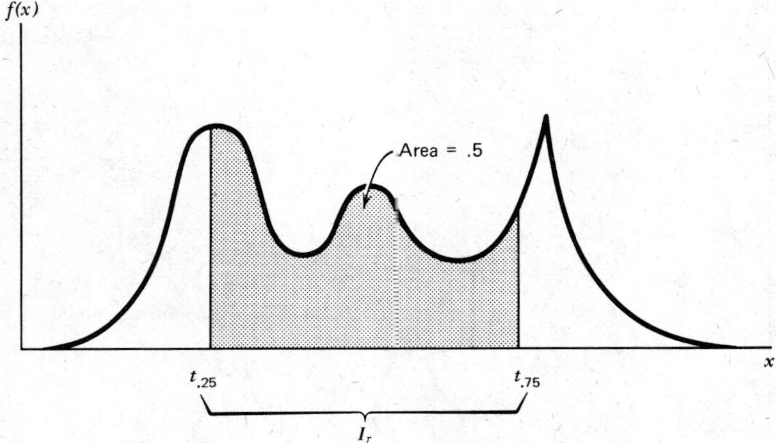

Figure 4.3.5 Interquartile range.

the left of m_X must be .5, the median is identical with the 50th percentile:

$$m_X = t_{.50}$$

Percentiles can be used to measure any desired aspects of the distribution of a random variable. The most frequently used measure of variability that is based on the percentiles is the *interquartile range* I_r (Figure 4.3.5). As with summarizations of data, discussed under descriptive statistics in Chapter 2, it is defined to be the difference between the 75th and the 25th percentiles:

$$I_r = t_{.75} - t_{.25}$$

I_r, then, gives the length of an interval, centered at the median m_X, which includes 50% of the total probability. Thus, as I_r gets bigger, one must take a larger interval about m_X to include 50% of the probability; as I_r gets smaller a shorter interval will include 50% of the probability. It is in this fairly direct way that I_r is a measure of variability.

Let us now evaluate the percentiles for the two general families of densities we have seen in this section, the uniform and the triangular. Suppose we are given a uniform random variable U with

range

$$R_U = \{u : a \leq u \leq b\}$$

and density

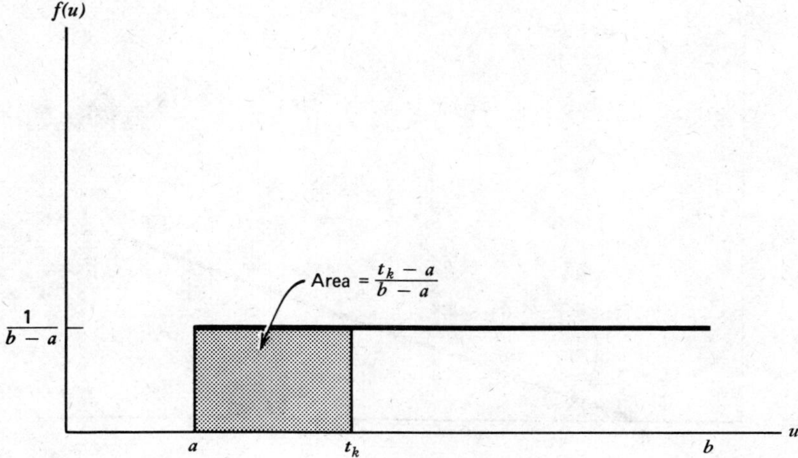

Figure 4.3.6   Percentiles for a uniform random variable.

$$f(u) = \frac{1}{b-a}$$

This density is constant over R_U (see Figure 4.3.6) and thus the area to the left of t_k will be a rectangle. The length of the base of the rectangle is $t_k - a$ and its height is $1/(b-a)$; thus the area to the left of t_k is $(t_k - a)/(b-a)$. If this is to equal k, then we must have

$$t_k = a + k(b-a) = a(1-k) + bk$$

The interquartile range for a uniform random variable is

$$I_r = t_{.75} - t_{.25}$$

$$= (.25a + .75b) - (.75a + .25b)$$

$$= .5\,(b-a)$$

A triangular random variable T has range $R_T = \{t: a \le t \le b\}$ and density

$$f(t) = \frac{2(t-a)}{(b-a)^2}$$

The area to the left of any point t_k is a triangle, using this density, whose base has length $t_k - a$ and whose height is equal to $2(t_k - a)/(b-a)^2$ (Figure 4.3.7). The area to the left of t_k, then, is

$$\frac{(t_k - a)^2}{(b-a)^2}$$

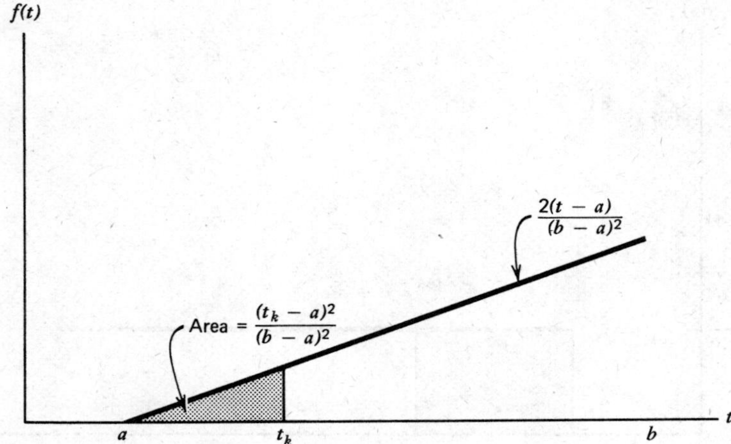

Figure 4.3.7   Percentiles for a triangular random variable.

If this area is to equal k, we must have

$$t_k = a + (b - a)\sqrt{k}$$
$$= a(1 - \sqrt{k}) + b\sqrt{k}$$

The interquartile range for a triangular random variable is

$$I_r = t_{.75} - t_{.25}$$
$$= [a(1 - \sqrt{.75}) + b\sqrt{.75}] - [a(1 - \sqrt{.25}) + b\sqrt{.25}]$$
$$= (\sqrt{.75} - \sqrt{.25})(b - a)$$
$$= .37(b - a)$$

Note that this is smaller than I_r for a uniform random variable over the same interval; thus, if we assume a triangular density for a random variable, instead of a uniform density on the same interval, we are assuming a density with less variability.

Example 4.3.4

Let us compute the percentiles and interquartile ranges for the two random variables V and T considered in Examples 4.3.1 and 4.3.2, respectively. V was uniform with $a = 0$ and $b = 12$ and, thus, its kth percentile is

$$t_k = 0(1 - k) + 12k = 12k$$

its interquartile range is

$$I_r = (.5)(12 - 0) = 6$$

The random variable T was triangular with $a = 5$ and $b = 15$. Its kth percentile, then, is

$$t_k = 5(1 - \sqrt{k}) + 15\sqrt{k}$$
$$= 5 + 10\sqrt{k}$$

The interquartile range for T is

$$I_r = .37(15 - 5) = 3.7$$

Exercise 4.3

1. Let Y be the height a two-year-old boy will attain as an adult. By examining the heights of his close relatives it is assumed that $R_Y = \{y\colon 69 \le y \le 72\}$, where the units of y are inches; it is also assumed that Y is uniform and thus its density is $f(y) = 1/3$.
   (a) What is the expected adult height of this boy?
   (b) What is the 90th percentile for this distribution?
   (c) Compute I_r.

2. Evaluate σ_Y and a_Y for the distribution given in question 1.

3. Assume that T is a triangular random variable on the interval from a to b. Verify that its median value is

$$m_T = a + \frac{b - a}{\sqrt{2}}$$

4. Assume V is a random variable with the density given in problem 4, Exercise 3.3. Find m_V.

5. The random variable V in problem 4 comes from a different trangular family than the one discussed in this section. The general case for this family is described by $R_V = \{v\colon a \le v \le b\}$ with density

$$f(v) = \frac{2(b - v)}{(b - a)^2} \quad a \le v \le b$$

   (a) What is μ_V for this general case? (Compare with the triangular density in the text).
   (b) What should σ_V be?
   (c) Find the formula for t_k for this density.

6. Example 3.3.3 discussed the point Y at which a 9-inch string will break when stretched. Using the density given in that example, what is the expected point at which the break will occur?

7. Assume that the number of miles, M, until a car will require an engine overhaul is given in Figure E.7, where $a = 50{,}000$, $b = 70{,}000$.
   (a) What is the value for c?
   (b) What is $E[M]$?

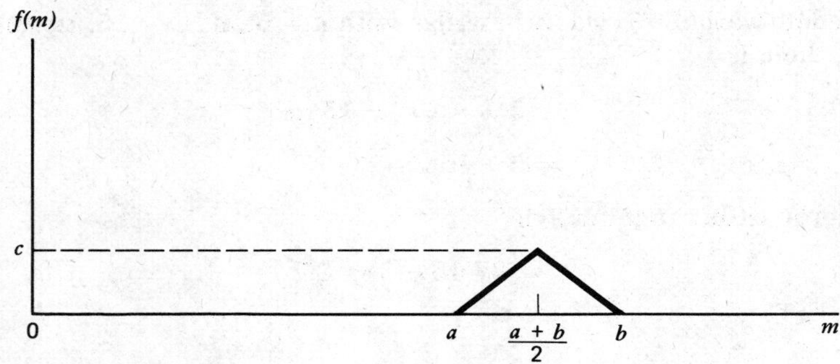

Figure E.7

8. Evaluate $t_{.25}$, $t_{.50}$, $t_{.75}$ for the distribution in question 7. What is the value for I_r for this distribution?

9. Doug is 20 years old and now weighs 170 pounds. Let W be his weight 10 years from now and assume $R_W = \{w: 150 \leq w \leq 190\}$; the density for W is shown in Figure E.9.
   (a) What is $E[W]$?
   (b) What is m_W?
   (c) Find I_r.

10. The *interdecile range* is defined to be the difference $t_{.90} - t_{.10}$. Evaluate this quantity
   (a) For the general uniform random variable on the interval from a to b.
   (b) For the triangular random variable on the interval from a to b.

4.4 SUMMARY

Chapter 3 was devoted to a discussion of the concept of a random variable, as well as of probability measures and their uses in describing the behavior of random variables. In this chapter we have studied the concept of expectation. Given a random variable X and its probability measure, the expected value of the random variable is the average of the values in R_X with respect to the given probability measure. This can be interpreted as the average value of the random variable X over a large number of repeated observations. It locates the balance point of the probability measure. The expected value of $h(X)$, a function of X, is the average value of $h(X)$, again with respect to the probability measure of X.

The expected value of X, μ_X, is the most commonly used measure

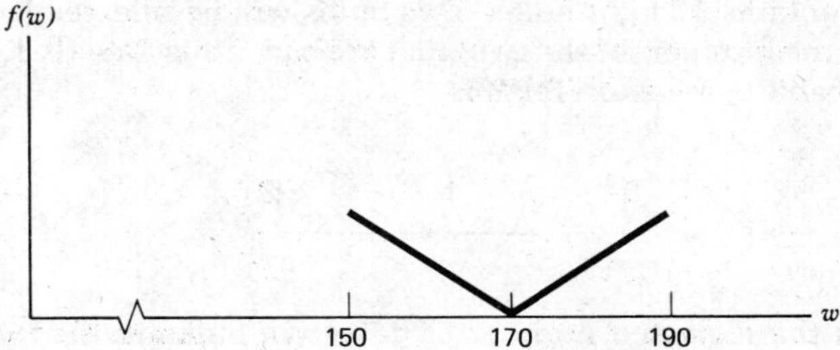

Figure E.9

of the middle of the probability distribution for X. Other measures of the middle of the distribution are the median, m_X, which splits the total probability of one into two equal parts, and the mode, the observed value that has the highest probability of occurrence.

The variance σ_X^2 of a random variable X is the expected value of the square of the difference between X and μ_X. The larger this quantity, the more variable X is about μ_X. The standard deviation σ_X is the positive square root of σ_X^2. It is measured in the same units as X itself. Another measure of variability that is occasionally used is the absolute deviation a_X; it is the expected value of the absolute difference between X and μ_X. For a continuous random variable X the $(100k)$th percentile t_k is defined by $P(X \leq t_k) = k$. Thus $(100)k\%$ of the time X is no larger than t_k. The interquartile range I_r is also used to measure variability; it is defined by

$$I_r = t_{.75} - t_{.25}$$

I_r is the length of an interval, centered at the median m_X, which contains 50% of the total probability.

Exercise 4.4

1. Assume you are going to play a game; let X be the amount you win and assume $R_X = \{x: x = -1, 1, 2, 3\}$. The probability measure for X is

$x =$	-1	1	2	3
Probability	.6	.2	.15	.05

   (a) Compute $E[X]$.
   (b) Would you like to play this game?

2. Compute σ_X and a_X for the random variable defined in question 1 above.

3. A box contains 10 light bulbs. Two bulbs will be selected from the box; let Y be the number (of the two) that are bad. Thus $R_Y = \{0, 1, 2\}$. Assume the probability measure for Y is

$y =$	0	1	2
Probability	2/9	5/9	2/9

What is the expected number of defective bulbs in the two selected?

4. Compute the values of a_Y and σ_Y for the distribution given in question 3.

5. A uniform random variable U has mean $\mu_U = 1$ and standard deviation $\sigma_U = 1$. What is R_U?

6. Assume that the clock time required for two college football teams to finish a game is a random variable T that has the triangular density discussed in Section 4.3 with $a = 100$ and $b = 150$ (minutes).
   (a) Find μ_T, the expected length of the game.
   (b) Evaluate I_r.

7. Evaluate I_r for the uniform distribution defined in question 5.

8. A random variable V has the triangular distribution given in Section 4.3; it is known that $\sigma_V = 1$. What is I_r for this random variable?

9. Assume you get out of bed each day between 7:00 and 7:15 A.M.; let W be the number of minutes past 7 at which you get up and assume W has the triangular distribution discussed in Section 4.3 with $a = 0$ and $b = 15$.
   (a) What is the expected time you get up each day?
   (b) What is the probability you get up before 7:10 A.M. tomorrow?

CHAPTER 5
SOME SPECIAL PROBABILITY MEASURES

In the previous two chapters we have studied the concepts of random variables, their ranges and probability measures, and the idea of expectation and some of its uses. Some probability measures are very important because they are frequently assumed in practical situations. We shall study several of these measures in this chapter, concentrating on what these probability measures are and when they may be appropriately used. The succeeding sections are devoted to discussions of these probability measures.

5.1   THE BINOMIAL PROBABILITY MEASURE

We have in fact already used a binomial probability measure in several cases in Chapters 3 and 4, without actually identifying it. Specifically, Example 3.2.2 considered the number, U, of football games a collegiate team will win out of 10 in a given year; four different probability measures for U were considered in the example. Measures 2, 3, and 4 discussed there are in fact particular cases of binomial probability measures. Several of the probability measures assumed in the exercises were also particular examples of binomial measures.

A binomial probability measure may be appropriate for a large number of different situations. Among those cases in which a binomial measure might prove satisfactory are the number of questions a student answers correctly in a 50-question true-false exam, the number of free throws a basketball player may put in the basket out of 10 attempts, the number of insurance policies on which an insurance company may have to pay, out of 1000 policies written, the number of gray offspring out of 200 produced from a mating of two flies, and so on. The binomial distribution, or probability measure, is one of those most frequently assumed in a wide variety of practical problems.

When is a binomial probability measure appropriate? What assumptions must be made to justify its use in a particular case? We shall examine these matters now. First, a second reading of the illustrations quoted above will reveal that in each case mentioned the random variable is a count of the number of times a certain event occurs (winning a game, getting the correct answer to a true-false question, making a basket from the free throw line, paying out on a policy, producing a gray offspring versus some other color). Thus the range of the random variable in each of these cases is a collection of integers or whole numbers; a binomial random variable is discrete and not continuous. It is counting the number of times an event occurs, which means the possible observed values are integers.

Second, common to all the cases mentioned is that each situation described can be regarded as a sequence of simple experiments or trials. Thus, for the football team the full season consists of 10 games, each of which will be won or will not be won (which includes the ties) by the team we are concerned with. Thus each game could be called a trial; the result of each trial in this case is either win or not win. Thus the random variable (number of games won) could be called the number of trials on which the event "win" will occur. Similarly, each true-false question could be called a trial (the answer is right or wrong), as could each attempt from the free throw line, each policy sold, and each offspring produced. Thus, in each of the situations described, a specified number of trials is performed and the random variable is counting the number of times a certain event is observed; on each trial the event of concern may occur or it may not occur. Thus, in each case, there were only two different possible outcomes that could be observed on each trial. All these things are necessary for the assumption of a binomial probability measure.

One remaining assumption is necessary for a binomial probability measure; this assumption involves the probability of observing the event of interest on each individual trial. For the binomial measure to be appropriate it is necessary to assume that this probability is a constant, say p, for each trial, regardless of the outcomes of the other trials. (We shall say the events occur independently.) Thus, in the football example, this requirement says that the probability that the team wins each game must be the same for each game, regardless of whether it won the previous week or not. For example, in the true-false examination it is necessary that the probability of answering the question correctly is the same for *each* question, and in the free throw example the probability the player makes a basket must be the same for each attempt. This assumption is not to be taken lightly and may or may not be justified in any particular case. This discussion is formalized in the following definitions and theorem.

> **Definition 5.1.** A *Bernoulli trial* is a simple experiment having only two possible outcomes, called success and failure.

As already indicated, each football game has two outcomes: win (success) or not win (failure); thus an individual game is an example of a Bernoulli trial. Similarly each true-false question is either answered correctly (success) or incorrectly (failure), so each such question answered is a Bernoulli trial. Each attempt by the basketball player from the free throw line either results in a basket (success) or not (failure); therefore each attempt is again a Bernoulli trial. The reader may easily verify that the other examples mentioned also concern Bernoulli trials.

> **Definition 5.2.** Assume n Bernoulli trials are to be performed and that the probability of the outcome success is a constant p for *each* trial, regardless of the outcomes for any other trials. The number X of successes that occur on the n trials is called a *binomial random variable* (meaning it has a binomial probability measure).

Notice immediately then that we may, at one extreme, observe 0 successes and, at the other, we can observe at most n successes; any number of successes in between could occur as the value for X. Thus, the range of a binomial random variable is

$$R_X = \{x : x = 0, 1, 2, \ldots, n\}$$

Assume we are willing to make the stated assumptions.

   (i)   The number of football games won out of the 10 played.
   (ii)   The number of questions answered correctly on the true-false examination.
   (iii)   The number of shots made from the free throw line.
   (iv)   The number of policies the insurance company must pay off.
   (v)   The number of gray offspring resulting from the mating of the two flies.

Each of these is a binomial random variable. What then is this binomial probability measure and how is it expressed?

Before answering this question let us first examine some simple notation that is very useful in expressing these binomial probability measures. The reader may be acquainted with factorial notation from a previous course in algebra. The factorial function is defined for the nonnegative integers; $n!$ is read n-factorial and is by definition the product of all the whole numbers from 1 to n inclusive. That is,

$$n! = n(n-1)(n-2)\ldots 3\cdot 2\cdot 1$$

so

$$4! = 4\cdot 3\cdot 2\cdot 1 = 24$$

$$5! = 5\cdot 4\cdot 3\cdot 2\cdot 1 = 120$$

and so on; as n increases, $n!$ increases at a very fast rate. The values of $n!$ for $n = 1, 2, \ldots, 10$ are given in the following table

n	$n!$
1	1
2	2
3	6
4	24
5	120
6	720
7	5,040
8	40,320
9	362,880
10	3,628,800

Each succeeding value for $n!$ is given by the product of the succeeding value of n with the preceding value for $n!$. Factorial notation is a necessity for expressing the solutions to "counting problems," which are discussed in the Appendix; it is also very useful in giving a formula for binomial probability measures and other standard measures.

A binomial coefficient is a ratio of two quantities involving factorials. It is defined by

$$\binom{n}{r} = \frac{n!}{r!\,(n-r)!};$$

both n and r must be nonnegative integers. Thus

$$\binom{5}{2} = \frac{5!}{2!\,3!} = \frac{120}{2(6)} = 10$$

$$\binom{7}{1} = \frac{7!}{1!\,6!} = \frac{5040}{720} = 7$$

$$\binom{6}{3} = \frac{6!}{3!\,3!} = \frac{720}{6(6)} = 20$$

The binomial coefficients are also useful in solving counting problems. It is useful, in terms of a unified notation for binomial probability measures, to define $0! = 1$. In doing so notice that, for example,

$$\binom{4}{0} = \frac{4!}{0!\,4!} = 1$$

$$\binom{8}{0} = \frac{8!}{0!\,8!} = 1$$

and for any n,

$$\binom{n}{0} = \frac{n!}{0!\,n!} = 1$$

The following theorem, whose proof is given in the Appendix, gives the formula for a binomial probability measure.

Theorem 5.1

Assume X is a binomial random variable as given in Definition 5.2. Then, letting $q = 1 - p$,

$$P(X = r) = \binom{n}{r} p^r\, q^{n-r} \qquad \text{for} \qquad r = 0, 1, 2, \ldots, n$$

In order to evaluate the probabilities assigned to the different possible observed values for X, two quantities must be known; these are n, the number of trials involved, and p, the probability of success on each trial. These quantities are called *parameters*; we shall say X is a binomial random variable with parameters n and p.

Example 5.1.1

Assume a professional baseball player has a .300 batting average and that he will be at bat four times in a given game. Let Y be the number of hits he will get in this game and $R_Y = \{0, 1, 2, 3, 4\}$. Making the assumptions given in Definition 5.2, Y is a binomial random variable with parameters $n = 4$ and $p = 3$. Thus, the probabilities of the various possible observed values are

$$P(Y = 0) = \binom{4}{0} (.3)^0 (.7)^4 = .2401$$

$$P(Y = 1) = \binom{4}{1} (.3)^1 (.7)^3 = .4116$$

$$P(Y = 2) = \binom{4}{2} (.3)^2 (.7)^2 = .2646$$

$$P(Y = 3) = \binom{4}{3} (.3)^3 (.7)^1 = .0756$$

$$P(Y = 4) = \binom{4}{4} (.3)^4 (.7)^0 = .0081$$

With these assumptions, the probability he gets at least 1 hit is

$$P(Y \geq 1) = 1 - P(Y = 0) = .7599$$

In slightly more than 3/4 of all games in which he bats 4 times a .300 hitter will get at least 1 hit, assuming the attempts are independent. In about 1/4 of such games he will get no hits and in less than 1% of such games will he get 4 hits.

Example 5.1.2

Assume a student must take an 8-question, true-false examination and that to pass the exam he must answer at least 6 questions correctly. Assume further that the student has, for one reason or another, no preparation and must guess in answering each question. What is the probability he answers half the questions correctly? What is the probability he passes the exam?

Let W be the number of questions he answers correctly. Then

$$R_W = \{w: w = 0, 1, 2, \ldots, 8\}$$

and the answers to the two questions posed are given by $P(W = 4)$ and $P(W \geq 6)$, respectively. If the student does guess at the answer to each question, he has a probability $p = .5$ of guessing the correct answer for each, regardless of whether he was correct on any others and W is a binomial random variable with parameters $n = 8$ and $p = .5$. We find

$$P(W = 4) = \binom{8}{4} (.5)^4 (.5)^4 = .273$$

and

$$P(\text{passes exam}) = P(W \geq 6)$$

$$= P(W = 6) + P(W = 7) + P(W = 8)$$

$$= \binom{8}{6} (.5)^6 (.5)^2 + \binom{8}{7} (.5)^7 (.5)^1 + \binom{8}{8} (.5)^8 (5)^0$$

$$= .145$$

Thus, 27.3% of all people that guess at the answers to 8 true-false questions will get (exactly) 4 correct; 14.5% of the people guessing will get at least 6 correct out of 8.

Tables of the binomial probability distribution, for various values of n and p, are readily available. Table A.1 in the Appendix gives the values of $P(X = r)$ for $n = 2, 3, \ldots, 20$ and for p in increments of .05 between .05 and .50. Thus, with $n = 10$, $p = .4$ we have $P(X = 5) = .2007$, $P(X = 4) = .2508$; with $n = 10$, $p = .2$ we have $P(X = 5) = .0264$, $P(X = 4) = .0881$. With $n = 15$, $p = .4$, we see $P(X = 5) = .1859$, $P(X = 4) = .1268$.

It is only necessary to arrange in a table the values of the binomial distribution for p between 0 and .5, because interchanging the two classes, success and failure, will replace p by $1 - p$. Thus, suppose $n = 8$, $p = .75$ and we want to evaluate $P(X = 5)$; we will observe exactly 5 successes if and only if we observe exactly 3 failures, where the probability of a failure is $1 - p = .25$ on each trial. Thus, if X is the number of successes and $Y = n - X$ is the number of failures,

$$P(X = r) = P(n - X = n - r) = P(Y = n - r)$$

For the case discussed above $n = 8$, $r = 5$, $p = .75$, and $Y = 8 - X$, and we find $P(X = 5) = P(Y = 3) = .2076$ as the entry with $1 - p = .25$, $n - r = 3$.

Examining table 1 will reveal that the binomial probability measures change with both n and p. With n fixed, X has higher probability of equaling values close to 0, as p becomes smaller. Conversely, the probabilities that X equals values close to n get larger, the closer that p is to 1. That is, changing p with n fixed will cause the probability to shift over the values in R_X. Similarly, if p is held fixed, then $P(X = r)$ will take on different values as n changes. Fig-

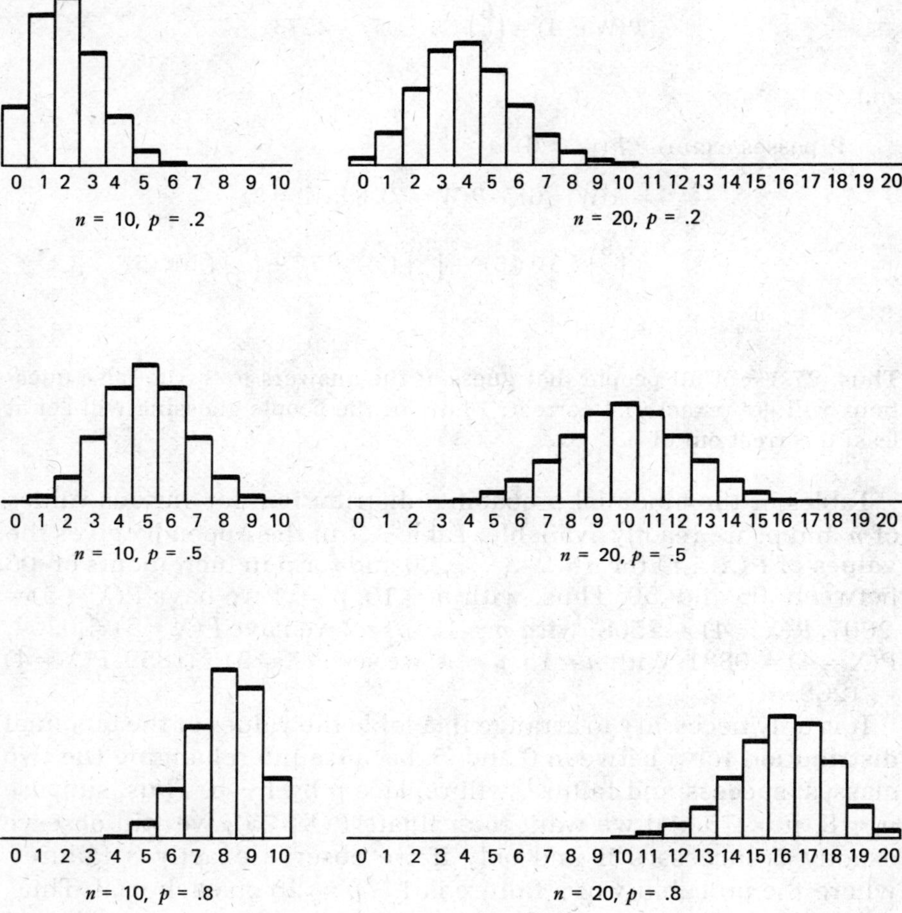

Figure 5.1.1 Binomial probability measures.

ure 5.1.1. gives the histograms for 6 different pairs of values for n and p. Note that as p goes from .2 to .8, the bulk of the histogram is shifted from lower values to higher values. With $p = .5$ the histogram is symmetric; if $p \neq .5$ the histogram is not symmetric but it is noticeably more symmetric with $n = 20$ than with $n = 10$. This trend continues; for any value of p the histogram gets more symmetric as n increases.

Example 5.1.3

A dairy that makes chocolate-covered ice cream bars on a stick has announced that 10% of the sticks have stars; anyone getting such a stick will receive a prize. Assume that Joe purchased 10 bars. What is the probability that he gets exactly one prize? At least one prize? Here we look at the bars he has purchased as being $n = 10$ independent trials with probability $p = 1/10$ of a success on each trial. If we let X be the number of bars with a star, that Joe purchased, then X is a binomial random variable with $n = 10$, $p = 1/10$. Its probability function is

$$P(X = r) = \binom{10}{r} (.1)^r (.9)^{10-r} \quad \text{for} \quad r = 0, 1, 2, \ldots, 10$$

The probability he gets exactly one star, then, is

$$P(X = 1) = \binom{10}{1} (.1)^1 (.9)^{10-1} = .387$$

from Table 1. The probability he gets at least one bar with a star is

$$\sum_{r=1}^{10} P(X = r) = 1 - P(X = 0)$$
$$= 1 - .349 = .651$$

We can evaluate this probability either by adding together the probabilities of 1 or of 2 or of 3, etc., or of 10 stars, or we can realize that the sum we want may be evaluated by subtracting $P(X = 0)$ from 1, as indicated.

Example 5.1.4

Assume that a high school graduate submits an application to attend each of five different colleges. Assuming that his probability of being accepted by each is .3, what is the probability he will be accepted by at least one? By more than one? This experiment consists of five trials (acceptance or nonacceptance by each college) with a probability of success $p = .3$ for each. If Y is the number of colleges that accept him, then Y is binomial with $n = 5$, $p = .3$. The probability he is accepted by at least one college is $P(Y \geq 1) = 1 - P(Y = 0) = 1 - .168 = .832$, from Table 1. The probability that he is accepted by more than one is $P(Y \geq 2) = 1 - .168 - .360 = .472$.

The binomial random variable has a *mean* and *variance* that are especially simple functions of its two parameters. If X is binomial with parameters n and p, then by definition

$$\mu_X = E[X] = \sum_{R_X} x P(X = x) = \sum_{x=0}^{n} x \binom{n}{x} p^x q^{n-x}$$

It can be shown that the value of this sum is np; that is, the *mean* of a binomial random variable is $\mu_X = np$, the product of its two parameters. Similarly, by definition, the *variance* of a binomial random variable is

$$\sigma_X^2 = E[(X - \mu)^2] = \sum_{R_X} (x - np)^2 \, P(X = x) = \sum_{x=0}^{n} (x - np)^2 \binom{n}{x} p^x q^{n-x}$$

The value of this sum can be shown to be npq; thus $\sigma_X^2 = npq$, and the *standard deviation* of a binomial random variable is $\sigma_X = \sqrt{npq}$.

Example 5.1.5

The random variable Y in Example 5.1.1 (number of hits in four attempts) has mean $\mu_Y = 4(.3) = 1.2$ and standard deviation $\sigma_Y = \sqrt{4(.3)\,(.7)} = \sqrt{.84} = .92$. The random variable W in Example 5.1.2 (number of true-false questions guessed correctly) has mean $\mu_W = 8(1/2) = 4$ and standard deviation $\sigma_W = \sqrt{8(1/2)\,(1/2)} = 1.41$. The random variable X in Example 5.1.3 (number of starred ice cream sticks) has mean $\mu_X = 10(1/10) = 1$ and standard deviation $\sigma_X = \sqrt{10(1/10)\,(9/10)} = .95$. The random variable Y in Example 5.1.4 (number of colleges accepting the high school graduate) has mean $\mu_Y = 5(.3) = 1.5$ and standard deviation $\sigma_Y = \sqrt{5(.3)\,(.7)} = 1.02$.

Once we have decided on a binomial probability measure for a random variable and we know the values for n and p, the mean and standard deviation for the random variable are immediately known, as the preceding example indicates. Example 5.1.3 above suggests a type of problem that is occasionally of interest: namely, how large should n be (how many trials should be performed) to have a known probability of getting at least one success? If X is a binomial random variable with parameters n and p,

$$P(X \geq 1) = \sum_{x=1}^{n} \binom{n}{x} p^x q^{n-x} = 1 - P(X = 0) = 1 - q^n$$

since we know the sum of the probability measure over the whole range is 1.

Thus, if we want to have at least probability γ of getting *at least* one success, we require

$$P(X \geq 1) \geq \gamma$$

$$1 - (1 - p)^n \geq \gamma$$

which gives

$$n \geq \frac{\log\,(1 - \gamma)}{\log\,(1 - p)}$$

If it happens that $\log(1 - \gamma)/\log(1 - p)$ is not an integer, we shall round to the next higher integer; in doing so, the probability of getting at least one success, is actually slightly greater than γ. If we rounded down instead of up, we would find the probability of at least one success, with that n, is slightly smaller than γ.

Example 5.1.6

Let us return to the starred ice cream stick example. The probability an individual stick has a star on it is 1/10. How many ice cream bars, n, must we buy to have at least probability .9 of getting at least one bar with a starred stick? Here we have $\gamma = .9$, $p = .1$ and, using logs to the base 10, we find

$$n \geq \frac{\log (.1)}{\log (.9)} = \frac{-1.00000}{-.04576} = 21.85 \sim 22$$

Thus, if we buy 22 bars the probability is (a little greater than) .9 that at least one of them has a starred stick. Another way of saying this is the following. Suppose we put these ice cream bars into boxes of 22. Then (slightly more than) 90% of the boxes will have at least one bar with a starred stick; (slightly less than) 10% of the boxes will contain no starred sticks.

Exercise 5.1

1. Assume that a college football team plays 10 games in a given year. If their probability of winning each game is .8, regardless of whether they win any others, what is the probability they will win at least 8 games out of the 10? Do you feel all the assumptions made here are reasonable?

2. Five fair dice are rolled one time. What is the distribution of the number of 6's that will occur? What is the distribution of the number of even numbers that will occur?

3. The probability is .1 that any person that enters a radio-television store will make a purchase. If 20 people enter the store one morning, what is the probability that the store makes at least four sales?

4. In Example 5.1.3 how many ice cream bars should you buy to be 99% sure you get at least one with a star?

5. A photographer assumes he has a probability of .7 of getting a good print of a particular scene, each time he takes a picture. If he takes three pictures of the same scene, what is the probability that he gets at least one good print of the scene?

6. Genetic theory predicts that 25% of the offspring of a certain mating should be gray in color and that the color of one offspring does not

affect any others. If there are 20 descendants from the mating, what is the distribution of the number of gray offspring in the 20?

7. Given that a binomial random variable has a mean of 10 and standard deviation of 3, can you completely specify the distribution (i.e., can you identify n and p)?

8. A binomial random variable has mean of 4.5 and variance of 2.25. Find n and p.

9. An insurance company writes $1000, one-year, term life insurance policies for each of 100 men, aged 52, in a given year. The probability a man this age will die before his next birthday is .01.
   (a) What is the probability that the insurance company will have to pay, on every policy?
   (b) What is the probability it will have to pay on none of the policies?
   (c) What is the expected number of policies it will have to pay?

10. A nurseryman plants flower seeds in flats, then sells the flats when the plants are about two inches high. He plants 20 seeds per flat; assume that the probability is .9 that each seed will germinate.
    (a) What is the probability that any given flat contains at least 15 plants when he sells it?
    (b) Assume he sells 10 flats. What is the probability that each of them has at least 15 plants in it?

11. The probability that a high jumper will clear the bar, when it is set at six feet, is .5. Assume he is allowed three attempts at this height during a track meet; if he misses on all three, then he is eliminated from competition. What is the probability he is not eliminated?

12. Assume the probability is .8 that a linotype operator will set a full newspaper page without any misprints. On a given day the newspaper contains 16 pages. What is the probability that there are no misprints in the newspaper?

13. In a given semester, a student takes five courses. The probability is .9 that he gets an A in each. What is the probability he gets all A's?

14. Twenty students are in a statistics class, and each is 20 years old. The probability is .5 that a 20-year-old will live to reach age 75.
    (a) What is the expected number of people in the class that will reach age 75?
    (b) What is the probability 10 of them reach age 75?
    (c) What is the probability at least 10 of them reach age 75?

15. During a semester, a statistics class meets 80 times. After every class, the same two students gamble to see who will pay for their coffee. Assuming the game is fair, what is the expected number of times each will have to pay?

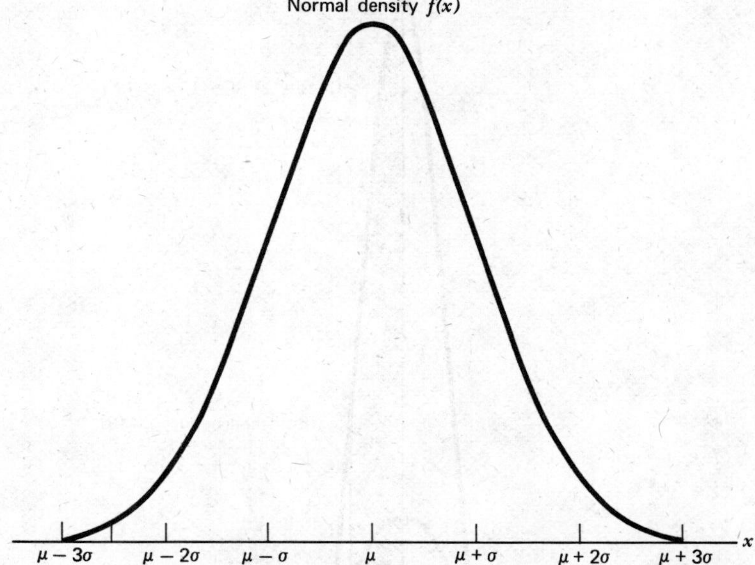

Figure 5.2.1   General normal density.

5.2   THE NORMAL PROBABILITY MEASURE

In Sections 3.3 and 4.3 we briefly studied continuous probability measures and some of their properties. It will be recalled that probabilities for continuous random variables are computed as areas under continuous curves, called density functions. We saw some specific examples of density functions that required computing areas under familiar geometric shapes; these areas gave the probability that the observed value of the random variable would lie in the corresponding interval.

By far the most important type of continuous probability measure is given by the *normal density* function. A general normal density is pictured in Figure 5.2.1. The equation of this curve is

$$f(x) = \frac{1}{\sigma\sqrt{2\pi}}\, e^{-(x-\mu)^2/2\sigma^2}$$

where $e = 2.718\ldots$ is a constant, the base of the natural logarithms, μ is any constant, and σ is a positive constant. Notice from the figure that the density is a symmetric, bell-shaped curve, with its middle located at μ. Because of the symmetry, the place at which the curve will balance is μ; thus, the constant μ is actually the mean or ex-

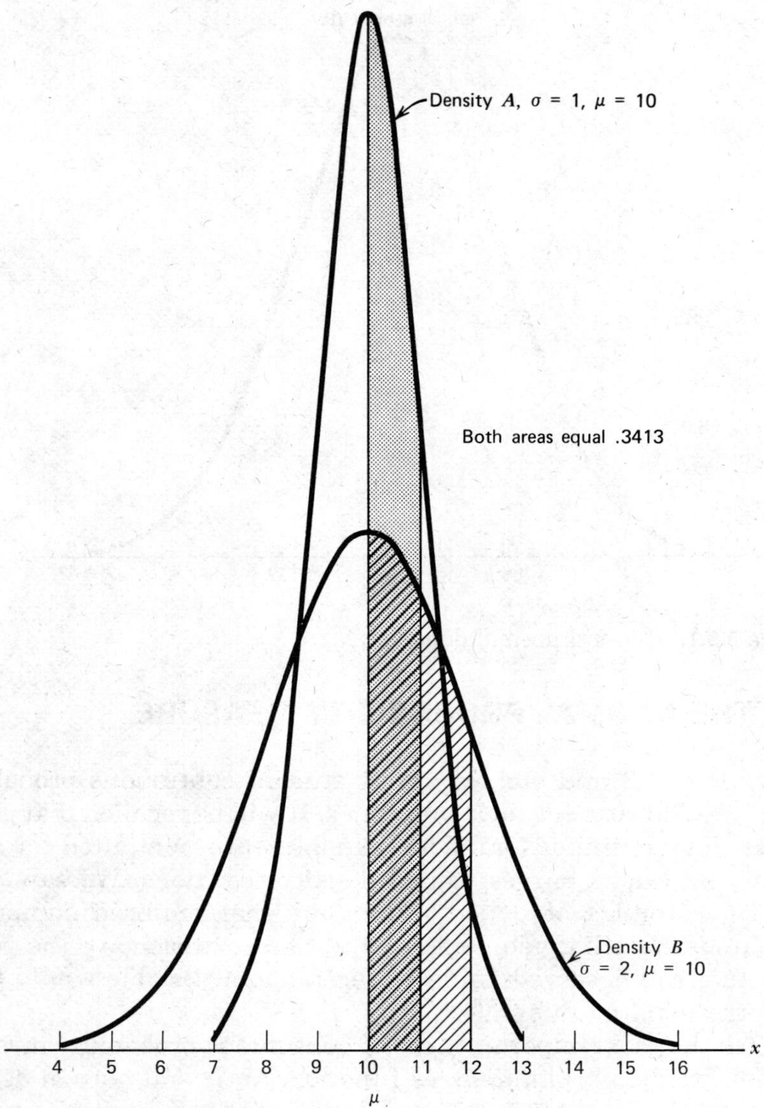

Figure 5.2.2   Two normal densities.
A. $\sigma = 1$, $\mu = 10$.   B. $\sigma = 2$, $\mu = 10$.

pected value μ_X of the normal random variable X; it is also the median, m_X, of the distribution. Since the constant μ can equal any number, positive or negative, we are able to have a normal distribution with its mean μ_X equal to any desired number. As the constant μ changes, the bell-shaped density is moved to different places.

The constant σ that occurs in the equation of the normal density $f(x)$ is, as might be guessed from the symbol used, the standard deviation σ_X of the normal random variable. It can be shown that the area under the normal density between μ and $\mu+\sigma$ is a constant (and equals .3413), no matter what the particular values of μ and σ. That is, suppose we have normal density A with $\mu = 10$ and $\sigma = 1$; then the area between 10 (μ) and 11 $(\mu + \sigma)$ is .3413. Assume also that normal density B has $\mu = 10$ and $\sigma = 2$; then the area under normal density B, between 10 (μ) and 12 $(\mu + \sigma)$ is also .3413. The only way this can be true is for normal density A to be more peaked and, thus, to have greater area close to 10 (the value for μ). This idea is pictured in Figure 5.2.2. Since both densities have $\mu = 10$, they are both symmetric about that point; however, density B with $\sigma = 2$ is flatter and more dispersed. Keeping σ constant and changing μ simply moves the bell shape; keeping μ constant and changing σ holds the center of the bell at the same point but changes the height of the bell and moves the sides so that the total area under the bell stays equal to 1.

As mentioned, the normal probability measure is the most important continuous distribution. This is true because many measurements made in practical applications have been shown to be well described by a normal distribution; such diverse measurements as scores on achievement examinations, weights of people of the same age and racial group, lifetimes of lightbulbs, time necessary for an airliner to travel from one city to another, lengths of fish of the same age and species, etc., may all be well described by a normal probability measure. Any sort of continuous measurement (and some that are not as we shall see) that we expect to have the highest probability of lying in a fixed-length interval centered at μ, and for which the probability trails off symmetrically as the interval center moves away from μ either way, may be well described by an appropriate normal probability measure.

Quite independently of whether any specific type of measurement itself is well described by a normal density function, the central limit theorem states that sums of random variables will be well described by normal probability measures. This fact alone assures the normal density a central and unique role in applied statistics. We shall examine this important theorem in the next chapter and see its use in the approximation of the probabilities of many different events.

As with any other continuous probability measure, the probability that the observed value of a normal variable will lie in any specific interval is given by the area under the density over the interval.

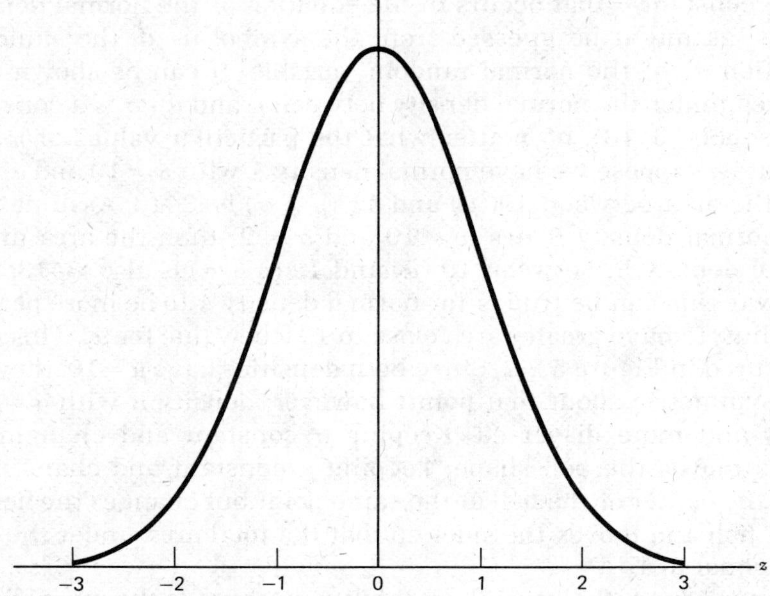

Figure 5.2.3 The standard normal density, $\mu = 0$, $\sigma = 1$.

Since the normal density over any fixed interval is not a familiar geometric shape, we shall not be able to use familiar formulas to compute areas, as we could for the density functions considered in the previous chapters. Fortunately, once we are able to evaluate probabilities for the *standard normal density,* it is quite straightforward to evaluate probability statements for any other normal density. The definition of a standard normal random variable is as follows.

> **Definition 5.3.** Assume Z is a continuous random variable whose density function is normal with $\mu = 0$ and $\sigma = 1$. We shall call Z a *standard normal random variable.*

The standard normal random variable, then, has a density that is centered at 0. Figure 5.2.3 shows this density. Table 2 in the Appendix presents the areas under this standard normal density in the following way. The probability that a standard normal random variable is less than or equal to any particular number z is denoted by $P(Z \le z)$; this probability, then, is the total area under the standard normal density to the left of z. Because of the symmetry of the normal density, we can see immediately that

$$P(Z \le 0) = .5$$

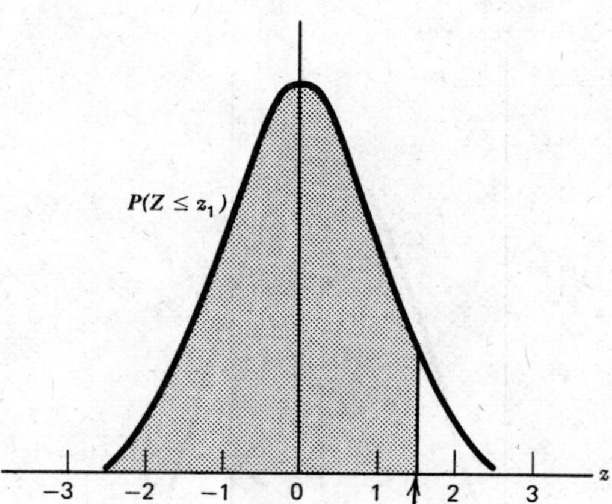

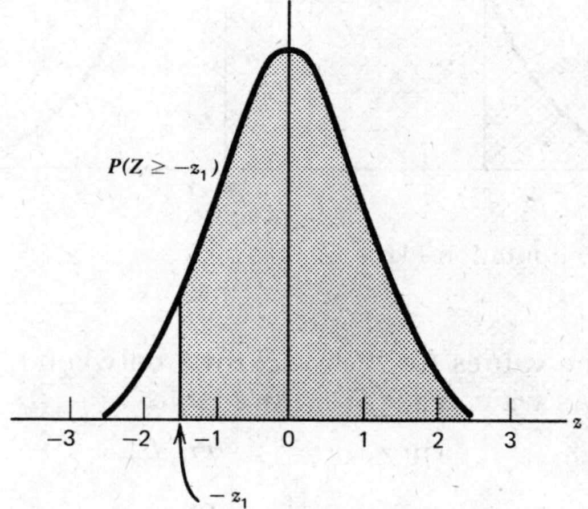

Figure 5.2.4 Equality of $P(Z \le z_1)$ and $P(Z \ge -z_1)$.

Also, because of the symmetry of the normal density, notice that if z_1 is any positive constant, then

$$p(Z \le z_1) = P(Z \ge -z_1)$$

$$= 1 - P(Z \le -z_1)$$

so that

$$P(Z \le -z_1) = 1 - P(Z \le z_1)$$

(see Figure 5.2.4). Because of these relationships, a table of values for $P(Z \le z)$, for $0 \le z \le 3$ can also be used to evaluate $P(Z \le z)$, where z is between 0 and -3, as we shall see. Table 2 in the Appen-

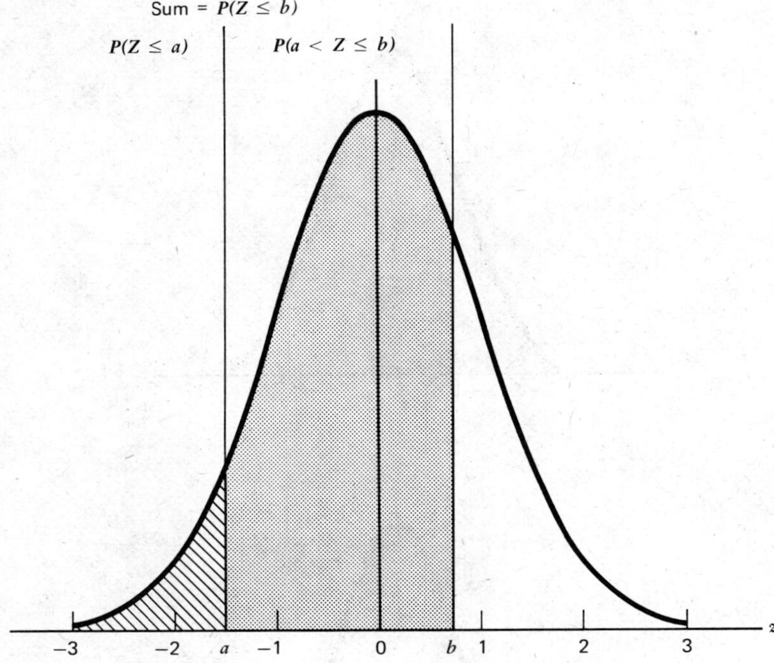

Figure 5.2.5 Computation $P(a < Z \leq b)$.

dix presents the values for $P(Z \leq z)$ for z between 0 and 3 in increments of .01. As we can see from the table

$$P(Z \leq .75) = .7734$$

$$P(Z \leq 2.10) = .9821$$

$$P(Z \leq 1.42) = .9222$$

Then, we know as well that

$$P(Z \geq -.75) = .7734$$

$$P(Z \geq -2.10) = .9821$$

$$P(Z \geq -1.42) = .9222$$

using the result pointed out above for z_1. As we can see from Figure 5.2.5, if we are given any two numbers a and b such that $a < b$,

$$P(Z \leq a) + P(a < Z \leq b) = P(Z \leq b)$$

Thus, subtracting $P(Z \leq a)$ from both sides, we have

$$P(a < Z \leq b) = P(Z \leq b) - P(Z \leq a)$$

which gives the way to compute the probability that the observed value of a standard normal random variable will lie between a and b. Using the tabular values mentioned above, we find

$$P(.75 < Z \le 2.10) = P(Z \le 2.10) - P(Z \le .75)$$

$$= .9821 - .7734$$

$$= .2087$$

$$P(-1.42 < Z \le .75) = P(Z \le .75) - P(Z \le -1.42)$$

$$= .7734 - (1 - .9222)$$

$$= .6956$$

$$P(-2.10 < Z \le 2.10) = P(Z \le 2.10) - P(Z \le -2.10)$$

$$= .9821 - (1 - .9821)$$

$$= .9642$$

This last numerical computation illustrates another formula of use in certain kinds of problems that we shall study. We know that $P(Z \le -a) = 1 - P(Z \le a)$; then

$$P(-a < Z \le a) = P(Z \le a) - P(Z \le -a)$$

$$= P(Z \le a) - [1 - P(Z \le a)]$$

$$= 2P(Z \le a) - 1$$

Since Z is a continuous random variable, we also have

$$P(-a \le Z \le a) = P(-a < Z < a) = 2P(Z \le a) - 1$$

This result is useful in certain inverse table look-up problems. Suppose we wanted to find the value of a such that

$$P(-a \le Z \le a) = .95$$

Then, from above, we need to find a such that

$$2P(Z \le a) - 1 = .95$$

$$P(Z \le a) = .975$$

This says we want to read the entries in the table until we find the area equal to .975; the corresponding value of z is the value a we want. Reading through the entries in the table we find that

$$P(Z \le 1.96) = .975$$

and, thus, $a = 1.96$.

Now that we have some familiarity with evaluating probability statements for a standard normal random variable, let us turn our attention to the evaluation of probability statements for arbitrary normal random variables, ones that do no necessarily have $\mu = 0$ and $\sigma = 1$. Rather than needing a special table for each possible pair of values for μ and σ, we can transform any arbitrary normal random variable into a standard normal random variable and then use the same table for the evaluation of probability statements. The following theorem gives the basic result required.

Theorem 5.2

Assume X is a normal random variable with mean μ_X and variance $\sigma_X{}^2$. Then

$$Z = \frac{X - \mu_X}{\sigma_X}$$

is a standard normal random variable.

We can in fact verify part of the result stated in the theorem. Theorem 4.1c shows how to compute the expected value of a linear function of a random variable. In fact,

$$E\left[\frac{X - \mu_X}{\sigma_X}\right] = \frac{E[X]}{\sigma_X} - \frac{\mu_X}{\sigma_X}$$

$$= 0$$

using this result; thus $(X - \mu_X)/\sigma_X$ has mean 0 as it must if it is to be a standard normal random variable. From Theorem 4.3 we see that the variance of $(X - \mu_X)/\sigma_X$ is

$$\sigma^2{}_{(X-\mu_X)/\sigma_X} = \frac{\sigma_X{}^2}{\sigma_X{}^2} = 1$$

as, again, it must be if Z is to be a standard normal random variable. The only part of the theorem, then, that we are not able to verify is that the density of Z must be normal in form; since to establish this requires calculus techniques, we shall simply accept that part of the theorem.

How, then, can we use Theorem 5.2, to evaluate probability statements about X, which is normal with arbitrary μ and σ? We note that the statement

$$X \leq t$$

is equivalent to

$$X - \mu_X \leq t - \mu_X$$

and to

$$\frac{X - \mu_X}{\sigma_X} \le \frac{t - \mu_X}{\sigma_X}$$

that is, it is equivalent to

$$Z \le \frac{t - \mu_X}{\sigma_X}$$

where $Z = (X - \mu_X)/\sigma_X$. Since the two statements are equivalent, their probabilities must be equal and

$$P(X \le t) = P\left(Z \le \frac{t - \mu_X}{\sigma_X}\right)$$

The probability on the right can be evaluated directly from Table 2. The two following examples illustrate the use of this result.

Example 5.2.1

Assume that the height that a college high jumper will attain on any given attempt is a normal random variable with mean $\mu_X = 76$ inches and standard deviation of one inch. What is the probability he will clear the bar if it is set at 6 feet 3 inches? What is the probability he will not clear the bar when it is set at 6 feet 6 inches? Define X to be the height he actually jumps on a given try. Then we are given that X is normal with mean 76 and standard deviation one and we want to evaluate $P(X \ge 75)$ and $P(X) < 78)$, translating the above numbers into inches. Then as described above,

$$P(X \ge 75) = P\left(Z \ge \frac{75 - 76}{1} = -1\right) = .8413$$

$$P(X < 78) = P\left(Z < \frac{78 - 76}{1} = 2\right) = .9773$$

Example 5.2.2

The number of ounces that a bottle-filling machine puts into a beer bottle is a normal random variable with mean 12.25 and standard deviation .12. The label on each bottle says it contains 12 ounces. What is the probability that any given bottle contains less than this amount? What proportion of the bottles filled by this machine will contain between 12 and 12.5 ounces? Again letting X represent the number of ounces that the machine puts into a bottle, we are given that X is normal with mean 12.25 and standard deviation .12; we want to evaluate $P(X < 12)$ and $P(12 \le X \le 12.5)$.

$$P(X < 12) = P\left(Z < \frac{12 - 12.25}{.12}\right) = P(Z < -2.08) = .0188$$

$$P(12 \le X \le 12.5) = P\left(\frac{12 - 12.25}{.12} \le Z \le \frac{12.5 - 12.25}{.12}\right)$$

$$= P(-2.08 \le Z \le 2.08) = .9624$$

The percentiles of an arbitrary normal random variable will be evaluated from Table 2. Because of their importance and frequency of usage, we shall use z_k (rather than t_k) to represent the $(100k)$th percentile of the standard normal distribution. Thus, z_k is defined by

$$P(Z \le z_k) = k$$

where Z is a standard normal random variable. The following table presents some of the commonly used percentiles of the standard normal distribution.

k	z_k
.01	-2.33
.025	-1.96
.05	-1.64
.10	-1.28
.25	$-.67$
.50	0
.75	$.67$
.90	1.28
.95	1.64
.975	1.96
.99	2.33

As has already been mentioned, the median for a standard normal random variable is

$$m_Z = z_{.50} = 0$$

the interquartile range for the standard normal is

$$I_r = z_{.75} - z_{.25} = .67 - (-.67) = 1.34$$

The percentiles of an arbitrary normal random variable will be denoted by t_k, the general notation introduced in Section 4.3 for any random variable. Thus t_k is defined by

$$P(X \le t_k) = k$$

but, as discussed above,

$$P(X \le t_k) = P\left(Z \le \frac{t_k - \mu_X}{\sigma_X}\right)$$

where Z is a standard normal random variable. The value that Z is less than or equal to, with probability k, is z_k. Thus, we have

$$\frac{t_k - \mu_X}{\sigma_X} = z_k$$

or

$$t_k = \mu_X + z_k \sigma_X$$

and we see that the percentiles of an arbritary normal random variable can be evaluated from knowledge of z_k, the standard normal percentiles.

Example 5.2.3

In Example 5.2.1 X was normal with $\mu_X = 76$ and $\sigma_X = 1$; thus, for example,

$$t_{.1} = 76 + (-1.28)1 = 74.72$$

$$t_{.95} = 76 + (1.64)1 = 77.64$$

The interquartile range for this random variable is

$$I_r = t_{.75} - t_{.25} = 76.67 - 75.33 = 1.34$$

In Example 5.2.2 X was normal with $\mu_X = 12.25$, $\sigma_X = .12$; thus, for this distribution

$$t_{.1} = 12.25 + (-1.28)(.12) = 12.10$$

$$t_{.90} = 12.25 + (1.28)(.12) = 12.40$$

and its interquartile range is

$$I_r = t_{.75} - t_{.25} = 12.33 - 12.17 = .16$$

Exercise 5.2

1. Use Table 2 to evaluate the following probabilities for a standard normal random variable.
    (a) $P(Z < .5)$
    (b) $P(Z < -.5)$
    (c) $P(|Z| < .5)$
    (d) $P(Z > 1.15)$
    (e) $P(|Z| < 1.15)$
    (f) $P(-1.96 < Z < 1.96)$

2. In each case, use Table 2 to evaluate the given constant
   (a) $P(Z > a) = .65$
   (b) $P(Z < a) = .65$
   (c) $P(|Z| < a) = .65$
   (d) $P(|Z| > a) = .75$

3. In "grading on the curve," an instructor assumes that the scores attained on a given exam are normally distributed (i.e., the proportions of scores that will fall in specific intervals are given by areas under a normal density) with known mean μ and standard deviation σ. Any score above $\mu + 1.5\sigma$ is an A, between $\mu + .5\sigma$ and $\mu + 1.5\sigma$ the grade is a B, between $\mu - .5\sigma$ and $\mu + .5\sigma$ the grade is a C, between $\mu - 1.5\sigma$ and $\mu - .5\sigma$ the grade is a D, and below $\mu - 1.5\sigma$ the grade is an F. What are the proportions of scores that will be given these various grades?

4. (See question 3). If an instructor gives an exam and assumes that the scores are normally distributed with $\mu = 75$ and $\sigma = 5$, what grades would he assign to the scores (a) 78, (b) 70, (c) 92, (d) 64, and (e) 82, given he grades on the curve?

5. In question 4, assume $\mu = 80$, $\sigma = 10$ and answer parts (a) to (e).

6. Assume that scores on the mathematics part of the SAT are normal with $\mu = 500$, $\sigma = 100$. If your score on this exam is 650, what proportion of all the scores is better than yours?

7. Assume that the lifetime of a fluorescent light bulb is a normal random variable with $\mu = 1000$, $\sigma = 150$. What is the probability that its lifetime will exceed 1200 hours?

8. Assume X is a normal random variable with mean μ_X and variance σ_X^2. Show that its interquartile range is $I_r = 1.34\sigma_X$.

9. A manufacturer has found that the length of time his product will operate satisfactorily without repair is a normal random variable with $\mu = 2.5$ years and $\sigma = .8$ years. He guarantees to pay all costs associated with repairs required in the first year of use. What fraction of his machines will require repair in its first year of use?

10. (Refer to question 9.) A second manufacturer of the same product finds that the length of time his product will operate satisfactorily is a normal random variable with $\mu = 2$ years and $\sigma = .4$ years. He also guarantees to pay all costs associated with repairs required in the first year of use. What fraction of this second manufacturer's machines will require repair in its first year of service? Which of the two products would you, as a buyer, prefer to purchase, assuming both sell for the same price?

11. What is the probability that any normal random variable will take on a value within two standard deviations of its mean?

12. Assume Z is a standard normal random variable. Find the constant a defined by each of the following:
    (a) $P(Z > a) = .3$
    (b) $P(Z < a) = .4$
    (c) $P(|Z| < a) = .1$
    (d) $P(|Z| > a) = .1$

13. Assume X is a normal random variable with mean $\mu = 500$ and standard deviation $\sigma = 100$. In each case find the value of the constant b.
    (a) $P(X < b) = .95$
    (b) $P(X > b) = .95$
    (c) $P(|X - 500| < b) = .95$

14. Suppose three of the bulbs described in question 7 are put into service at the same time. What is the probability that all three last at least 1200 hours? That none last this long? That (exactly) one lasts this long?

5.3 THE POISSON PROBABILITY MEASURE

The *Poisson random variable* is a *discrete* random variable that is quite useful in some applications. We discussed the binomial random variable in Section 5.1; recall that it is the number of successes that will be observed in repeated trials, each of which may result in the occurrence or nonoccurrence of a success. The Poisson random variable also equals the number of successes observed, but rather than being observed with respect to well-defined trials, the successes (or occurrences) recorded by the Poisson random variable are observed in a continuous interval of time, distance, or area, for instance. An example is given by the number of electrons emitted by a piece of uranium in, say, a one-hour period. Each emission is essentially instantaneous, and it is quite feasible to record the number that do occur in a fixed interval of time. However, the one-hour period consists of an infinite number of instants, not some finite number n of instants, at each of which we may or may not observe an emission. Thus the number of electrons emitted certainly is discrete but not a binomial random variable. Other examples of this type of random variable are the number of cars that arrive at a shopping center parking lot in a two-hour period, the number of telephone calls that arrive at an industrial switchboard between 9 A.M. and noon, and the number of bacteria in a cubic centimeter of water. In each of these cases the discrete variable is counting the number of occurrences of a phenomenon within a continuum.

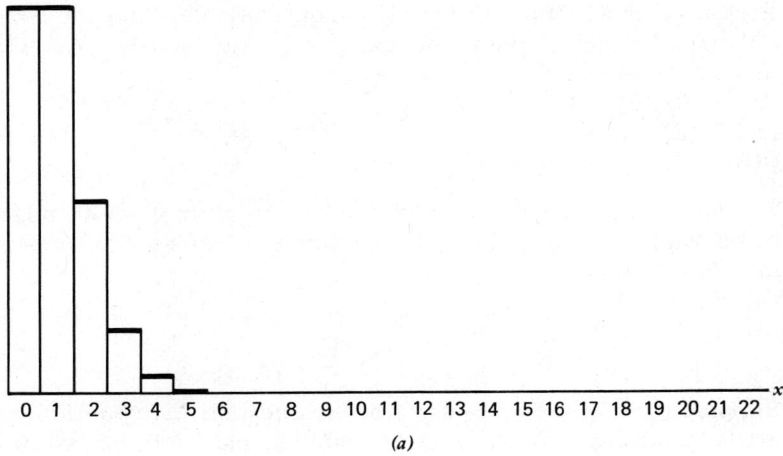

(a)

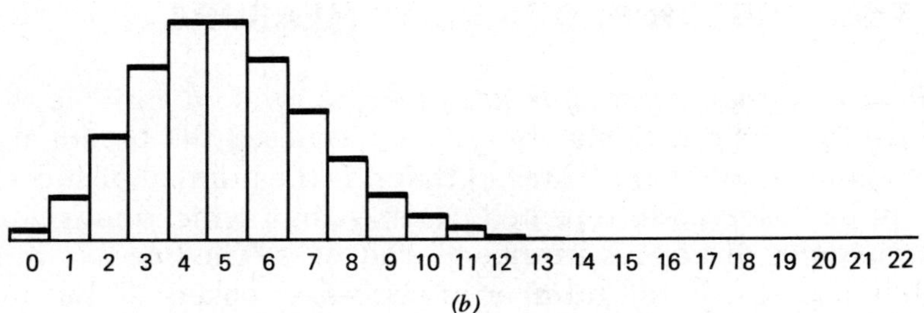

(b)

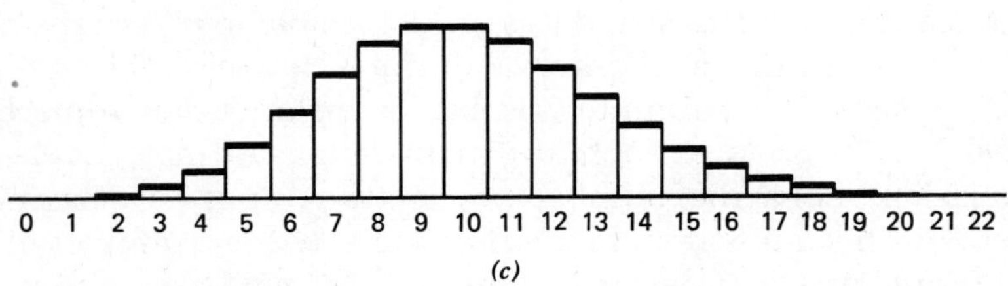

(c)

Figure 5.3.1 Poisson probability measures.
(a) $\lambda t = 1$. (b) $\lambda t = 5$. (c) $\lambda t = 10$.

Let us briefly review the assumptions behind the Poisson random variable. The phenomenon studied is observed over an interval of total length t; it is called a *Poisson process*. It must be possible to subdivide the continuous interval of observation into sufficiently small intervals of length h such that

1. The probability of exactly one occurrence in each small interval is proportional to h, regardless of whether previous intervals have contained occurrences or not.
2. The probability of more than one occurrence in the same small interval is approximately zero.

Then, if the Poisson process is observed for an interval of fixed length t and X is the number of occurrences in this interval, X is the *Poisson random variable*; its range is the infinite set of non-negative integers $R_X = \{0, 1, 2, 3, \ldots\}$ and its probability measure is

$$P(X = x) = \frac{(\lambda t)^x}{x!}\, e^{-\lambda t} \qquad \text{for} \qquad x \text{ in } R_X$$

e again is the base of the natural logarithms, the same constant that occurs in the definition of the normal density. It can be shown that both the *mean* μ_X and the variance σ_X^2 of the Poisson random variable are equal to λt. Note then that if we count the number of occurrences in an interval of length $1(t = 1)$, the mean number of occurrences is λ. Thus λ is called the *rate parameter* of the process; it describes the average number of occurrences per unit and, thus, must be a positive number.

Figure 5.3.1 presents the histograms for three different Poisson distributions ($\lambda t = 1, 5$ and 10). In doing this plotting, each probability has been rounded to the nearest hundredth; the range in each case is infinite, but those individual probabilities smaller than $.005$ are too small to be given a bar with nonzero height. Note that the histograms become markedly more symmetric as λt increases. As we shall see, this tendency continues as λt increases indefinitely.

Table A.3 in the Appendix presents the Poisson probability function for various values of λt. With $\lambda t = 1$, note that $P(X = 0) = P(X = 1) = .368$, $P(X = 4) = .015$; with $\lambda t = 3$, $P(X = 0) = .050$, $P(X = 1) = .149$, $P(X = 4) = .168$. The range of the random variable, in the table, is restricted to values that have nonzero probability to three decimals. Let us now examine some examples.

Example 5.3.1

Assume that telephone calls arrive at a large industrial switchboard at a rate of two per minute, in accordance with a Poisson process. In a five-minute period, what is the probability there will be exactly four calls? Exactly eight calls? Ten or more calls? To answer these questions, we let X be the number of calls arriving in a $t = 5$ minute period. We are given that the calls arrive at an average rate of $\lambda = 2$ calls per minute. Thus X is Poisson with parameter $\lambda t = 2(5) = 10$, and we can answer all three questions by referring to Table 3 with $\lambda t = 10$. We find there

$$P(X = 4) = .019 \qquad P(X = 8) = .113$$

$$P(X \geq 10) = P(X = 10) + P(X = 11) + P(X = 12) + \ldots$$

$$= .125 + .114 + .095 + .073 + .052$$

$$+ .035 + .022 + .013 + .007 + .004$$

$$+ .002 + .001$$

$$= .543$$

Example 5.3.2

Assume that the sales made by a used car salesman occur like events in a Poisson process with parameter $\lambda = 1$ per week. In a two-week period, what is the probability that he makes no more than two sales? That he makes exactly three sales? What is the probability of three two-week periods in a row with no sales? To answer the first two questions posed, let X be the number of sales he makes in a two-week period. Then X is a Poisson random variable with parameter $\lambda t = 1(2) = 2$. Thus, from Table A.3,

$$P(X \leq 2) = P(X = 0) + P(X = 1) + P(X = 2)$$

$$= .677$$
$$P(X = 3) = .180$$

To answer the last question posed above, let Y be the number of sales he will make in a six-week period; Y then is a Poisson random variable with parameter $\lambda t = 1(6) = 6$. The salesman will have three two-week periods in a row with no sales if and only if he makes no sales in the six-week period. Thus, the final question is answered by the probability that $Y = 0$. From Table A.3 we find

$$P(Y = 0) = .002$$

The final question posed in Example 5.3.2 above could be answered in a second, equivalent manner as follows. Suppose, as above, that we let X be the number of sales he makes in a two-week period. Then X is Poisson with parameter $\lambda t = 2$, and we find from Table 3

$$P(X = 0) = .135$$

Each two-week period can be taken to be a Bernoulli trial; let us define 0 sales to be a failure and 1 or more sales to be a success in the two-week period. Thus, in each two-week period we either observe a failure (with probability .135) or a success (with probability .865); furthermore, a rereading of the assumptions for the Poisson process will show that these probabilities are appropriate for each two-week period, regardless of whether he made a sale or not in any other, nonoverlapping period. Then, if we let W be the number of successes in the 3 two-week periods, W is a binomial random variable with parameters $n = 3$ and $p = .865$. The probability of 3 two-week periods in a row with no sales is

$$P(W = 0) = \binom{3}{0} (.865)^0 (.135)^{3-0}$$

$$= (.135)^3 = .002$$

the same answer derived in the example in a different way.

The Poisson probability measure gives a good approximation to binomial probabilities if n is large and p is small. The approximation is quite good if n is at least 100 and p is no larger than .1; the larger n is, or the smaller p is, or both, the better the approximation will be. To apply the approximation, we simply equate the Poisson mean to the binomial mean, that is, we set

$$\lambda t = np$$

The following example illustrates the use of the approximation.

Example 5.3.3

Assume that the probability any baby is born with some physical defect is .001. In the next 100 children born at a city hospital, what is the probability that none have physical defects? That exactly one has a physical defect? If we let U be the number of children with physical defects in the next 100 born, then U is a binomial random variable with parameters $n = 100$, $p = .001$. The probabilities asked for then are, using the exact binomial formula,

$$P(U = 0) = \binom{100}{0} (.001)^0 (.999)^{100}$$

$$= (.999)^{100} = .9048$$

$$P(U = 1) = \binom{100}{1} (.001)^1 (.999)^{99}$$

$$= .0906$$

Using the approximation suggested above, we should use a Poisson random variable V, say, with parameter

$$\lambda t = np = 100(.001) = .1$$

We find, then, from Table A.3.

$$P(V = 0) = .905$$

$$P(V = 1) = .091$$

The approximation of $P(U = u)$ by $P(V = u)$, for $u = 0, 1$, is quite good since n is large and p is quite small.

Exercise 5.3

1. If X is a Poisson random variable with parameter $\lambda t = 3$, use Table A.3 to evaluate
    (a) $P(X = 0)$
    (b) $P(X = 2)$
    (c) $P(X = 3)$
    (d) $P(X \geq 4)$

2. A piece of radioactive material emits pulses on an average of four per second. In a $\frac{1}{2}$-second period, what is the probability of there being no pulses emitted?

3. Assume that a printed page in a book consists of 40 lines, with 75 positions per line (each of which may be left blank or be filled with a letter or a grammatical symbol). Thus each page has $(40 \times 75) = 3000$ positions on it. Assume that a typesetter, on the average, makes one error per 6000 positions, and that these occur like events in a Poisson process. Let X be the number of misprints per page made by this type-setter; what is the distribution for X? What is the probability any given page has no errors? That it has exactly one error? If a chapter con-sists of 15 pages, what is the probability there are no misprints in the chapter?

4. The locations of bacteria in a volume of water occur like events in a Poisson process with rate 20 per cubic centimeter. If 1/10 of a cubic centimeter of this liquid is examined for this type of bacteria, what is the probability that one or more are found?

5. Commercially made chocolate chip cookies are prepared and baked in batches of 1000 cookies. A total of 3000 chocolate chips are placed in the batter for each batch and then are well mixed throughout the batter. What is the probability an individual cookie has no chips in it? That it has four or more?

6. Assume that accidents of various kinds (e.g., falls and bumps) occur to a given person at a rate of one per year. Compute, in two different ways, the probability he has exactly one such accident in a three-year period.

7. Customers arrive at the checkout station(s) of a large supermarket at the rate of two per minute, in accordance with a Poisson process.
   (a) In a five-minute period what is the probability that at least 12 persons arrive at the checkout stations?
   (b) What is the probability that fewer than four arrive (during the five-minute period)?

8. At a dangerous intersection, automobile collisions occur at a rate of one per week, in accord with a Poisson process. What is the probability there are no accidents in a two-week period? That there are exactly two?

9. Let X be a Poisson random variable with parameter $\lambda t = 9$. What is the probability that X takes on a value within two standard deviations of its mean?

10. Note, in Table 3, that if X is a Poisson random variable with parameter λt equaling a positive integer, then $P(X = \lambda t - 1) = P(X = \lambda t)$. Show algebraically that this must be the case. (*Hint.* Look at their ratio.)

11. Assume that the probability a college professor is late to class is .005 and that being late any single time has no effect on the probability that he is late any other time. Approximate the probability he is late at least one time in his next 200 classes.

12. Assume the probability that there is a crash of a passenger airplane, somewhere in the world, is .008 for any particular 24-hour period; assume also that this probability remains unchanged, no matter what occurs in any other 24-hour period. What is the (approximate) probability there are no crashes in a 90-day period?

5.4 SUMMARY

In this chapter we have studied some frequently occurring probability measures, often used in practical problems. A binomial random variable is discrete; the binomial probability measure depends on two parameters, n and p. n measures the number of trials and p is the probability of observing a success on each trial; the binomial random variable X, then, is the total number of successes observed. It has mean $\mu_X = np$ and variance $\sigma_X{}^2 = npq$. Table A.1 in the Appendix presents the binomial probability measures for several values of n and p.

A normal random variable is continuous and thus is described by a density function. The normal density function is bell-shaped, symmetric, with its middle located at the mean μ_X of the random variable; the standard deviation σ_X controls how peaked or flat the density function is. Table A.2 in the Appendix presents areas under the standard normal density, the one with mean 0 and variance 1. This same table can be used to evaluate probabilities for any normal random variable.

The Poisson random variable is discrete and counts the number of occurrences in a Poisson process in a fixed interval of length t. A Poisson random variable with parameter λt has mean $\mu_X = \lambda t$ and variance $\sigma_X^2 = \lambda t$. Table A.3 in the Appendix presents Poisson probability measures for several different values of λt. The Poisson probability measure, with $\lambda t = np$, gives a good approximation to the binomial probability measure if n is large and p is small.

Exercise 5.4

1. In a certain town, half of the residences put lights outside their home at Christmas time. Given a block with eight homes on it, what is the probability that all eight put lights outside their home? What is the probability that half the homes on the block put lights outside?

2. Ten percent of all the cars on the highway in California have some defect that should be corrected. The highway patrol periodically sets up stations by the roadside and stops and inspects cars for such defects. Assume such a station was set up and that 10 cars were stopped in the first half hour.
   (a)   What is the probability none of them had defects?
   (b)   What is the expected number with defects?

3. In extruding continuous nylon filament, defects occur along the length at a rate of one per 200 feet, in accord with a Poisson process. Given a 100-foot segment
   (a)   What is the probability it has no defects?
   (b)   What is the probability it has two or more defects?

4. Westinghouse 40-watt bulbs commonly claim on their packages "2000 Avg. Hours," meaning the distribution of lifetimes of this sort of bulb has a mean of 2000 hours (in a "typical" usage). The package does not give the standard deviation of the lifetimes. Assume the standard deviation is 100 hours and that the lifetimes are normal.
   (a)   What is the probability that one of these bulbs will last at least 1900 hours?
   (b)   What is the probability that one of these bulbs will last no more than 2300 hours?
   (c)   What is the probability that one of them lasts exactly 2000 hours?
   (d)   What is the probability that one of them lasts between 1999 hours and 2001 hours?

5. The elapsed time from the instant a commercial airliner leaves the ground at San Francisco International Airport until its wheels touch down at Kennedy Airport in New York is a normal random variable with mean of 275 (minutes) and standard deviation of 10 minutes, assuming the flight is safely concluded.
   (a) On any given day, what is the probability that the flight takes no more than 270 minutes?
   (b) If the flight is made seven times in one week, at the same time of day, what is the probability that it takes less than 270 minutes each of these seven days?

6. Twenty percent of all kindergarten children have difficulty in pronouncing the letter "l." In a class of 20
   (a) What is the expected number of children who have difficulty pronouncing this letter?
   (b) What is the probability none of them have difficulty in pronouncing this letter?

7. Vacuum tubes used in a home radio will fail (cease to work) at instants that occur like events in a Poisson process, with a rate of one each two years.
   (a) What is the probability that any tube of this type will last at least two years?
   (b) Given that you have a radio that contains four such tubes (and nothing else that could fail), what is the probability the radio performs without servicing for at least two years? (Assume that the tubes fail independently of each other.)

8. Suppose you have 15 lamps in your home, in each of which you put one of the bulbs described in question 4. What is the expected number of the 15 that will burn out in less than 2128 hours?

9. What is the probability that at least 14 of them burn out in less than 2128 hours? (See question 8.)

10. In a large city, suicides occur at a rate of 10 per week, in accord with events in a Poisson process. What is the probability there are at least five suicides in any given week in this city?

11. The Olympic record for the javelin throw is 295 feet, $7\frac{1}{4}$ inches. The distance a new contender will toss the javelin on each attempt is a normal random variable with mean of 285 feet and standard deviation of 5 feet. What is the probability this person will break the Olympic record on any given attempt?

12. It is assumed that 5% of the fish in a certain lake have a detectable amount of mercury in their tissue. How many fish should be taken from the lake to have probability at least .99 of getting at least one fish with a detectable amount of mercury in its tissue?

13. Assume that the weights of albacore in a certain fishing area are normally distributed with mean of 30 (pounds) and standard deviation of 5 pounds. What proportion of the albacore caught in this area will exceed 40 pounds in weight?

14. A student is given a true-false exam with 40 questions on it. He must answer at least 30 correctly to pass. He finds he knows the answers to 25 of the questions. What is the probability that he passes the exam?

CHAPTER 6
POPULATIONS AND SAMPLES

We have studied the idea of a random variable and the concept of a probability measure to describe the behavior of a random variable. In this chapter we shall begin our study of statistical inference and will see how random variables occur and are used to answer practical questions. As has been mentioned previously, essentially all statistical procedures can be cast in the framework of sampling from a population and using the sample to make inferences about certain properties of the population that are of interest. Thus, we shall first study the idea of a population and, subsequently, that of sampling from the population.

6.1 POPULATIONS AND DISTRIBUTIONS

The word *population* is commonly used in everyday English in much the same sense that we shall use it in our study of statistics. A *population* is a collection of things or objects that have something in common. Examples are the population of the United States, the population of fish in Lake Erie, the population of holly trees in the state of California, and the population of 1972 cars made by General Motors. Populations are extremely diverse in nature.

As has been mentioned previously, we shall be concerned with techniques for making *inferences* about a population, based upon an examination of only a *sample* or portion of the population. Generally, we shall be concerned only with making inferences about populations of numbers. These are generated or occur as the numerical value of some attribute attached to the members of the population. For example, we might want to make inferences about the annual incomes of United States citizens. Then, rather than being concerned with the population of people living in the United States, we would only be really concerned with the population of numbers that represents their incomes. Similarly, if we were interested in the weights or lengths of fish in Lake Erie, rather than being concerned with the population of fish, per se, it is the population of numbers that would be of interest. Thus, in general, we shall find we are normally interested in investigating, or making inferences about, populations of numbers. These are examples of *real* populations, in the sense that they were all concerned with populations that really exist, the numbers being measurements of some aspect of the population members that was of interest. A conceptual population, on the other hand, is a collection of numbers that does not and may never exist, yet it is frequently of interest to make inferences about it. For example, suppose a light bulb manufacturer has a research division that develops a new, cheaper method of making light bulbs. The company could conceive of making any number of bulbs by this new process and would undoubtedly be interested in the lengths of life such bulbs would have in specific uses. This, then, would be a *conceptual* population of light bulb lifetimes; the company would most likely want to investigate features of this population, such as the mean or average lifetime, to be able to decide whether the new procedure should be adopted for manufacturing in the future. If they decide not to adopt the new procedure, the conceptual population will never come into existence. It is very useful to be able to refer to either the real or conceptual type of population in making inferences. The word population will always refer to a collection of numbers, when used in this book.

Populations can be of two distinct types: *real* and *conceptual*.

Whether a population is real or conceptual, it may be *discrete* or *continuous*. If the population consists of only integers (or if the values in the population form any discrete set), it is discrete. Some examples of discrete populations are: the number of hits per game made by a professional baseball player over his career, the number of children per family for all families in the United States on July 1, 1980, the number of auto accidents per day at a given intersection over a three-year period, and so on. Generally discrete populations that we shall be concerned with will be generated by counting occurrences of some kind and thus will be populations of integers.

On the other hand, populations that consist of measurements such as height, weight, length, area, and time, will be taken to be continuous. For example, the population of heights of all individuals in the United States, the weights of all fish in Lake Erie, and the lengths of time you would take to drive to work with a 1976 Chevrolet, over a five-year period, will all be taken to be continuous populations. Each such population is in fact truly discrete because there is a finite number of individuals in each population (person, fish, working days), and each has associated with it a specific value of the variable concerned. Furthermore, the variable defined (height, weight, driving time) is theoretically continuous, but any measurements made of it will necessarily be discrete since any measuring instrument available can only measure to the nearest .1 or .01 or .001, or whatever, of a unit. Thus the actual measured numbers could take on values only at specific discrete points of a continuous interval. When dealing with variables of this type, the number of distinct values that could occur is quite large and it is then convenient to make the idealization that the measurements are in fact continuous. The continuous techniques are much simpler to apply; studies have demonstrated that this type of idealization does not lead to significant errors in inference.

Let us now define a *population random variable*. Clearly, if we are trying to investigate a population of numbers, the population values will not, in general, all be equal; some values will be this number, some that. We shall say that the population values are *distributed* over the real line in some way. If the population is discrete, the population values will be distributed over a set of integers (for our purposes); if the population is continuous the values will be distributed over a continuous interval. In either case we assume there is a probability measure describing the distribution of population values. If the population is discrete, the population probability measure gives the proportion of population values equal to the possible (discrete) integer values; if the population is continuous the population probability measure is described by a density function. The area under the density, over any specific interval, gives the proportion of population measurements that fall in that interval. Figure 6.1.1 illustrates a discrete population probability measure; for the case pictured, the population members each have one of the numbers $1, 2, 3, 4$ associated with them. The heights of the bars at these integers give the *proportions* of the population members that have those values. The random variable described by this measure is called the *population random variable*. Figure 6.1.2 illustrates a continuous population probability measure; for the case pictured the population values all lie between 1 and 10. Areas under the density give the proportions of population values in specific intervals. As with discrete populations, the random

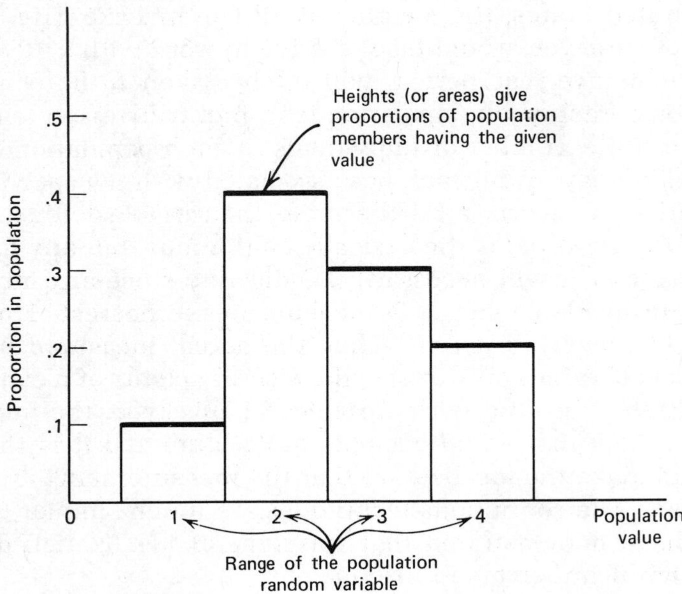

Figure 6.1.1 Graphic presentation of a discrete population.

variable described by the continuous density function is called the *population random variable.*

Example 6.1.1

Let us first examine a simple example of a conceptual population and the corresponding population random variable. Given a fair die, we could conceive of rolling this die any number of times and recording the number that occurs each time. We would then generate a discrete population of numbers, each of which has value 1 or 2 or 3 or 4 or 5 or 6. Granted that the die is fair, the conceptual population should consist of equal numbers of each of these values. Thus the random variable X with probability measure

$$P(X = x) = \frac{1}{6} \qquad x = 1, 2, 3, 4, 5, 6$$

would be the population random variable.

Example 6.1.2

Assume that every housewife in the United States either does or does not use brand A laundry detergent. If we assign 1 as the value for each house-

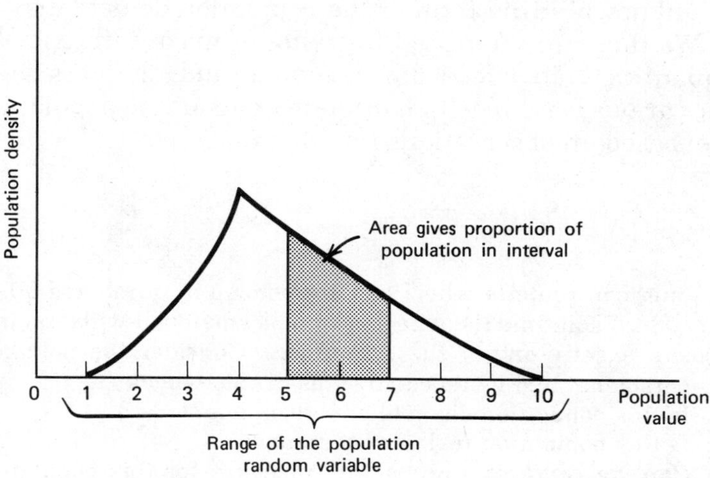

Figure 6.1.2   Graphic presentation of a continuous population.

wife that uses brand A and 0 for each housewife that does not, we have a discrete population of 0's and 1's. Let p be the proportion of 1's and $q = 1 - p$ be the proportion of 0's. Then if X is the random variable with probability measure

$$P(X = 1) = p$$

$$P(X = 0) = 1 - p = q$$

X is the population random variable.

Example 6.1.3

Consider the weights of United States adult males. This population of numbers, then, consists of a finite (very large) number of values of a continuous variable; as discussed above the population will be idealized and taken to be continuous. We are then assuming there exists some density function that describes the proportions of weights in the population which fall into various intervals. It would seem likely that the weights in the population may be distributed symmetrically, and we might then assume a normal density for the population random variable, with suitable parameters μ_X and σ_X.

In general, of course, we do not have perfect knowledge of the population; if we did, then there would be no need for making inferences about the population. Thus the population parameters (parameters of the probability measure of the population random variable) may be unknown or the form of the population density may be unknown. We then are interested in trying to make inferences about these quantities. In succeeding sections and chapters we shall consider various ways in which inferences about the population may be made, based on observations found in a sample.

Exercise 6.1

1. An American roulette wheel has 38 spots on it, numbered 00, 0, 1, 2, 3, ... , 36. Each time the wheel is spun, a small ball will land in a hole opposite exactly one of these numbers. Consider the population of numbers generated by repeated spins of this wheel.
   (a) Is this population discrete or continuous?
   (b) Is this population real or conceptual?
   (c) Can you suggest a probability measure for this population random variable?

2. The California Department of Highways is considering replacing an existing two-lane highway with a four-lane highway. Consider the population consisting of the number of collisions per day on this 10-mile stretch of road, after it is changed to four lanes.
   (a) Is this population real or conceptual?
   (b) Is this population discrete or continuous?

3. Consider the population of driving times required by motorists to complete the 10-mile stretch of four-lane highway referred to in question 2 above.
   (a) Is this population real or conceptual?
   (b) Is this population discrete or continuous?

4. Assume a new drug has been developed, and it is to be tested for its effect on a skin infection. An unlimited number of people with the same skin infection are available for testing. Each person tested will be classified as cured (assigned value 1) or not cured (assigned value 0). We can then refer to the population of numbers that would be generated by applying the drug to people with this infection.
   (a) Is this population real or conceptual?
   (b) Is this population discrete or continuous?

5. Draw the histogram for the population random variable discussed in Example 6.1.1.

6. Draw the histogram for the population random variable discussed in Example 6.1.2.

7. Consider the number of years of formal education for every person over age 21 living in the United States.
   (a) Is this population real or conceptual?
   (b) Is this population discrete or continuous?

8. Consider the population of times at which you get out of bed in the morning. Is this population discrete or continuous?

9. Consider the population of incomes of heads of households in the United States.
   (a) Is this population real or conceptual?
   (b) Is this population discrete or continuous?

10. For the case discussed in Example 6.1.3, assume $\mu_X = 160$ and $\sigma_X = 20$. If one adult male is selected at random from the population, compute the probability that his weight
    (a) Is at least 170 pounds.
    (b) Is between 120 and 140 pounds.

6.2 SAMPLES

In the last section we defined a population, the population random variable and the probability measure for the population (or population random variable). As noted, if the population and its probability measure were completely known, then there would be no need for making inferences about the population. We would know, or be able to derive, any and all aspects about the population, including its mean or average value, its variance, its range, and its 40th percentile. Any question regarding the population could be definitively answered, given this perfect knowledge.

In practice, of course, we do not have perfect knowledge of the population; that is why we want and need to rely on inferences about the population. In Example 6.1.1 we considered the population of numbers generated by repeated rolls of the same die. It is easy to see that this population is discrete and that every number in the population must be a 1 or a 2 or a 3 or a 4 or a 5 or a 6. It is not necessary, though, that all of these numbers occur with equal frequency. That is, the population random variable does not necessarily have probability 1/6 of equaling each of these numbers and, for any specific die, we do not truly know the probability measure for the population. Granted we do not know the exact probability measure for the population, we do not know its mean, variance or any other quantity that depends on the probability measure. We could, of course, roll the die 10 times, or 100 times, or 1000 times, and observe what happens on those rolls. Surely the results of the rolls we make contain some information, for example, about the population probability measure, the population mean, or population variance. Conclusions we might draw about the population, based on the sample of 1000 rolls (or any other number), are in fact inferences about the conceptual population. The numbers we observe in rolling the die are a sample from the conceptual population.

In Example 6.1.2 we considered the population of housewives in the United States; each housewife that uses brand A laundry detergent was assigned the value 1 and each one that does not was assigned the value 0. Thus every number in the population is a 1 or a 0; we let p represent the proportion of 1's in the population (and $q = 1 - p$, then, is the proportion of 0's). Since we do not know the value of p, again we do not know the probability measure of the population. To get information about p (and thus about the population probability measure), we could select a sample of 10, 100, or 1000 housewives, recording a 1 for each housewife that does use brand A and a zero for each one that does not. Again, then, this would be a sample from the population; the values we get in the sample should give us information about the population. Based on the sample numbers observed, we could make inferences about the population.

These two examples are meant to illustrate two basic facts:

1. For any practical case, we do not know the population in full, including all the specific details of its probability measure.
2. Granted the ignorance stated in (1), we should be able to get information about the population by selecting and examining a sample from the population.

Sample results enable us to make inferences about the population probability measure (or about derived quantities depending on that measure).

How should the sample be selected? If we are to use the sample to make inferences about the population, we would like the sample to be representative of the population, to mirror the population as closely as possible. How, then, can we be assured that we have a representative sample?

There are many ways of trying to make the sample representative of the population, depending on what prior knowledge, if any, is already available about the population. Courses in sampling techniques or sample survey theory discuss many different practical situations that have been studied and techniques that are known to assure representative samples. Granted that we do not know anything about the population, a sample selected randomly or haphazardly from the population can be expected to give the most accurate inferences about the population, from many points of view.

In this book, we shall consider only situations in which inferences are based upon a random sample selected from the population.

What, then, do we mean by a sample selected at random from the population? A value is selected *at random* from the population if *each* population member has equal probability of being selected; thus if there are twice as many 3's in the population as there are 1's, for example, then the value selected is twice as likely to be a 3 as it is to be a 1. In short, defining X_1 (a random variable) to be the first value selected from the population, the probability measure for X_1 is identical with the probability measure for the population random variable, if that first value is selected at random from the population. The same is true for all additional sample values. Thus, if we let X_2 be the second value selected, then the probability measure for X_2 again is identical with the probability measure for the population random variable, regardless of which value X_1 equaled; letting X_1, $X_2, \ldots, X_n$ be the values to be selected for a sample of size n, the probability measure for *each* X_i is identical with the population probability measure, independent of the values of the other elements in the sample. If we satisfy this requirement, then, we shall say that we have a random sample of the population random variable. All of the techniques we shall discuss assume that the sampling has been done in this way. (The random variables X_1, $X_2, \ldots, X_n$ are said to be *independent,* if we satisfy the above requirement; the interested reader is urged to consult the Appendix for more information about the important concept of independence.)

Example 6.2.1

Assume, as in Example 6.1.3, that the weights of adult United States males are normally distributed with mean μ and variance σ^2. (This says that if X is the population random variable for this population of weights, then X is normal with $\mu_X = \mu$ and $\sigma_X = \sigma$.) Assume further that $\mu = 160$ and $\sigma = 20$ and that we select a random sample of 3 weights from this population. We shall compute the probabilities that all 3 weights exceed 170 pounds, that none of them exceed 170 pounds, and that exactly 1 of the 3 exceeds 170 pounds (and thus the other 2 do not). Let X_1 be the first weight selected, X_2 the second, and X_3 the third. Then, since these are a random sample from the population, the probability measure for each individual X_i, $i = 1, 2, 3$, is the same as the population random variable X, namely normal with mean 160 and standard deviation 20, regardless of what values we observe for the other weights. Thus, the probability that each individual sample value exceeds 170 pounds is

$$P(X > 170) = P\left(Z > \frac{170 - 160}{20}\right)$$

$$= P\left(Z > \frac{1}{2}\right)$$

$$= .3085$$

where Z is the standard normal random variable. Then to evaluate the probabilities mentioned above we can reason as follows: we call the random selection of an individual from the population a Bernoulli trial with two possible outcomes — either the weight selected exceeds 170 pounds or it does not. Defining a success to be a weight in excess of 170 pounds, we see then that our selection of 3 weights from the population is the same as the performance of $n = 3$ Bernoulli trials with probability of success on each trial equal to $p = .3085$. Thus,

$$P(\text{all 3 exceed 170 pounds}) = P(\text{3 successes})$$

$$= \binom{3}{3} (.3085)^3 (.6915)^0$$

$$= .0294$$

$$P(\text{none exceed 170 pounds}) = P(\text{3 failures})$$

$$= \binom{3}{0} (.3085)^0 (.6915)^3$$

$$= .3307$$

$$P(\text{exactly 1 exceeds 170 pounds}) = P(\text{1 success})$$

$$= \binom{3}{1} (.3085)^1 (.6915)^2$$

$$= .4425$$

Note that we are using the binomial probability measure here to answer the question posed, because our random sample from the population can be looked at as repeated Bernoulli trials with the same probability of success for each trial.

Example 6.2.2

As in Example 3.3.2, assume that you drive to work five mornings a week, always taking the same route that passes through one intersection with a stoplight. The time at which you arrive at the stoplight is a random variable W with density as shown in Figure 3.3.2. (See page 60.) We can define the conceptual population of all the days you will be driving to work this way and the conceptual population of times at which you arrive at the stoplight;

W then is the population random variable for this conceptual population of arrival times at the stoplight. Your arrival times at this light during one 5-day work week can be looked at as a random sample of size 5 from this conceptual population, so long as it is reasonable to assume that your arrival time on any particular day is unaffected by the time you arrived the other days. As computed in Example 3.3.2, the probability that you have to stop at the light any particular day is .329. Thus we can look at your 5 arrival times at the light during one week as being 5 repeated Bernoulli trials with probability of success given by $p = .329$ for each (calling success the outcome that you must stop at the light). It follows, then, that

$$P(\text{you must stop every day}) = P(5 \text{ successes})$$

$$= \binom{5}{5} (.329)^5 (.671)^0 = .0039$$

$$P(\text{you don't stop during the week}) = P(0 \text{ successes})$$

$$= \binom{5}{0} (.329)^0 (.671)^5$$

$$= .1360$$

$$P(\text{you must stop 3 times}) = P(3 \text{ successes})$$

$$= \binom{5}{3} (.329)^3 (.671)^2$$

$$= .1603, \text{ etc.}$$

Again, probabilities of certain types of events happening in the random sample from the population are evaluated by using the binomial probability measure.

If we have a finite population, there are two different ways in which the sampling may be accomplished, either *with replacement* or *without replacement*. By sampling with replacement we mean that the first element selected for the sample is replaced in the population and is eligible to be selected a second (or further) time in the same sample. Sampling without replacement says the first, or any other element, selected for the sample is *not* replaced in the population before the subsequent sample elements are chosen. Thus, with this type of sampling any particular population member can occur at most once in the same sample. The following two examples illustrate the differences between these two methods.

Example 6.2.3

Assume a bowl contains 10 identical pieces of paper, except that the pieces are numbered from 1 to 10. Considering the contents of this bowl to be a population and letting X be the population random variable, we have $P(X = x) = 1/10$, $x = 1, 2, \ldots, 10$. Suppose we draw a random sample of size 2 from the bowl *with replacement*. Thus, we will randomly draw one slip, record the number on it, then replace that slip to the bowl and again choose one slip randomly from the 10 in the bowl to determine the second number for the sample. Let X_1 be the first number selected and let X_2 be the second number selected. Then, since the first slip is drawn at random from the 10, it is clear that the probability measure for X_1 is

$$P(X_1 = x) = \frac{1}{10} \qquad x = 1, 2, \ldots, 10$$

Regardless of which slip was drawn for X_1 it will be replaced in the bowl and the second slip is drawn at random from the 10. Thus the probability measure for X_2 is

$$P(X_2 = x) = \frac{1}{10} \qquad x = 1, 2, \ldots, 10$$

regardless of the value for X_1. Notice that in this case the probability measures for X_1 and X_2 are each identical with the probability measure for X, regardless of the value the other variable equals. When sampling at random with replacement from a finite population, the elements of the sample will satisfy the definition given above for a random sample of the population random variable. This is not the case if the sampling is done without replacement, as the following example illustrates.

Example 6.2.4

Assume the same situation described in Example 6.2.3 and assume two slips are selected at random from the bowl, without replacement. Thus one slip will be selected at random from the 10 in the bowl to determine the value for X_1, the first number drawn; this first slip is *not* replaced in the bowl before the second slip is drawn. Thus, the second slip will be drawn at random from the nine remaining in the bowl, to determine the value of X_2, the second number drawn. Then, it is easy to see that the probability measure for X_1 is

$$P(X_1 = x) = \frac{1}{10} \qquad x = 1, 2, \ldots, 10$$

the same as the probability measure of the population random variable. The probability measure for X_2, however, depends on which nine numbers remain in the bowl when the second slip is drawn. If the first slip drawn has a 1 on it, the probability measure for X_2 is

$$P(X_2 = x) = \frac{1}{9} \qquad \text{for} \qquad x = 2, 3, \ldots, 10, \text{ given } X_1 = 1$$

On the other hand, if the first slip drawn has a 10 on it, the probability measure for X_2 is

$$P(X_2 = x) = \frac{1}{9} \qquad \text{for} \qquad x = 1, 2, \ldots, 9, \text{ given } X_1 = 10$$

Thus the probability measure appropriate for X_2 depends on the particular value X_1 was equal to (since the same number cannot occur twice). Thus, if we sample without replacement from a finite population, the sample values do not satisfy the requirements given above for a random sample of the population random variable.

As was mentioned above, all of the techniques for statistical inference that we will discuss are appropriate for random samples of the population random variable. If the population is continuous or conceptual (or both), selecting each sample value at random will give rise to a random sample of the population random variable and the techniques of this book apply directly. The same is true if the population is discrete and infinite (or equivalently, the population is discrete, finite, and the sampling is done with replacement). Only in the case of a finite discrete population, and sampling without replacement, as illustrated in Example 6.2.4, do we end up with the random variables in the sample not giving a random sample of the population random variable. Even in this latter case, though, if the sample size is small compared to the population size (the sample is no more than 10% of the population), the techniques we shall study are appropriate and inferences made should be subject to a negligible error.

Exercise 6.2

1. Assume that the amount of beer that an automatic bottle-filling machine puts into a beer bottle is a normal random variable with $\mu = 11.1$ (ounces), $\sigma = .1$ (ounce). (Thus the conceptual population random variable X is normal with $\mu_X = 11.1$ and $\sigma_X = .1$.)
    (a) If you buy a six-pack of this beer, what is the probability that each of the six bottles you buy contains at least 11 ounces?
    (b) What is the probability that all of your bottles contain less than 11 ounces?

2. Assume that the height a college high jumper will clear on each try is a normal random variable with $\mu = 74$ (inches) and $\sigma = 2$ (inches). (The conceptual population random variable X is normal with $\mu_X = 74$ and $\sigma_X = 2$.) At an intercollegiate track meet he has three chances to clear the bar set at 6 feet 4 inches, before he is eliminated. What is the proba-

bility he is not eliminated at this height? (Here is an equivalent question. Given a random sample of size 3 of X, what is the probability that at least one value exceeds 76 inches?)

3. Assume that the time T for you to drive from your residence to your job in the morning has the triangular density described in Example 4.3.2 and shown in Figure 4.3.3. (We then are assuming a triangular population, and T is the population random variable.) Assume the five times required for you to drive to work in the same week constitute a random sample of T.
    (a)   What is the probability that it takes you at least 10 minutes to get to work each day in the week?
    (b)   What is the probability that you get to work in less than eight minutes on two of the five days (and eight or more minutes are required the other three days)?

4. Assume a continuous population random variable U with median m_U. Given a random sample of size n of U, what is the probability that all n observations are smaller than m_U?

5. Assume a continuous population random variable V with 90th percentile $t_{.9}$. Given a random sample of size n of V, what is the probability that none of the sample observations is smaller than $t_{.9}$?

6. Consider the population of baseball games, in each of which player A is at bat exactly four times. Assume that the number of hits that A gets in any one of these games is a binomial random variable with parameters $n = 4$, $p = .25$.
    (a)   What is the expected number of hits he will make in one of these games?
    (b)   What is the probability he makes (exactly) one hit in one of these games?
    (c)   In one month he plays in 10 games in which he is at bat four times. What is the probability he gets exactly 1 hit in at least three of these games?

7. Assume that the number of automobile accidents on a 10-mile-long two-lane highway, per week, is a Poisson random variable with parameter $\lambda t = 2$ (the conceptual population random variable X is Poisson with parameter $\lambda t = 2$).
    (a)   What is the probability of at least two automobile accidents in a week?
    (b)   What is the probability of at least two accidents per week for five weeks in a row?

8. Assume that the scores made by high school seniors on a mathematics achievement test are normally distributed with $\mu = 500$, $\sigma = 100$ (the conceptual population random variable X has a normal distribution

with $\mu_X = 500$ and $\sigma_X = 100$). A random sample of 10 seniors is selected and each takes the test.

(a) What is the probability that the first student's score lies between 400 and 600?

(b) What is the probability that exactly five of these students score between 400 and 600?

6.3 SUMS OF RANDOM VARIABLES

In the last section we considered the idea of sampling at random from a population and the concept of a random sample of the population random variable. As we saw there, given a random sample of the population random variable, we can then look at each individual sample value as a Bernoulli trial; this enables us to employ the binomial probability measure to compute probabilities of occurrence of different possible numbers of successes that we might observe (e.g., number of weights that exceed 170 pounds, number of mornings you must stop at a stoplight, and a number of bottles of beer that contain at least 11 ounces). While this type of question does occur frequently, it is certainly not the only type of computation that is of interest, given a random sample of a population random variable.

In many problems of inference we will want to add together the values of the random variables in the sample and to make probability statements about this sum (or a constant times this sum). For example, most one-pound packages of butter consist of four individually wrapped sections. The weight of the butter in the package, then, is equal to the sum of the weights of the four sections. Granted we know the probability measure of the population of sections, what can we say about the weights of the packages? This sort of question cannot be answered by solely considering the numbers of sections (out of four in a package) that do or do not exceed 1/4 pound (or some other value) in weight; we need to consider the actual weights of the sections and their sum to be able to answer questions about the weights of the packages.

Suppose we assume that weights of United States adult males are normally distributed but that the mean, μ, of the population is unknown; we want to estimate or guess at its value based on the numbers we observe in a random sample of the population random variable. One way we might use the sample weights to get a guess of the value of μ is to compute the average of the numbers we observe in the sample; this sample average, then, could be our guess for the unknown value of μ. Since the sample average consists of adding together the weights in the sample and dividing by n, the size of the sample, we would be interested again in the distribution of the sum of the sample values in studying whether this sample average seems a good rule to use in making a guess about the unknown value for μ. (See Chapter 7 for more information about guessing values for population parameters.)

A basic result, which is usually proved in courses in mathematical statistics, is given in the following theorem. Two random variables, X and Y, are *independent* if the probability measure for Y is the same, no matter which observed value X may equal and vice versa.

Theorem 6.1

Assume X and Y are independent random variables. Define the new random variable $W = aX + bY$, where a and b are any constants. Then the mean of W is $\mu_W = a\mu_X + b\mu_Y$ and the variance of W is $\sigma_W^2 = a^2\sigma_X^2 + b^2\sigma_Y^2$.

Example 6.3.1

Suppose we take $a = b = 1$ in the above theorem, and W then is the sum of X and Y; the mean of the sum is

$$\mu_W = \mu_X + \mu_Y$$

and the variance of the sum is

$$\sigma_W^2 = \sigma_X^2 + \sigma_Y^2$$

The standard deviation is $\sigma_W = \sqrt{\sigma_X^2 + \sigma_Y^2}$, *not* the sum of the two standard deviations. If we take $a = 1$ and $b = -1$, then W is the difference, $X - Y$; the mean of the difference is

$$\mu_W = \mu_X - \mu_Y$$

and the variance of the difference is

$$\sigma_W^2 = (1)^2\sigma_X^2 + (-1)^2\sigma_Y^2 = \sigma_X^2 + \sigma_Y^2$$

the same as the variance of the sum. At first glance the fact that the variance of the sum and the variance of the difference are equal may seem strange. Remembering that the variance of a random variable is the average of the square of the difference between the value of the variable and its mean, this result is simply saying that the sum varies equally about its mean as does the difference, so long as X and Y are independent. Put in this way, the result may seem more natural. In both cases the variance is the sum of the variances of X about its mean and of Y about its mean. The reader is cautioned not to make the common error of forgetting Theorem 6.1 and using $\sigma_X^2 - \sigma_Y^2$ as the variance of $X - Y$.

The following theorem, which we shall not prove, gives a frequently used result about the sum of the values observed in drawing a random sample of a population random variable. It is a generalization of Theorem 6.1.

Theorem 6.2

Assume we have a population random variable X with mean μ_X and variance σ_X^2; $X_1, X_2, \ldots, X_n$ is a random sample of X. Then the random variable

$$V = \sum_{i=1}^{n} X_i$$

(which is the sum of the sample random variables) has mean $\mu_V = n\mu_X$ and variance $\sigma_V^2 = n\sigma_X^2$.

Notice immediately, then, that the mean of the sum is the sum of the means and that the variance of the sum is the sum of the variances (this result holds true even if the X_i's added together do not have the same means and variances, so long as they are independent); the standard deviation of the sum then is

$$\sigma_V = \sqrt{\sigma_V^2} = \sigma_X \sqrt{n}$$

not the sum of the standard deviations.

Example 6.3.2

A dairy uses quarter-pound molds to form quarter-pound sections of butter; four quarter-pound sections are then placed in the same package and sold as one pound of butter. Because of variations in the specific gravity of the butter used, as well as minute variations in how full the mold is from one section to another, the "exact" weights of all sections is not a quarter pound. Assume that the population of weights of sections is normal with mean $\mu = .255$ (pounds) and standard deviation $\sigma = .01$ (pounds). (Thus the population random variable, X, is normal with mean $\mu_X = .255$ and standard deviation $\sigma_X = .01$.) Four of these sections will be placed together in the same package. What can we say about the distribution of weights of the packages? Let X_1, X_2, X_3, and X_4 be the weights of the four sections in the same package and assume that these four random variables are a random sample of the population random variable X. The weight of the package is

$$V = X_1 + X_2 + X_3 + X_4$$

From Theorem 6.1

$$\mu_V = 4\mu_X = 4(.255) = 1.02 \text{ (pounds)}$$

and

$$\sigma_V^2 = 4\sigma_X^2 = 4(.01)^2 = .0004$$

The standard deviation of the weights of the packages is

$$\sigma_V = \sqrt{.0004} = .02$$

In this example we started with one population, the weights of the quarter-pound sections. From this population we derived the mean and variance of a second, derived population which is, in a sense, one-fourth as large: the population of weights of packages, each of which contains four sections. If sections are selected at random, placed in sets of four, then the population of weights of the packages necessarily has mean 1.02 pounds and standard deviation .02 pounds (granted the initial assumptions about the weights of the sections).

It was mentioned above that we might logically want to consider the average of the numbers in a sample as a guess for the value of the population mean μ, if that quantity were unknown. As we shall see in subsequent chapters, there are many good reasons supporting this type of reasoning. Because of its importance we shall reserve a special symbol for the sample mean. Assume we have a random sample of size n of a population random variable X. As above, the random variables in the sample will be denoted by X_1, X_2, ..., X_n, representing the first value selected, second value selected, etc. The sample mean (or average), then, is obtained by dividing the total of the sample by n, the size of the sample. We shall denote the sample mean by the symbol for the population random variable, with a bar above it. Thus for this case

$$\overline{X} = \frac{1}{n} \sum_{i=1}^{n} X_i$$

(read X-bar) is the way we will denote the sample mean. If we had used Y for the population random variable, then the sample mean would be, for example,

$$\overline{Y} = \frac{1}{n} \sum_{i=1}^{n} Y_i$$

This sample mean is itself a random variable; as the particular observed values in the sample change, so does the observed value for $\overline{X}$. The following corollary to Theorem 6.2 gives the mean, variance, and standard deviation of the distribution for $\overline{X}$, computed from random samples of the population random variable X; the result follows quite easily from Theorem 6.2 and Theorems 4.1 and 4.3, as you are asked to verify in Exercise 10 below.

Corollary 6.2.1

Assume we have a population random variable X with mean μ_X and variance σ_X^2; $X_1, X_2, \ldots, X_n$ is a random sample of X. Then the sample mean

$$\overline{X} = \frac{1}{n} \sum_{i=1}^{n} X_i$$

has mean $\mu_{\overline{X}} = \mu_X$ and variance $\sigma_{\overline{X}}^2 = \sigma_X^2/n$.

The amazing thing about this result is the fact that the average of a random sample of the population random variable varies less about μ_X than do the individual population members. Suppose we are given a population random variable X with mean μ_X and standard deviation $\sigma_X = 1$. If we have a random sample of n observations of X, and average them together, than X has mean $\mu_{\overline{X}} = \mu_X$, the same as the population random variable; its standard deviation, though, is $\sigma_{\overline{X}} = 1/\sqrt{n}$. Thus, for example, if $n = 4$, $\sigma_{\overline{X}} = 1/2$ and if $n = 100$, $\sigma_{\overline{X}} = 1/10$. The larger n becomes, the smaller the standard deviation of $\overline{X}$ and, thus, the less that $\overline{X}$ varies about $\mu_{\overline{X}} = \mu_X$. These statements refer to the *sampling distribution* for $\overline{X}$; given repeated samples of size n of X, we are describing the way in which $\overline{X}$ will vary from one sample to another.

Theorem 6.2 gives results about the mean and variance of the sum of values in a random sample of a population random variable. We need still more information before we could compute the values of probability statements for such a sum; we would need to know the appropriate probability measure for such a sum, rather than just the mean and variance of the probability measure. The following theorems state the probability measure of sums of two of the types of random variables we have discussed in Chapter 5, the normal and Poisson.

Theorem 6.3

Assume $X_1, X_2, \ldots X_n$ is a random sample of a normal random variable X with mean μ_X and variance σ_X^2. Then the probability measure for

$$V = \sum_{i=1}^{n} X_i$$

is *normal* with mean $\mu_V = n\mu_X$ and variance $\sigma_V^2 = n\sigma_X^2$. The probability measure for

$$\overline{X} = \frac{1}{n} \sum_{i=1}^{n} X_i$$

is also normal with mean $\mu_{\overline{X}} = \mu_X$ and variance $\sigma_{\overline{X}}^2 = \sigma_X^2/n$.

Theorem 6.4

Assume $X_1, X_2, \ldots, X_n$ is a random sample of a Poisson random variable X with parameter λt. Then the probability measure for

$$V = \sum_{i=1}^{n} X_i$$

is also *Poisson* with parameter $n\lambda t$.

The following examples illustrate some uses of these results.

Example 6.3.3

Assume, as in Example 6.3.1, that the quarter-pound sections of butter produced by a dairy have a normal population with mean $\mu_X = .255$ pounds and standard deviation $\sigma_X = .01$ pounds. Four of these sections are put into the same package. What proportion of the resulting packages will weigh less than one pound? As discussed in the previous example, if we let X_1, X_2, X_3, X_4 represent the weights of the four sections in the same package, then the weight of the package is

$$V = \sum_{i=1}^{4} X_i.$$

Assuming the four weights in the same package are a random sample from the population, Theorem 6.2 says that the probability measure for V is normal with $\mu_V = 4\mu_X = 1.02$ and $\sigma_V^2 = 4\sigma_X^2 = .0004$; thus $\sigma_V = \sqrt{.0004} = .02$. The proportion of packages that weigh less than one pound is given by $P(V < 1)$. Since the probability measure for V is normal,

$$P(V < 1) = P\left(Z < \frac{1 - 1.02}{.02}\right)$$

$$= P(Z < -1)$$

$$= .1587$$

where Z is the standard normal random variable. About 16% of the packages of butter sold by this dairy will weigh less than one pound, with the given assumptions.

Example 6.3.4

Assume that patients with broken bones arrive at a university hospital like occurrences in a Poisson process at a rate of two per week. In a four-week period what is the probability the university is called on to treat 10 cases of broken limbs? If we let X_1, X_2, X_3, X_4, respectively, be the number of cases that the university receives in the four individual weeks, then the total number they receive in the four-week period is

$$V = \sum_{i=1}^{4} X_i$$

Each X_i has a Poisson probability measure with parameter $\lambda t = 2$. From Theorem 6.4 the probability measure for V is again Poisson with parameter $4 \cdot 2 = 8$. The probability they must treat 10 cases of broken limbs in the four-week period, then, is

$$P(V = 10) = .1186$$

as we see from Table A.3.

Exercise 6.3

1. Assume that the amount of soda that an automatic bottling machine puts into a seven-ounce bottle is a normal random variable X with mean $\mu_X = 6.98$ ounces and standard deviation $\sigma_X = .05$ ounces. Bottles are placed at random into six-pack containers. Assume you buy one six-pack of this type of soda.
    (a) What is the distribution of the total amount of soda in your six-pack?
    (b) What is the probability there is less than 42 ounces, in total, in the six-pack you purchase?

2.   Refer to the sections of butter discussed in Examples 6.3.2 and 6.3.3, and assume that I buy three one-pound packages of this dairy's butter. What is the probability each of the packages weighs at least one pound?

3.   Again assume I buy three packages of butter, as discussed in question 2 above.
     (a)   What is the probability that the total weight of the butter I get exceeds three pounds?
     (b)   Why is this probability different from the answer to question 2 above?

4.   Assume that molecules of a rare gas occur at a rate of one per cubic foot of air at 70°F at sea level. Five samples of air, each of one cubic foot, are examined. What is the probability that at least one molecule of the gas is found?

5.   Assume that the heights of United States adult males are normally distributed with mean μ (unknown) and standard deviation $\sigma = 4$ inches.
     (a)   What proportion of heights in the population are within two inches of μ?
     (b)   What proportion of means of random samples of size 16 are within two inches of μ?

6.   A random sample of $n = 20$ packages of frozen fish, each from the same manufacturer, are weighed. If we assume the population of weights of these packages has a mean $\mu = 11.9$ (ounces) and a standard deviation $\sigma = .2$ (ounces), what is the mean and standard deviation of the total of the 20 weights?

7.   If we assume the package weights in question 6 are normally distributed, what is the probability that the sum of the 20 weights in the sample would equal 237 ounces or less?

8.   Assume that X is a uniform random variable on the interval from 0 to 12 (see Example 4.3.1 and discussion following) and that Y is triangular on the interval from five to 15 (see Example 4.3.2 and discussion following). X and Y are independent.
     (a)   What is the mean and variance of $X + 2Y$?
     (b)   What is the mean and variance of $2X + Y$?

9.   The result given in Theorem 6.1 (as well as the others in this section) are exact only if the random variables are independent. Suppose we have a random variable X with some specified probability measure with $\mu_X = 10$ and $\sigma_X = 2$, and we let $Y = -X$. (By this we mean the value Y equals is always the negative of the value X equals; for example, if $X = 1$, then $Y = -1$ and if $X = 2.64$, $Y = -2.64$.) Since the value of Y is determined by the value of X, the X and Y are certainly not independent; the probability measure for Y is *not* the same for every possible observed value for X. Let $V = X + Y$.
     (a)   What are the values for μ_V and σ_V?
     (b)   What are the values for μ_Y and σ_Y?

(c)   What would Theorem 6.1 give as the values for μ_V and σ_V?

10.   Use Theorems 6.2, 4.1, and 4.3 to show that the result given in Corollary 6.2.1 is true.

6.4   THE CENTRAL LIMIT THEOREM; APPROXIMATIONS

In this section, we shall discuss one of the most important theorems for applications of statistical methods; many people agree that it is *the* most important single result. The result is called the central limit theorem and is concerned with the probability measure for sums of random variables. There are many versions of this theorem, differing from each other in terms of the assumptions made about the random variables involved. Although surprisingly few prior assumptions can be shown to be required, the conclusion of the theorem still follows. The assumptions we shall state can in fact be weakened in several ways; the statements of these weaker assumptions involve more mathematics and probability theory than we have studed.

Theorem 6.5   Central Limit Theorem

Assume $X_1, X_2, \ldots, X_n$ is a random sample of a population random variable X with mean μ_X and variance σ_X^2. Then, for a sufficiently large n, the probability measure of the total,

$$Y_n = \sum_{i=1}^{n} X_i$$

is well approximated by the *normal* probability measure with $\mu_{Y_n} = n\mu_X$ and $\sigma_{Y_n}^2 = n\sigma_X^2$. The probability measure of the sample mean

$$\overline{X} = \frac{1}{n} Y_n = \frac{1}{n} \Sigma X_i$$

is well approximated by the *normal* probability measure with

$$\mu_{\overline{X}} = \mu_X, \sigma_{\overline{X}}^2 = \frac{\sigma_X^2}{n}$$

The amazing thing about this result is that no assumption is necessary about the probability measure for the population (except that μ_X and σ_X^2 exist); still the probability measures for the sample total and sample mean are well approximated by *normal* probability measures. The population measure can be discrete or continuous, symmetric or not; still totals and means approximately have normal distributions. This is one of the reasons that the normal probability measures are so important, and so frequently used, in problems in statistical inference. If we observe the value of a random variable, say V, and are able to look at the value of V as the sum of other random variables, then Theorem 6.5 implies that the probability measure for V can be well approximated by a normal measure.

To illustrate the results of this theorem, let us consider the discrete conceptual population of numbers we will observe if a fair die is rolled repeatedly. The population random variable X, then, has probability measure

$$P(X = x) = \frac{1}{6} \quad \text{for} \quad x = 1, 2, 3, 4, 5, 6$$

The histogram for this probability measure is given in Figure 6.4.1a. If we roll this die $n = 2$ times and let X_1 be the first number, X_2 the second, and define the total of the two to be

$$Y_2 = X_1 + X_2$$

methods of mathematical statistics can be used to show that the exact probability measure for the average of the two numbers, $\overline{X} = Y_2/2 = (X_1 + X_2)/2$, is as given in Figure 6.4.1b. Figure 6.4.1c presents the histogram of the exact probability measure of $\overline{X}$ based on $n = 3$ rolls of this die, as well as the approximating normal probability measure. It can be verified that the mean and variance of the population random variable are $\mu_X = 3.5$ and $\sigma_X^2 = 35/12$; thus the mean and variance of $\overline{X}$ based on $n = 3$ rolls, are $\mu_{\overline{X}} = \mu_X = 3.5$ and $\sigma_{\overline{X}}^2 = \sigma_X^2/3 = 35/36$. These are the values of the parameters of the normal curve drawn in Figure 6.4.1c.

In Section 3.2 we discussed histograms for discrete random variables. It will be recalled that the area of a bar is equal to the probability that the random variable equals the value at which the bar is

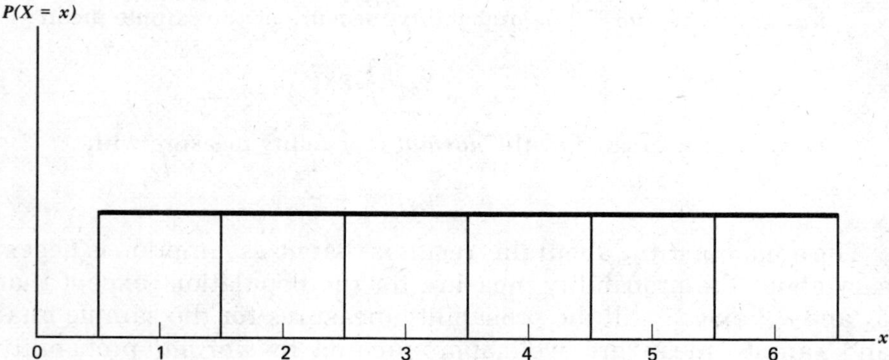

Figure 6.4.1a Population measure.

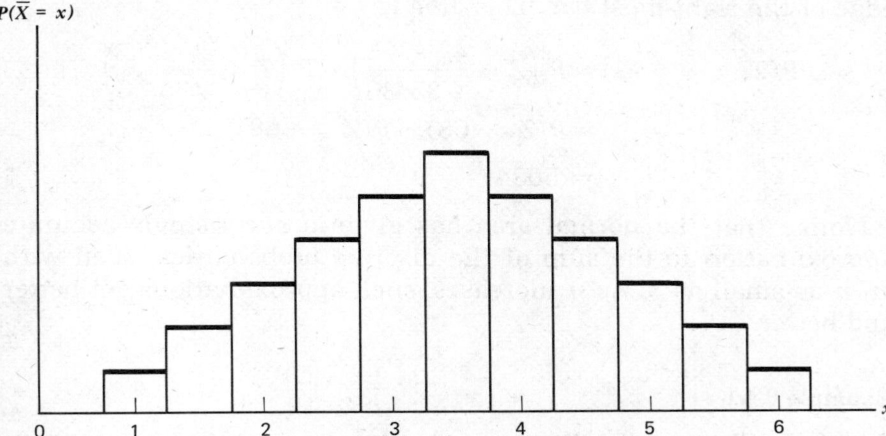

Figure 6.4.1*b*   Probability measure for $\overline{X}$, $n = 2$.

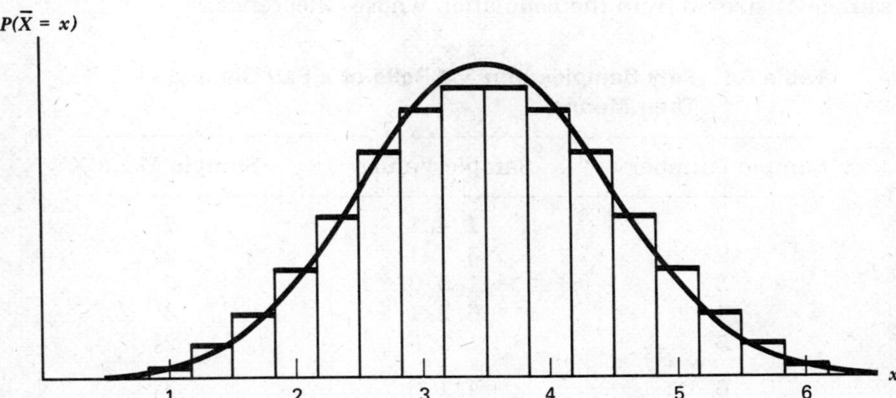

Figure 6.4.1*c*   Probability measure for $\overline{X}$, $n = 3$ and approximating normal measure, $\mu = 3.5$, $\sigma = .986$.

centered; the probability that the discrete random variable equals some value in an interval is given by the sum of the areas of all bars centered in the interval. In Figure 6.4.1c the exact probability that $\overline{X}$ lies between 3 and 4, inclusive, then is given by the sum of the areas of four bars, those centered at 3, $3\frac{1}{3}$, $3\frac{2}{3}$, and 4. (The sum of these four areas is .481.) From the figure we can see that this area should be well approximated by the area under the normal density between $2\frac{5}{6}$ and $4\frac{1}{6}$, so that we have started from the left-hand edge of the left-most bar and included all the area up to the right-hand edge of the right-most bar. This area is

$$P(2\tfrac{5}{6} \le \overline{X} \le 4\tfrac{1}{6}) = P\left(Z \le \frac{4\tfrac{1}{6} - 3.5}{\sqrt{35/36}}\right) - P\left(Z \le \frac{2\tfrac{5}{6} - 3.5}{\sqrt{35/36}}\right)$$

$$= P(Z \le .68) - P(Z \le -68)$$

$$= .5034$$

Notice that the normal area has given a surprisingly accurate approximation to the sum of the discrete probabilities, even with an n as small as 3. As n increases, such approximations get better and better.

Example 6.4.1

To further illustrate the discussion above, a die was rolled 3 times, then another 3 times, etc., for a grand total of 50 repetitions of 3 rolls. The numbers that occurred and the averages of the numbers are given in Table 6.1. The values of $\overline{X}$ that occurred in these 50 sets of rolls constitute a random sample of size 50 from the population whose (theoretical)

Table 6.1   Fifty Samples of $n = 3$ Rolls of a Fair Die and Their Means

Sample Number	Sample Values	Sample Mean $\overline{X}$
1	1, 4, 1	2
2	5, 1, 1	$2\tfrac{1}{3}$
3	1, 5, 3	3
4	6, 2, 1	3
5	1, 1, 6	$2\tfrac{2}{3}$
6	4, 1, 6	$3\tfrac{2}{3}$
7	4, 6, 6	$5\tfrac{1}{3}$
8	1, 1, 4	2
9	1, 4, 6	$3\tfrac{2}{3}$
10	2, 5, 3	$3\tfrac{1}{3}$
11	6, 2, 2	$3\tfrac{1}{3}$
12	6, 5, 1	4
13	2, 1, 2	$1\tfrac{2}{3}$
14	2, 3, 6	$3\tfrac{2}{3}$
15	4, 6, 5	5
16	3, 2, 5	$3\tfrac{1}{3}$
17	5, 5, 5	5
18	3, 3, 3	3
19	5, 1, 2	$2\tfrac{2}{3}$
20	2, 5, 3	$3\tfrac{1}{3}$

**Table 6.1   Fifty Samples of $n = 3$ Rolls of a Fair Die and
Their Means (Continued)**

Sample Number	Sample Values	Sample Mean $\overline{X}$
21	4, 2, 3	3
22	6, 5, 3	$4\frac{2}{3}$
23	4, 5, 5	$4\frac{2}{3}$
24	2, 6, 6	$4\frac{2}{3}$
25	6, 1, 1	$2\frac{2}{3}$
26	4, 3, 6	$4\frac{1}{3}$
27	1, 5, 5	$3\frac{2}{3}$
28	5, 2, 4	$3\frac{2}{3}$
29	1, 3, 4	$2\frac{2}{3}$
30	3, 6, 2	$3\frac{2}{3}$
31	2, 5, 2	3
32	2, 6, 5	$4\frac{1}{3}$
33	6, 3, 3	4
34	6, 4, 1	$3\frac{2}{3}$
35	5, 1, 2	$2\frac{2}{3}$
36	4, 3, 4	$3\frac{2}{3}$
37	3, 4, 3	$3\frac{1}{3}$
38	2, 1, 3	2
39	6, 1, 6	$4\frac{1}{3}$
40	5, 5, 6	$5\frac{1}{3}$
41	3, 3, 4	$3\frac{1}{3}$
42	5, 6, 2	$4\frac{1}{3}$
43	4, 2, 1	$2\frac{1}{3}$
44	6, 3, 3	4
45	6, 2, 4	4
46	6, 5, 6	$5\frac{2}{3}$
47	1, 6, 3	$3\frac{1}{3}$
48	6, 3, 4	$4\frac{1}{3}$
49	4, 5, 3	4
50	3, 6, 1	$3\frac{1}{3}$

histogram is given in Figure 6.4.1c. Thus, if we summarize these 50 values
of $\overline{X}$, we would expect the resulting histogram to be similar in shape to
that in Figure 6.4.1c. Table 6.2 summarizes the 50 values of $\overline{X}$ given in
Table 6.1 and Figure 6.4.2 gives the histogram of these 50 observed values;
notice that there is a marked similarity between this histogram and that
given in Figure 6.4.1c.

Table 6.2 Distribution of the Sample Means of Fifty Samples of Three Rolls of a Fair Die

Value	Frequency
$1\frac{2}{3}$	1
2	3
$2\frac{1}{3}$	2
$2\frac{2}{3}$	5
3	5
$3\frac{1}{3}$	8
$3\frac{2}{3}$	8
4	5
$4\frac{1}{3}$	5
$4\frac{2}{3}$	3
5	2
$5\frac{1}{3}$	2
$5\frac{2}{3}$	1
Total	50

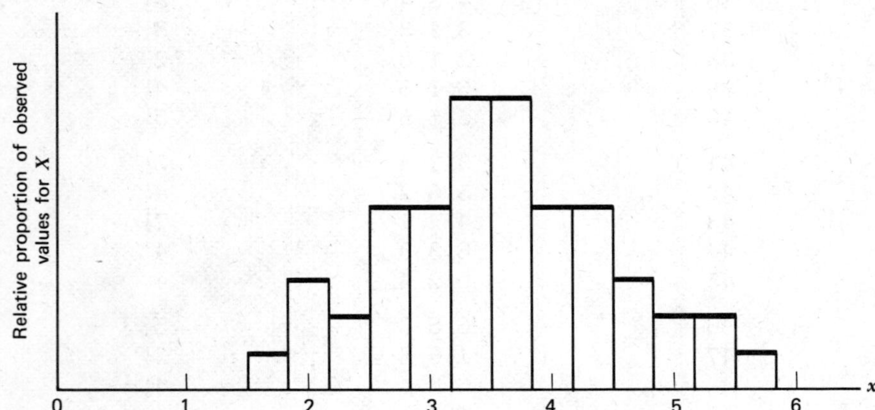

Figure 6.4.2 Observed values of $\overline{X}$ for 50 samples of three rolls.

Note that in these 50 rolls, the observed value for $\overline{X}$ was between 3 and 4, inclusive, 26 times; the relative proportion of the time it was in this interval is $26/50 = .52$

The central limit theorem (Theorem 6.5) is the basis for approximating binomial and Poisson probability measures, in appropriate cases (as well as for the approximation of many other measures).

We shall first discuss the normal approximation for binomial probability measures. It will be recalled that the binomial random variable Y with parameters n and p is the total number of successes observed in n repeated independent Bernoulli trials, where the probability of success on each trial is p. Suppose we consider this binomial experiment, consisting of n repeated independent trails and define n random variables, $X_1, X_2, \ldots, X_n$. We define X_1 to be 1 if we get a success on the first trial and 0 otherwise. X_2 equals 1 if we get a success on the second trial and 0, otherwise. Similarly, X_n equals 1 only if we get a success on the nth trial and is 0, otherwise. Then, the probability is p for each X_i to equal 1 and is $q = 1 - p$ for each X_i to equal 0, $i = 1, 2, \ldots, n$, The total number of successes in the n trials, then, is $\sum\limits_{i=1}^{n} X_i$ that is,

$$Y = \sum_{i=1}^{n} X_i$$

and we can see that our binomial random variable Y is the sum of the n individual random variables $X_1, X_2, \ldots, X_n$. The mean of each X_i is easily found to be $\mu_{X_i} = p$ and the variance is $\sigma_{X_i}^2 = pq$; then, as we know from Theorem 6.2, the mean of the sum is np and the vari

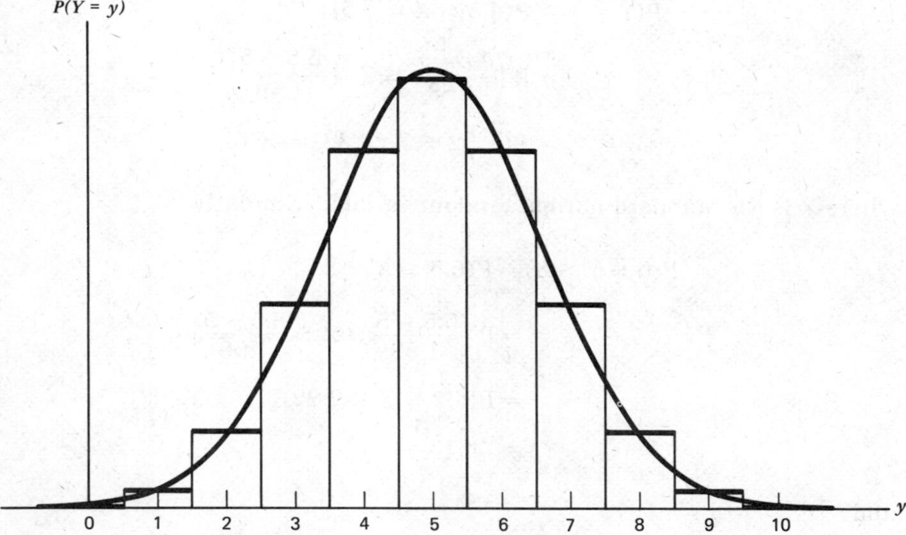

Figure 6.4.3   Exact histogram for binomial $n = 10, p = .5$ and approximating normal density $\mu = 5, \sigma = 1.58$.

ance of the sum is npq, the appropriate values of these quantities for the binomial random variable Y. The central limit theorem then says that the exact binomial histogram should be well approximated by a normal density with $\mu = np$ and $\sigma^2 = npq$, for n sufficiently large. The following two examples illustrate the use of this approximation.

Example 6.4.2

Let us first examine the approximation for a case in which we know the exact values for the probabilities (from Table A.1); we can then compare these with the normal approximate areas. Assume Y is binomial with $n = 10$ and $p = .5$; we shall evaluate $P(Y = 5)$, $P(6 \leq Y \leq 8)$, and $P(Y \leq 2)$. From Table 1 we find

$$P(Y = 5) = .246$$

$$P(6 \leq Y \leq 8) = .205 + .117 + .044 = .366$$

$$P(Y \leq 2) = .001 + .010 + .044 = .055$$

The histogram for these probabilities is given in Figure 6.4.3, as is the approximating normal density with $\mu = np = 10(.5) = 5$ and $\sigma = \sqrt{npq} = \sqrt{10(.5)(.5)} = 1.58$. We can see that the area under the smooth curve between 4.5 and 5.5 should give a good approximation to the area of the bar centered at 5 [which is $P(Y = 5)$]. Letting X be the normal random variable whose density is given in Figure 6.4.3, then,

$$P(Y = 5) \doteq P(4.5 \leq X \leq 5.5)$$

$$= P\left(\frac{4.5 - 5}{1.58} \leq Z \leq \frac{5.5 - 5}{1.58}\right)$$

$$= P(-.32 \leq Z \leq .32) = .2510$$

where Z is the standard normal random variable. Similarly,

$$P(6 \leq Y \leq 8) \doteq P(5.5 \leq X \leq 8.5)$$

$$= P\left(\frac{5.5 - 5}{1.58} \leq Z \leq \frac{8.5 - 5}{1.58}\right)$$

$$= P(.32 \leq Z \leq 2.22)$$

$$= .3613$$

and

$$P(Y \leq 2) \doteq P(X \leq 2.5)$$

$$= P\left(Z \leq \frac{2.5 - 5}{1.58}\right)$$

$$= P(Z \leq -1.58) = .0571$$

Each of these approximations can be seen to be quite good. As n gets larger they get better and better. Note that in the last case we approximated the sum of the areas of the bars centered at 0, 1, and 2, but we did *not* stop the continuous area at $-1/2$, the leftmost edge of the 0 bar. It can be shown that the approximation is generally better if we do not end these continuous areas, but include all the area in the tail.

The normal approximation in Example 6.4.2 is good, with n as small as 10, because we assumed $p = .5$, in which case the exact histogram is symmetric; as p differs from $1/2$, either toward 0 or toward 1, the exact histogram gets less symmetric and would not be well approximated by the normal density. No matter what the value of p, though, for n sufficiently large the approximation will be excellent; the closer p is to 0 or 1, the larger n has to be to give the same degree of approximation. A good rule of thumb is that the normal approximation should prove quite adequate if

$$9 \le (n + 9)\, p \le n$$

that is if

$$\frac{9}{n + 9} \le p \le \frac{n}{n + 9}$$

Example 6.4.3

Assume that the probability a person with cancer is cured by an operation is .75 (this is called the *success rate*). If 100 patients have this operation, approximate the probability that at least 80 are cured. Our exact binomial variable in this case has $n = 100$, $p = .75$ and we would approximate its probabilities with a normal density with mean $np = 75$ and standard deviation $\sqrt{npq} = 4.33$. Then

$$P(Y \ge 80) \doteq P(X \ge 79.5)$$

$$= P\left(Z \ge \frac{79.5 - 75}{4.33}\right)$$

$$= P(Z \ge 1.04)$$

$$= .1492$$

The probability that 65 or fewer are cured is approximately

$$P(Y \le 65) = P(X \le 65.5)$$

$$= P\left(Z \le \frac{65.5 - 75}{4.33}\right)$$

$$= P(Z \le -2.19) = .0143$$

The probability that exactly 75 are cured is approximately

$$P(Y = 75) = P(74.5 \leq X \leq 75.5)$$

$$= P\left(\frac{74.5 - 75}{4.33} \leq Z \leq \frac{75.5 - 75}{4.33}\right)$$

$$= P(-.12 \leq Z \leq .12)$$

$$= .0956$$

Poisson probability measures can also be well approximated by normal densities, in certain cases. It will be recalled, from Theorem 6.4, that if X_1, X_2, ..., X_n are independent Poisson random variables, each with parameter λt, then

$$Y = \sum_{i=1}^{n} X_i$$

is again a Poisson random variable with parameter $n \lambda t$. The central limit theorem says that the probability measure for Y should be well approximated by a normal density, as n gets large. Since the parameter of the Poisson probability measure for the sum is $n \lambda t$, this implies that the Poisson probability measure is well approximated

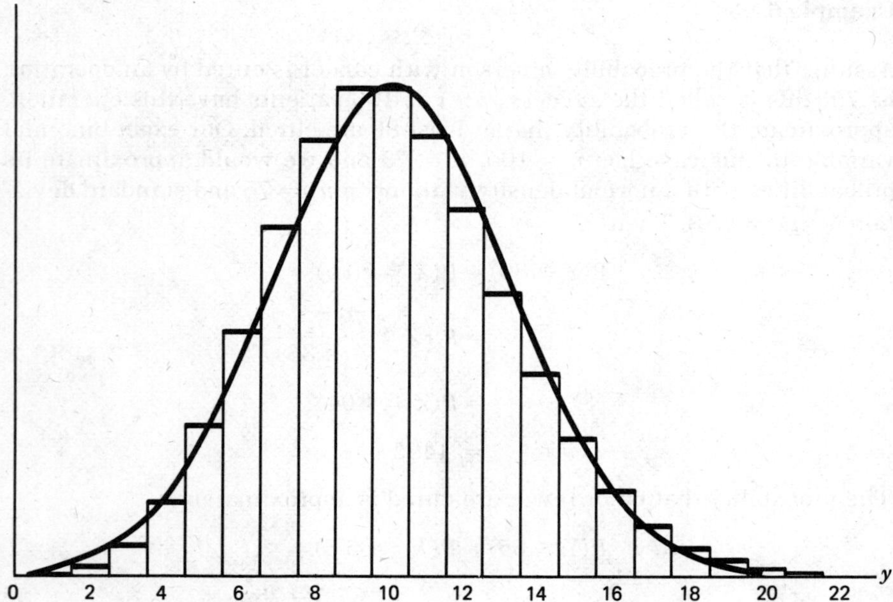

Figure 6.4.4   Poisson probability measure, $\lambda t = 10$ and approximating normal density, $\mu = 10$, $\sigma = 3.16$.

by a normal density as the value of the parameter becomes larger. Thus, if Y is a Poisson random variable with parameter γ, say, then its probability measure is well approximated by a normal density with parameters $\mu = \gamma$, $\sigma = \sqrt{\gamma}$, if γ is large. In practice, if the parameter is at least 10 the normal approximation should be quite good; it gets better the larger the value for the parameter.

Example 6.4.4

Let us assume Y is a Poisson random variable with parameter 10 and compare the exact probability values for Y with the approximate values given by a normal density with $\mu = 10$ and $\sigma = \sqrt{10} = 3.16$. Figure 6.4.4 plots the exact Poisson probability measure, as well as the approximating normal density. From Table A.3, the following are exact Poisson probabilities:

$$P(Y = 15) = .035$$

$$P(4 \leq Y \leq 8) = .322$$

$$P(Y \geq 11) = .417$$

Again letting X be the normal random variable whose density is used to approximate probabilities for Y, we find the following approximations to the above values:

$$P(Y = 15) \doteq P(14.5 \leq X \leq 15.5)$$

$$= P\left(\frac{14.5 - 10}{3.16} \leq Z \leq \frac{15.5 - 10}{3.16}\right)$$

$$= P(1.42 \leq Z \leq 1.74)$$

$$= .0369$$

And

$$P(4 \leq Y \leq 8) \doteq P(3.5 \leq X \leq 8.5)$$

$$= P\left(\frac{3.5 - 10}{3.16} \leq Z \leq \frac{8.5 - 10}{3.16}\right)$$

$$= P(-2.06 \leq Z \leq -.47)$$

$$= .2995$$

And

$$P(Y \geq 11) \doteq P(X \geq 10.5)$$

$$= P\left(Z \geq \frac{10.5 - 10}{3.16}\right)$$

$$= P(Z \geq .16)$$

$$= .4364$$

As the Poisson parameter gets larger, the approximate probabilities get closer to the exact values.

Example 6.4.5

Assume that traffic deaths occur in the United States over a three-day holiday weekend at a rate of 175 per day. Approximate the probability that there will be between 500 and 600 traffic deaths over a three-day United States holiday weekend. Again let Y be the exact Poisson random variable and let X be the normal random variable whose density will be used to approximate the probability measure for Y. Then Y is Poisson with parameter $\lambda t = 525$; X is normal with mean 525 and standard deviation $\sqrt{525} = 22.92$; and

$$P(500 \leq Y \leq 600) \doteq P(499.5 \leq X \leq 600.5)$$

$$= P\left(\frac{499.5 - 525}{22.92} \leq Z \leq \frac{600.5 - 525}{22.92}\right)$$

$$= P(-1.11 \leq Z \leq 3.30)$$

$$= .8665$$

The probability of exactly 525 deaths is approximately

$$P(Y = 525) = P(524.5 \leq X \leq 525.5)$$

$$= P\left(\frac{524.5 - 525}{22.92} \leq Z \leq \frac{525.5 - 525}{22.92}\right)$$

$$= P(-.02 \leq Z \leq .02)$$

$$= .016$$

Exercise 6.4

1. Roll a die 150 times yourself, giving 50 sets of three rolls. Compute the value of $\overline{X}$ for each set of three and compare the distribution of your 50 values of $\overline{X}$ with Figures 6.4.1c and 6.4.2. How close is your average of all 150 numbers to 3.5, the (theoretical) mean of the population you have sampled from?

2. Assume that the yearly income per capita for residents of a large town has a mean of $8000 and a standard deviation of $2000. If a random sample of 100 residents is selected, what is the probability that the sample average will lie between $7500 and $8500?

3. A sample of 10 light bulbs is selected from a population of bulbs whose lifetimes are normally distributed with mean 1000 hours and standard deviation 100 hours. What is the probability that the lifetime

of the third bulb lies between 900 and 1100 hours? What is the probability that the average of the 10 lifetimes lies in this interval?

4. Assume that if a diet is rigorously followed for two weeks, the distribution of amount of weight lost has a mean of five pounds and a standard deviation of five pounds. If a doctor has 25 patients following this diet at the same time, what is the probability that the average weight loss per patient is at least four pounds? At least seven pounds?

5. Assume Y is a binomial random variable with $n = 10$, $p = .3$. Compare the exact and approximate values for $P(Y = 3)$, $P(Y \le 1)$, and $P(5 \le Y \le 7)$. Note the approximation is not as good in this case as it was in Example 6.4.2; the exact histogram for this case is not very close to being symmetric.

6. A large industrial plant receives light bulbs in 2000 bulb shipments. The supplier has a 5% defective rate (i.e., the probability is .05 that any single bulb will not work). What is the (approximate) probability there are fewer than 95 defective bulbs in a 2000 bulb shipment?

7. Assume that you and a friend gamble to decide who will pay for the coffee every morning after class. If you do this for 150 days and the game is fair, how many times would you expect to pay? What is the (approximate) probability you would have to pay between 70 and 80 times, inclusive?

8. Assume you are given a 40-question multiple choice exam; each question lists four possible answers. How many questions would you expect to answer correctly if you are guessing? If you are guessing on each one, what is the probability of getting at least 15 correct? At least 20?

9. In question 8 assume the answers presented are not well chosen and you can eliminate two possibilities on each question; thus you only have to guess between the remaining two. Answer the questions posed in question 8 in this case.

10. In a large city suicides occur at a rate of three per day during hot, humid weather. What is the (approximate) probability there would be fewer than 50 suicides during a 20-day period of hot, humid weather?

11. In assembly line painting of automobiles, assume that surface imperfections occur like events in a Poisson process at a rate of one per 17 square feet. If a car has 204 square feet of painted surface, what is the (approximate) probability that a single car has 10 or fewer surface imperfections in its paint?

12. (Refer to question 11.) Assume that the assembly line paints 40 cars per day, all of the same make and model. What is the distribution of the number of cars per day with 10 or fewer surface paint imperfections? Compute the (approximate) probability that at least half of the cars painted each day have 10 or fewer surface imperfections.

6.5 SUMMARY

A population is a collection of things having something in common; in general, we shall be concerned with populations of numbers and with making inferences about such populations. The random variable whose probability measure describes the distribution of values in the population is called the population random variable. The probability measure for a value selected at random from the population is identical with the probability measure of the population random variable. The techniques we shall discuss are appropriate given a random sample of the population random variable; for a random sample of a random variable, the probability measure for each element in the sample is identical with the population measure, regardless of the other values that occur in the same sample.

The total of a random sample (of size n) is itself a random variable; its mean or expected value is n times the population mean and its variance is n times as large as the population variance. The sample mean $\overline{X}$ has expected value equal to the population mean and variance equal to $1/n$ times the population variance. If the population has a normal probability measure, then both the sample total and the sample mean have normal probability measures. If the population measure is Poisson, the sample total also has a Poisson probability measure.

Regardless of the population probability measure, the total or the mean of a random sample has a probability measure that is well approximated by a normal probability measure, for a sufficiently large sample size. This result allows normal approximations of both binomial and Poisson probabilities in appropriate situations.

Exercise 6.5

1. A large delivery truck can carry 100 boxes. The boxes are all the same size and shape; the weight of each box is a normal random variable with mean 110 pounds and standard deviation 3 pounds. What is the probability measure for the total weight of 100 boxes loaded on the truck?

2. The probability a Fuller brush man will make a sale, in stopping at any given residence is .1. If he stops at 200 residences in a week, what is the approximate distribution of the number of sales he makes?

3. Assume you flipped a new coin 200 times and observed 130 heads, 70 tails. Would you suspect the fairness of the coin?

4. An airline has found through experience that the probability a person who reserves a seat on one of their flights will actually show up for the flight is .9. To avoid running this flight with a large number of vacant seats, they will "reserve" more seats on the flight than there are on the plane. Assume a Boeing 707 has 130 seats. The airline will take reservations for 140 seats. Assuming the probability is .9 that each person reserving a seat actually shows up for the flight, and that all these people act independently, what is the approximate probability that everyone that shows up for the flight will get a seat?

5. Answer question 4 if everything is as described there, except the airline only takes 135 reservations. What if they take only 130 reservations?

6. Assume that weights of United States citizens are normally distributed with mean $\mu = 160$ pounds and standard deviation $\sigma = 30$ pounds. Six United States citizens (selected at random) enter an elevator to ride up one floor. The elevator has a plaque that says capacity: 6 persons, weight 990 pounds. What is the probability that the weight allowance is exceeded by the people entering the elevator?

7. Assume seven people squeeze into the elevator mentioned in problem 6. What is the probability that they exceed the weight allowance?

8. Assume that 60% of the electorate favor a school bond proposition. A random sample of 100 voters is contacted. What is the (approximate) probability that a majority of the people in the sample favor the proposition?

9. The female fly, drosophila, frequently lays 200 or more eggs at one time. Assume that a particular female has exactly 200 offspring; genetic theory predicts that the probability is 1/4 that each will bear a certain characteristic, such as being gray in color, and that the offspring are independent in terms of possessing the characteristic. Assuming the theory is correct, what is the (approximate) probability that 60 or more of the offspring (out of the 200) will be gray in color?

10. A packager of frozen foods finds that the weights of its packages of one type of food are normally distributed with mean equal to 12.05 ounces and standard deviation .05 ounces. A government agency is interested in checking the claim made on the package: "Net Weight 12 ounces." It will select 10 packages (at random) and will weigh each of them. What is the probability that the average of the 10 will be less than 12 ounces?

11. A manufacturer of light bulbs prints "Average life 2100 hours" on the packages of its bulbs. If the distribution of lifetimes is normal with $\mu = 2120$ and $\sigma = 100$, what is the probability that the average lifetime of a random sample of 25 bulbs will exceed 2100 hours?

12. Assume that the distribution of the number of days between earth-

quakes of magnitude 4 or more has mean 14 and standard deviation 14. What is the probability that the average number of days between the next 100 earthquakes will be less than 15?

13. An airliner has 250 seats. Assume that the distribution of the weights of passengers has a mean of 150 pounds and a standard deviation of 30 pounds. If all the seats are filled, what is the approximate distribution of the total weight of all the passengers? What is the probability the total passenger weight exceeds 19 tons?

14. For a normal population, how long is the interval, centered at μ, which includes 95% of the total distribution? If we take a sample of n values from this population, how long is the interval, centered at μ, which includes 95% of the distribution for $\overline{X}$?

15. Assume that earthquakes of magnitude 4 or more occur at a rate of once each 14 days (somewhere in the world) like events in a Poisson process. What is the (approximate) probability there will be at least 20 recorded in a 52-week period?

16. A large new car dealer sells new cars at a rate of three per week. What is the (approximate) probability this dealer will sell at least 150 cars in a year?

17. Assume that you bet $1 on the occurrence of a red number when a roulette wheel is spun, for each of 100 successive spins. Thus the probability you win on each spin is 9/19. What is the (approximate) probability you come out ahead after the 100 spins? (*Hint.* If X is the number of times you win out of 100, you come out ahead if and only if $X \geq 51$.)

18. What is your expected winnings over the 100 plays in question 17?

CHAPTER 7
STATISTICAL INFERENCE: ESTIMATION

We shall now turn our attention to problems of statistical inference. We have studied the concept of a population of numbers, the population random variable, and the probability measure for a population random variable. Granted a population, and its probability measure, we have seen how to describe various aspects of a population. We can compute the population mean, or median, as a measure of a "typical" population value, or we can evaluate the population standard deviation as a measure of the variability in the population. Granted the population probability measure, we can compute the probability that a single number selected at random from the population will lie in any specific interval, or the probability (at least approximately) that the total or average of n numbers selected at random will lie in any specific interval.

All of the computations referred to above can be carried out if we know the population probability measure. But suppose we do not know the population probability measure. This is generally the situation in practical problems and the reason that methods of inference are needed.

Specifically, consider a medicine, say tetracycline, and its use in curing a skin infection. We can conceive of any number of people, each with the same type of skin infection and can also conceive of each of these people being treated with the same dosage of tetracycline. To keep the example simple, suppose we can classify each person as belonging to one of two classes, after the treatment with tetracycline: cured (1) or not cured (0). This, then, is a discrete conceptual population of 0's and 1's. Letting p be the proportion of 1's in the population (and $q = 1 - p$ is the proportion of 0's), the population probability measure would be completely specified if we knew the value of p. But do we know p? Unfortunately, as long as the value of p is unknown, we do not completely know the population probability measure.

As a second example, consider the occurrences of fatal automobile accidents (in each of which one or more people are killed) per day on a 20-mile stretch of two-lane highway. Here again we have a discrete population of numbers since the number of fatal accidents on any day must be a whole number. What are the relative proportions of 0's, 1's, or 2's, for example, in this population? If we assume that fatal accidents occur on this road at a constant rate per month, that the probability is zero of there being two fatal accidents exactly simultaneously, and that the occurrences are independent (the occurrence or nonoccurrence of a fatal accident at 9 A.M. on Monday has no effect on the probability of another one at 10 A.M. Monday or any other time), then we have made the assumptions for a Poisson process (see Section 5.3). The relative proportions of 0's, 1's, 2's, etc., in the population, then, would be described by a Poisson probability measure with parameter $\lambda \cdot 1 = \lambda$, where λ is the average rate of fatal accidents per month. That is, with these assumptions about the way in which fatal accidents occur, the proportion of 0's in the population must be

$$\frac{(\lambda)^0}{0!} \, e^{-\lambda} = e^{-\lambda}$$

the proportion of 1's must be

$$\frac{(\lambda)^1}{1!} \, e^{-\lambda} = \lambda e^{-\lambda}$$

the proportion of 2's must be

$$\frac{(\lambda)^2}{2!} \, e^{-\lambda} = \frac{\lambda^2}{2} \, e^{-\lambda}$$

and so on. These values are determined by the Poisson probability measure. Do we know λ? Again, since we do not know the value for λ, we do not know the probabilities of occurrence (relative proportions in the population) of these different discrete values.

The two populations just discussed are examples selected from a large number of practical situations. We do not in general know the probability measures of populations. By reasoning we may be able to convince ourselves that the probability measure must be of a specific form (such as the Poisson above), but we do not generally know the values of the parameters of the probability law.

Statistical methods are concerned with using the numbers observed in a sample from the population to make inferences about the population or, more specifically, the probability measure of the population. There are two broad categories that encompass essentially all statistical methods: *estimation,* which we shall study in this chapter, and *tests* of *hypotheses,* which we shall study in the next chapter.

Problems of estimation are concerned with manipulations of the numbers that occur in a sample to guess or estimate the values of unknown parameters of the population probability law. In the example about tetracycline above we might want to treat some number of people, say 100, with the tetracycline. What can we do with the resulting number of people cured to estimate p, the proportion of cures in the population? Or, in the example of fatal traffic accidents, what should we do with the numbers of accidents observed over, say, 10 months to estimate λ, the unknown parameter of the population probability measure? Or how could we estimate the probability there will be no fatal accidents on a given day, which is a function of λ? These are all problems of estimation.

A hypothesis, on the other hand, is a statement about the probability measure of the population. To test a hypothesis means to select a sample from the population and, on the basis of the numbers observed in the sample, to decide whether to accept or reject the hypothesis. For example, the drug firm marketing tetracycline may claim that its product is effective in curing at least 80% of all people with a given skin infection; this statement says, then, that $p \geq .8$, where p is the proportion of 1's in the population. Given the observed values in a sample from the population, should this statement be accepted or rejected? We shall study the methodology appropriate to this type of problem in the next chapter.

7.1 POINT ESTIMATES

We have seen that parameters of populations (p in the binomial, μ and σ^2 in the normal, λ in the Poisson) are numbers. If the value of a parameter is unknown, that means we do not know what its specific numerical value is. To estimate its value, we could use any number we like; if we can in some way base the number we choose on the results observed in selecting a random sample from the population, then we certainly should expect to do better than if we simply select an arbitrary number as our guess for the value of the parameter. The following two subsections discuss the topic of point estimation, what to do with the numbers observed in a sample to estimate the value of an unknown parameter.

7.1.1. Parameter Estimates

Let us begin our study of estimation by considering the tetracycline-skin infection population discussed earlier. We have an infinite conceptual population of people, each with the same skin infection; each person is (conceptually) treated with the same dosage of tetracycline and p is the proportion that are cured. Suppose we select a sample of $n = 100$ people from this population. By this we mean we have 100 people with this infection, and each of them is given the same dosage of tetracycline. Let us suppose, furthermore, that 70 of the 100 are cured (and thus 30 were not). Based on this sample result what should we use as our guess, or estimate, of the unknown value of p?

Let us briefly examine the methodology that is generally applied in answering such a question. If we can assume that the 100 people in the sample constitute 100 independent repeated trials each with two possible outcomes, success (cured) and failure (not cured), then we know that the total number of successes we will observe in the 100 trials, Y, is a binomial random variable with parameters $n = 100$ and p, the unknown probability a person will be cured by this dosage of tetracycline (see Section 5.1). Thus, the probability measure for Y is

$$P(Y = y) = \binom{100}{y} p^y (1 - p)^{100-y} \qquad y = 0, 1, 2, \ldots, 100$$

The probability we would observe 70 cures in the 100 patients is

$$P(Y = 70) = \binom{100}{70} p^{70} (1 - p)^{30}$$

which involves the unknown parameter p. We might then ask what value of p makes this probability (that Y will equal 70) as large as possible? The answer is that the value $\hat{p} = 70/100$ maximizes $P(Y = 70)$ for this problem; the proof of this statement is beyond the mathematics assumed in this text, but the general result is certainly appealing on intuitive grounds. If we observe that a binomial random variable, Y, equals a particular value y, then the ratio $\hat{p} = y/n$ is the value of p that maximizes $P(Y = y)$. Thus, it would seem logical to use y/n as a guess for the unknown value of p (and for the particular numbers given earlier to use $\hat{p} = .7$ as our guess for the unknown probability that a person with the given skin infection will be cured by tetracycline). This is an example of the use of the *method of maximum likelihood*, a procedure that is generally applicable for choosing point estimates of unknown parameters. (See question 12 below.)

It proves very useful to draw a distinction between the formula used to estimate a parameter and the particular value that the formula gives when the sample values are observed. The following two definitions draw this distinction.

> **Definition 7.1.** A (point) *estimate* of an unknown parameter is a number, computed from observed sample values, that is used as our guess for the value of the unknown parameter.

> **Definition 7.2.** A (point) *estimator* for an unknown parameter is the mathematical formula (or function of the sample values) that is used to compute the estimate.

Notice, then, that estimates and estimators occur together; the formula used is called the estimator and the particular value the formula equals, given specific sample values, is called the estimate (this is the number computed from the sample values). In the tetracycline example just discussed, the estimator of p was Y/n, the estimate was $70/100 = .7$ (for the sample results assumed). In general, the estimator will be a function of the random variable(s) occurring in the sample (such as Y/n) and, thus, will itself be a random variable; the estimate is simply a number, neither random nor variable once the sample values are known. Thus estimators will be denoted by capital letters and estimates by lower case letters.

Example 7.1.1

Assume a production line is used to produce items. Each item that comes off the line either is or is not defective in some way; defectives occur independently, from item to item, and the probability any item is defective is p (unknown). Suppose we select a random sample of $n = 30$ of these items. Then again we could use the estimator Y/n, where now Y is the number of defectives in the sample; this is the value that maximizes the probability of occurrence of the observed sample results, as discussed earlier. Thus, for example, if we find one defective in the 30 the estimate of p would be $\hat{p} = 1/30$, if we observe two defectives the estimate would be $\hat{p} = 2/30$, and if we observe three defectives the estimate would be $3/30$. In each case the same formula (estimator) is used.

The important thing to grasp from this example is the fact that *estimates* of unknown parameters will *vary* from one sample to another; the *estimator* or formula used to compute the estimate will remain *fixed* but, since the particular sample values that occur will vary from one drawing or sampling to another, the *value* for the estimator, or the estimates will also vary. Because of this inherent variability in any sampling problem, we would certainly not expect the value of the estimate, for any given sample, to be identical with the value of the unknown parameter being estimated.

If we have a sample from a normal population, there are two parameters, μ and σ^2 (or σ), to be estimated; with a sample of a Poisson random variable there is only one parameter (λ) to be estimated. In either of these cases we can use the same reasoning mentioned above to estimate p: the estimators are those functions of the sample values that maximize the probability of occurrence of the sample (again called the maximum likelihood method). Table 7.1 gives the maximum likelihood estimators of these parameters, based on a random sample from the population (The estimator given for σ^2 in the normal case differs slightly from the maximum likelihood estimator; the divisor should be n instead of $n - 1$ to truly maximize the probability of the sample. If n is large, there is very little difference between dividing by $n - 1$ instead of n.)

Example 7.1.2

Assume that the population of weights of United States male citizens, age 18 to 26, is normal in form and that the values of the mean μ and variance σ^2 for this population are unknown. Suppose we decide to take a random sample of $n = 5$ members of this population and record their weights. If we denote the five sample weights by X_1, X_2, X_3, X_4, X_5, then $\overline{X}$ is the *estimator* for μ and S^2 is the *estimator* for σ^2. If the five weights in the sample were

Table 7.1 Estimators Based on Samples of Size *n*

Population	Unknown Parameter(s)	Estimator(s)
Bernoulli	p	$\hat{p} = Y/n$ (Y is total number of successes.)
Poisson	λ	$\hat{\lambda} = \overline{X} = \dfrac{1}{n} \Sigma X_i$
Normal	μ, σ^2	$\hat{\mu} = \overline{X} = \dfrac{1}{n} \Sigma X_i$
		$\hat{\sigma}^2 = S^2 = \dfrac{1}{n-1} \Sigma (X_i - \overline{X})^2$

175, 203, 199, 164, and 210, then the *estimate* for μ for this sample is

$$\bar{x} = \tfrac{1}{5}(175 + 203 + 199 + 164 + 210) = 190.2$$

the *estimate* for σ^2 is

$$s^2 = \tfrac{1}{4}\{[(175)^2 + (203)^2 + (199)^2 + (164)^2 + (210)^2] - 5(190.2)^2\}$$

$$= 387.7$$

These are the numbers we would use to estimate μ and σ^2, given these observed sample values. If, on the other hand, the five sample weights had been 170, 159, 221, 200, and 187, we would still use the estimators $\bar{X}$ and S^2, but their values would be different. In fact, for these sample values we have

$$\bar{x} = \tfrac{1}{5}(170 + 159 + 221 + 200 + 187) = 187.4$$

$$s^2 = \tfrac{1}{4}\{[(170)^2 + (159)^2 + (221)^2 + (200)^2 + (187)^2] - 5(187.4)^2\}$$

$$= 599.3$$

The *estimates* for the standard deviation, computed from the two sets of sample outcomes, are $s = \sqrt{387.7} = 19.7$ and $s = \sqrt{599.3} = 24.5$, respectively.

In addition to estimating the parameters of a distribution, we are also able to estimate functions of those parameters. For example, suppose we were interested in estimating the 60th percentile of the weights of United States male citizens, age 18 to 26. By referring to Table A.2 in the Appendix, we see that the 60th percentile of a normal distribution is given (approximately) by $t_{.60} = \mu + .25\sigma$. We can estimate this quantity by $\hat{t}_{.60} = \bar{x} + .25s$, where $\bar{x}$ and s are the estimates of μ and σ. Thus, given a sample that yielded $\bar{x} = 190.2$, $s = 19.7$ we would estimate $t_{.60}$ to be

$$\hat{t}_{.60} = 190.2 + (.25)(19.7) = 195.1$$

Given a sample that gave $\bar{x} = 187.4$, $s = 24.5$, the estimate would be

$$\hat{t}_{.60} = 187.4 + (.25)(24.5) = 193.5$$

Again, as with all estimates, as the values observed in the sample change, so does the estimate of $t_{.60}$

7.1.2. Properties of Point Estimators

Having recognized that the sample variability will lead to variability in the estimates of parameters, we can now briefly discuss *properties* of estimators. What sort of behavior of an estimator, from sample to sample, is desirable and what is not? In Example 7.1.2 above we used the sample mean $\overline{X}$ as the estimator for μ, the population mean. There are many other estimators (rules or formulas for computation) that might also seem reasonable to use in computing the value of an estimate. In particular, to estimate the mean μ of a normal population, in addition to $\overline{X}$ we might consider, for example, the median of the sample to estimate μ or we might take the average of the largest and smallest sample values, ignoring the rest, to estimate μ. In fact, for the same sample, we could determine the value of $\overline{X}$, the value of the median, and the average of the two sample extremes and, then, average these three quantities together to estimate μ. Each of these is simply describing a different estimator, a different formula that could be employed to determine the estimate. Why is it that $\overline{X}$ in particular is taken as the usual estimator for μ in the case described, rather than one of these others?

Different possible estimators for the same parameter, such as those just described, are compared on the basis of their properties from sample to sample. The numbers that occur in a sample from a population are observed values of random variables, since they are chosen at random from the population. Different possible estimators of an unknown parameter, which then are simply different possible functions of the sample random variables, are themselves random variables; the values they take on, for specific samples, are controlled by the chance mechanism that chooses the values for the sample. Thus, if samples of the same size are repeatedly chosen, any particular estimator will take on different values, as we have seen; there then exists a probability distribution that describes the probabilities of occurrence of the various values in the range of the estimator. If we could find, for every possible sample, an estimator whose value was equal to the unknown parameter it is supposed to estimate then we would clearly prefer to use it instead of any other; it would have probability 1 of equaling the unknown parameter. Unfortunately, when·sampling from a population with an unknown parameter, it is impossible to find such a perfect estimator. Thus, it is necessary to compare the probability distributions of imperfect estimators.

There are many different ways in which the probability distributions of different estimators can be compared. Courses in mathematical statistics generally consider this type of problem in some detail. We shall content ourselves with discussing briefly two aspects of the sampling distribution of an estimator: its mean and its variance.

First, if the mean of the sampling distribution of an estimator is equal to the unknown parameter being estimated, then the estimator is *unbiased*. Thus, if we repeatedly take samples of the same size, and for each sample we compute the value of an unbiased estimator of an unknown parameter, the average of these estimates over the repeated samples is equal to the unknown parameter value. For any particular sample the estimate may or may not be close to the parameter value, but over an infinite number of repeated samples the *average* of the estimates equals the unknown parameter value, if the estimator is unbiased.

It certainly is desirable, then, to use unbiased estimates for unknown parameters. It can be shown that each of the estimators of the parameters listed in Table 7.1 are in fact unbiased. Many other unbiased estimators of those same parameters could be quoted. In fact, in estimating the mean μ of a normal population, the sample median and the average of the largest and smallest sample values, mentioned above, are also unbiased estimators for μ. Why, then, should we prefer $\overline{X}$ to these other two estimators?

When considering several unbiased estimators for the same parameter, it would seem desirable to choose the unbiased estimator with the highest probability of giving an estimate close to the unknown parameter value. As we have seen for random variables in general, the variance or standard deviation of the probability measure gives an indication of how close the observed value of a random variable may be to its mean; that is, for an unbiased estimator, the variance or standard deviation will give an idea of how close the estimate should be to the mean of the distribution, that is, to the unknown parameter value when the estimator is unbiased. Thus we would prefer the unbiased estimator with the smallest variance (or, equivalently, the smallest standard deviation). On this basis, in estimating μ, the sample mean, $\overline{X}$, is clearly to be preferred to either the sample median or the average of the largest and smallest sample values, when sampling from a normal population. As we know, for samples of size 5, the variance of $\overline{X}$ is $\sigma^2/5 = .2\sigma^2$, where σ^2 is the variance of the normal population; it can be shown that the variance of the sample median is $.287\sigma^2$, for samples of size 5, and that the variance of the average of the largest and smallest observations is $.261\sigma^2$, for samples of size 5. Thus, even though all three are unbiased estimators for μ, we can expect the observed value of the sample mean to be closer to μ than either of the other two since it varies less about μ than they do. The densities for these three estimators are plotted in Figure 7.1.1. Note that the density for $\overline{X}$ has more area in any interval centered at μ than do the other two; thus the probability of $\overline{X}$ taking on a value within a given distance of μ is greater than the same probability for the other two estimators.

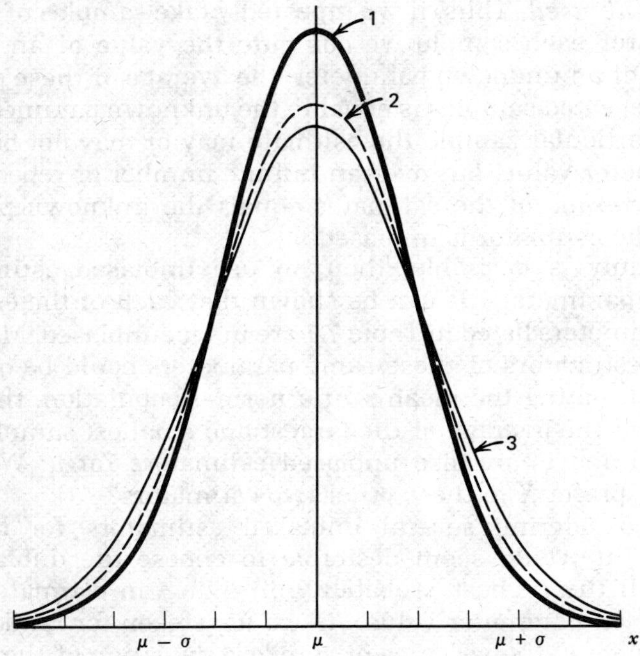

Figure 7.1.1   Densities for 1. $\overline{X}$.   2. Average of largest and smallest sample values.   3. Sample median.

In all areas of scientific research and experimentation, it is usual to give both the estimate of the unknown parameter of interest and also its estimated standard deviation. This is done to indicate the reliability of the estimate. Were the experiment to be repeated, how likely is it that the new estimate, computed from the new observations, will be somewhere close to the estimate already observed? The larger the standard deviation of the estimator, the more variable it is over repeated samples and the more likely that a repetition of the experiment might yield a quite different estimate. This standard deviation of an estimate is frequently called the *standard error of estimate*. The following example illustrates the computation of the standard errors of the estimates in two examples previously discussed.

Example 7.1.3

In Example 7.1.2 we discussed samples of size 5 from a normal population of weights. For the first set of 5 sample values we found $\overline{x} = 190.2$, $s = 19.7$.

The standard deviation of the estimator $\overline{X}$ we know is $\sigma/\sqrt{n}$; thus the standard error of $\overline{x}$ (estimated standard deviation of $\overline{X}$) is $19.7/\sqrt{5} = 8.81$. For the second set of five values from the same population we found $\overline{x} = 187.4$, $s = 24.5$; thus the standard error of this second estimate is $24.5/\sqrt{5} = 10.96$.

In Example 7.1.1 we discussed samples of size 30 of a production process with unknown probability p of producing a defective. The estimator for p was $Y/30$, where Y is the number of defectives found in the sample of size 30; Y is binomial with parameters $n = 30$ and p. Thus the variance of Y is $\sigma_Y^2 = 30\,pq$ (see Section 5.1) and the variance of the estimator $Y/30$ is $30pq/(30)^2 = pq/30$; the standard deviation of the estimator is $\sqrt{pq}/\sqrt{30}$. Thus, if we find one defective in the sample of 30, the estimate of p is $1/30 = .033$; its standard error is $\sqrt{(.033)(.967)}/\sqrt{30} = .0328$. If we find two defectives in the sample of 30, the estimate of p is $2/30 = .067$; its standard error is $\sqrt{(.067)(.933)}/\sqrt{30} = .046$.

Example 7.1.4

The number of fatal accidents on a 20-mile two-lane highway per month, for the 12 months of the same calendar year, were $0,1,1,0,0,2,0,0,1,0,1,0$, respectively. We assume that fatal accidents occur on this road like events in a Poisson process; thus the above 12 numbers are a random sample of size 12 of a Poisson random variable with parameter $\lambda \cdot 1 = \lambda$, where the unit is a month. From Table 7.1 we see that the estimator of λ is $\overline{X}$, the sample mean. The variance of this estimator is again σ^2/n, where σ^2 is the population variance. Recalling from Section 5.3 that the variance of a Poisson population is again λ, the same as the mean, the variance of $\overline{X}$ then is λ/n and its standard deviation is $\sqrt{\lambda}/\sqrt{n}$. Thus, for the sample values given above, the estimate of λ is $\overline{x} = (0 + 1 + 1 + \ldots + 1 + 0)/12 = .5$; the standard error of this estimate is $\sqrt{.5}/\sqrt{12} = .204$.

In the following section we shall discuss interval estimates of unknown parameters. These techniques combine, in a sense, the two steps in which the value of the estimate and its standard error are given, resulting in an interval of values with a known probability of including the unknown parameter value.

Exercise 7.1

1. It is assumed that the scores made by high school seniors on a mathematics ability test are normally distributed with unknown mean μ and unknown variance σ^2. Ten high school seniors were selected at random to take the test; their scores were 604, 427, 515, 570, 483, 592, 653, 526, 460, and 499.
    (a)   Use this data to estimate the population mean and variance.
    (b)   Compute the standard error of the estimate of μ.

2. The net weights of five frozen packages of fish were 12.1, 10.9, 11.2, 11.7, and 11.5 ounces.
   (a) Assuming these are a random sample of weights of packages put out by this supplier, and that the weights of all his packages are normally distributed, estimate the population mean and variance.
   (b) Compute the standard error of your estimate of the mean.
   (c) Each package says the net contents weighs 12 ounces. Does this seem truthful, given these sample results?

3. An assembly line turns out items, each of which is either defective or not defective. A random sample of 100 items is inspected and 10 are found to be defective.
   (a) Given this sample, what value would you estimate for p, the probability an item turned out by this line is defective?
   (b) What is the standard error of your estimate?

4. In 15 times at bat a baseball player made four hits.
   (a) Assuming these results to be a random sample of his performances, what would you estimate is his probability of getting a hit each time he is at bat?
   (b) What is the standard error of your estimate?

5. Accidents occur at an industrial plant like events in a Poisson process. The numbers of accidents per month, over a 12-month period, were 0, 2, 1, 1, 3, 0, 2, 2, 1, 0, 1, 3.
   (a) What would you estimate as the value of λ, the number of accidents per month for this plant?
   (b) What is the standard error of your estimate?

6. For the data given in question 1, what would you estimate as the value for the 50th percentile in the population? The 90th? The 25th?

7. Again, for the data in question 1, what proportion of the population scores would you estimate to exceed a score of 600? To be less than a score of 500?

8. For the assembly line discussed in question 3, what would you estimate as the value for the probability of getting zero defectives, if five additional items are selected? What would you estimate as the probability of getting exactly one defective if five more items are selected?

9. For the industrial accidents described in question 5, estimate the probability there will be no accidents in a given month. What would you estimate as the probability there will be no accidents for three months in a row?

10. In a study of extrasensory perception, a student sits opposite a laboratory assistant at a table. The assistant has a deck of eight cards, four of which have green faces and four of which have red faces; all have identical backs. The cards are well shuffled and the assistant then looks at the face of each in turn. Each time the assistant looks at a card the student attempts to guess its color, without being allowed to see it.

The student knows there are four cards of each color and is not told whether he is right or wrong on individual cards. This experiment is repeated 10 times; the number of cards that the student was right on were 4, 6, 6, 2, 4, 8, 6, 8, 6, and 2. What would you estimate as his probability of correctly guessing the color of a card? Why do only even numbers occur as the number correct?

11. A large university uses a computer to print the grade reports for all students. During one quarter a total of 58,967 individual course grades were processed; a check revealed that 312 of these were in error. What would you estimate as the value for p, the probability an individual grade will be reported erroneously? If you took five courses during this quarter, what would you estimate as the value of the probability that all your grades are reported correctly?

12. The estimators listed in Table 7.1, those recommended for the unknown parameters of the distributions we have studied, are all examples of *maximum likelihood* estimators. (See the discussion at the start of Section 7.1.1.) The maximum likelihood method, which can be applied to any probability law, gives estimators that have good properties in general. It (essentially) says that we should take the parameter value that maximizes the probability of observing the particular sample values, as the estimate of the unknown parameter; this, in a sense, then gives the most likely possible value, based on the given sample, as our guess for its value. Suppose we perform three independent Bernoulli trials, each with unknown probability of success p; assume we observe two successes in the three trials. This number 2 then is the observed value of a binomial random variable with parameters $n = 3$ and p, which is unknown. As we recall from Section 5.1, the probability that this binomial random variable is equal to 2 is $\binom{3}{2} p^2 (1 - p) = 3p^2 - 3p^3$. This is the quantity to be maximized, with respect to p, to get the value of the estimate for this sample. In general, calculus techniques are easily applied to such a problem. However, in this particular case, we can see that

$$3p^2 - 3p^3 = \frac{4}{9} - 3\left(p - \frac{2}{3}\right)^2 \left(p + \frac{1}{3}\right)$$

by expanding the right-hand side. Notice that $3(p - 2/3)^2 (p + 1/3) \geq 0$ for all $0 \leq p \leq 1$ and in particular $3(p - 2/3)^2 (p + 1/3) = 0$ only for $p = 2/3$ or $p = -1/3$. Thus,

$$3p^2 - 3p^3 = \frac{4}{9}$$

if $p = 2/3$, and

$$3p^2 - 3p^3 < \frac{4}{9}$$

for all other values of p between 0 and 1; that is, if we take $\hat{p} = 2/3$ as our guess for the value of the unknown p, we maximize the probability of observing two successes in three trials. Thus it is the *maximum likelihood estimate*. In general, the maximizing value of p is given by the ratio of the number of successes over the number of trials. Verify that taking $p = 1/3$ maximizes the probability of one success in three trials.

13. A random sample of 100 high school seniors made an average score of $\bar{x} = 450$ on the mathematics portion of the SAT tests. The standard deviation of their scores was $s = 90$. Assume their scores are a random sample from a normal population. What would you estimate as the proportion of all seniors that would have a score over 500 on this test?

14. For the sample results described in question 13, what would you estimate to be the 80th population percentile?

15. Assume the baseball player described in question 4 is at bat three times in a game. Using the information given there, estimate the probability that in the game he gets
    (a) No hits.
    (b) Exactly one hit.
    (c) Exactly two hits.
    (d) Three hits.

16. As of January 15, a college basketball team had played 16 games. In four of these games, the team's star scored at least 30 points.
    (a) Estimate the probability that this player will make at least 30 points in a game.
    (b) Assume that the team has eight remaining scheduled games. Estimate the probability that the star will make at least 30 points in exactly two of the remaining games.
    (c) Estimate the probability that the star will make at least 30 points in at least two of the eight remaining games.

17. A radioactive substance emits electrons at instants of time like events in a Poisson process. During a one-minute period, 600 electrons were emitted.
    (a) Estimate λ, the number of electrons emitted per second.
    (b) Estimate the probability there will be no electrons emitted during a period of length .1 seconds.
    (c) Estimate the probability that exactly one electron will be emitted in a period of length .1 seconds.

18. Assume that, nationwide, 20% of all adults are smokers. A doctor's practice includes 500 adults. Assuming these people are a random sample from the whole population, estimate the probability that at least 110 of his adult patients are smokers.

19. The height a college high jumper will in fact jump on any given try is a normal random variable with unknown mean μ and unknown standard deviation σ. He makes 10 attempts at clearing the bar, when

it is set at six feet, and is successful on eight of these. Then he makes 10 attempts to clear the bar, at height six feet, six inches, only three of which are successful. Based on this information, can you find a way to estimate μ and σ?

20. The council for religious freedom ran a national survey to estimate the proportion of people that attended church regularly in various sections of the country. They contacted random samples of 200 people in the northeast, 160 in the southeast, 250 in the southwest and 120 in the northwest. The numbers that said they regularly attended church were, respectively, 123, 120, 100, and 65, in these four regions.
    (a) Use these results to estimate the proportions of the population that regularly attend church in these four sections.
    (b) Calculate the standard errors of your estimates.

7.2 INTERVAL ESTIMATES

In the last section we studied point estimation, procedures that lead to the computation of a single number as our estimate of the value of the parameter. Additionally, quoting the standard error of the estimate indicates how variable the values of the estimator may be from one sample to another; this gives a rough idea of how likely it is that the estimate from the given sample may be close to the true unknown parameter value. It makes the choice of estimators with small standard deviations seem more desirable.

A more quantitative indication of how sure we are of the value of an unknown parameter is provided by an interval estimate for the parameter. In using interval estimates, as the name implies, we give an interval of values rather than a single number as our estimate for the unknown parameter. Attached to the interval is a probability statement that tells how sure we are that the interval includes the value of the unknown parameter. We are free to choose the value of the probability; thus we shall be able to derive 80% confidence intervals, or 90, 95, or 99% confidence intervals. As the probability that the interval covers the parameter increases, so does the length of the interval; in order to be 99% sure the interval includes the parameter, we use a longer interval than if we were only 80% sure the interval covers the parameter. The definition of a confidence interval is given below.

> **Definition 7.3.** Given a random sample from a population with unknown parameter θ, the interval from L to U is called a $100(1 - \alpha)\%$ *confidence interval* for θ if
>
> $$P(L \leq \theta \leq U) = 1 - \alpha,$$
>
> where L and U are functions of the sample values.

We are free to choose α to equal any number between 0 and 1. Thus, for example, if we choose $\alpha = .1$, the probability that the interval from L to U covers θ is $1 - .1 = .9$; by using $\alpha = .05$ we get a 95% interval. As α changes, so will L and U; they get further apart as α becomes smaller. The quantity $1 - \alpha$ is called the *confidence coefficient*. The following subsections consider the computations of confidence intervals for the parameters and distributions we have studied earlier.

7.2.1 Confidence Intervals for μ and σ in the Normal Distribution

We shall first examine an unrealistic case to see the typical logic employed in finding a confidence interval for an unknown parameter. Assume that the amount of mercury contamination, per fish, in a large lake is a normal random variable with unknown mean μ and *known* σ (this is the unrealistic assumption; generally if one parameter is unknown, so is the other). Further assume that a random sample of n fish is selected from this lake and that the amount of mercury is determined for each (in parts per million). Then, letting $\overline{X}$ be the average contamination of these n fish, we know from Section 6.3 that $\overline{X}$ has a normal probability measure with mean μ and variance σ^2/n. We also know then that

$$\frac{\overline{X} - \mu}{\sigma/\sqrt{n}}$$

is a standard normal random variable. From Table A.2 in the Appendix, we can find the value, $z_{1-\alpha/2}$, of the $100(1 - \alpha/2)$ percentile of the standard normal distribution; this quantity has the property, then, that

$$P\left(\frac{\overline{X} - \mu}{\sigma/\sqrt{n}} \le z_{1-\alpha/2}\right) = 1 - \frac{\alpha}{2}$$

Because of the symmetry of the normal density function, the area to the left of $-z_{1-\alpha/2}$ is also $\alpha/2$. Thus

$$P\left(-z_{1-\alpha/2} \le \frac{\overline{X} - \mu}{\sigma/\sqrt{n}} \le z_{1-\alpha/2}\right) = 1 - \alpha$$

(see Figure 7.2.1). This statement says that for any given sample, the probability is $1 - \alpha$ that

$$-z_{1-\alpha/2} \le \frac{\overline{X} - \mu}{\sigma/\sqrt{n}} \le z_{1-\alpha/2}$$

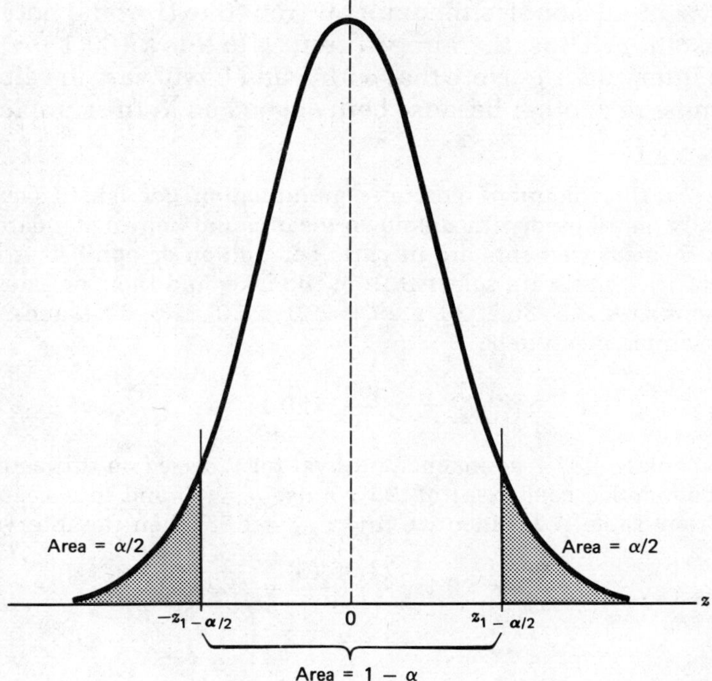

Figure 7.2.1   Area between $-z_{1-\alpha/2}$ and $z_{1-\alpha/2}$ for the standard normal measure.

The probability statement also means that in $100(1 - \alpha)\%$ of all samples of size n selected from this normal population we should find the value of $(\overline{X} - \mu)/(\sigma/\sqrt{n})$ between $-z_{1-\alpha/2}$ and $z_{1-\alpha/2}$ and in $100\alpha\%$ of these samples it would not lie between these limits. Algebraically it can be shown that

$$-z_{1-\alpha/2} \le \frac{\overline{X} - \mu}{\sigma/\sqrt{n}} \le z_{1-\alpha/2}$$

is equivalent to

$$\overline{X} - \frac{\sigma}{\sqrt{n}}\, z_{1-\alpha/2} \le \mu \le \overline{X} + \frac{\sigma}{\sqrt{n}}\, z_{1-\alpha/2}$$

Thus, in $100(1 - \alpha)\%$ of all samples of size n, we would find μ lying between

$$L = \overline{X} - \frac{\sigma}{\sqrt{n}}\, z_{1-\alpha/2} \qquad \text{and} \qquad U = \overline{X} + \frac{\sigma}{\sqrt{n}}\, z_{1-\alpha/2}$$

In $100\alpha\%$ of all samples the interval from L to U would not include the constant μ. Thus, the interval from L to U is a $100(1-\alpha)\%$ confidence interval for μ. Note that both L and U will vary in value from one sample to another because both depend on $\overline{X}$, the sample mean.

Example 7.2.1

Assume that the amount of mercury contamination, per fish, in a large lake is normally distributed with unknown mean μ and known standard deviation $\sigma = 5$ (measurements are in parts per million or ppm). If a random sample of $n = 9$ fish are selected from the lake, and their measured contaminations are 28.5, 30.2, 36.8, 29.4, 22.1, 25.0, 29.9, 33.4, and 26.7, we find the sample mean to be

$$\overline{x} = \frac{1}{9}\sum_{i=1}^{9} x_i = \frac{262}{9} = 29.1$$

Let us calculate a 95% confidence interval for μ, based on this sample. To have a confidence coefficient of .95 we use $\alpha = .05$ and thus require $z_{.975}$ ($z_{1-\alpha/2}$) from Table A.2. Since we find $z_{.975} = 1.96$, then the interval from

$$\overline{x} - \frac{1.96\sigma}{\sqrt{n}} = 29.1 - \frac{(1.96)(5)}{3} = 25.83$$

to

$$\overline{x} + \frac{1.96\sigma}{\sqrt{n}} = 29.1 + \frac{(1.96)(5)}{3} = 32.37$$

is a 95% confidence interval for μ.

The probability statement used in a confidence interval refers to the behavior of L and U *over repeated samples* from the population. Thus, if as in Example 7.2.1 a 95% confidence interval is used, over repeated samples of the same size from the population, it would be found that 95% of the computed intervals did cover the unknown parameter and 5% did not. In taking only one sample and computing the interval, either the interval does contain the unknown parameter or it does not; unfortunately we do not know which of these is the true situation. However, *if* repeated samples were taken and the same procedure followed to compute an interval for each, it would be found that 95% of the computed intervals included the unknown parameter; thus we are 95% sure or 95% confident that the single interval we compute from our sample includes this unknown value.

The confidence interval just discussed for μ depends on two important assumptions:

1. The population is normal.
2. The population standard deviation σ is known.

The second assumption is generally not warranted. It would be very unusual indeed if we had a normal population whose standard deviation was known but whose mean was not.

Let us now examine the more realistic case of a random sample from a normal population with *both* μ and σ *unknown*. It is still true that $(\overline{X} - \mu)/(\sigma/\sqrt{n})$ has a standard normal distribution, but this fact does not aid us in computing a confidence interval for μ since, as we have just seen, the end points of the resulting interval depend on the value of σ; since it is not known, we could not compute L and U as derived above. The natural thing to do would be to replace σ by its sample estimate S. Surprisingly, $(\overline{X} - \mu)/(S/\sqrt{n})$ has a distribution that is independent of σ and depends only on the sample size n; its distribution, rather than being standard normal, is called the *t distribution* with ν (ν is the Greek letter nu, pronounced noo) *degrees of freedom,* its only parameter. We shall see that this distribution occurs and is useful in different contexts. For the problems under discussion here, a confidence interval for the mean of a normal population based on a sample of size n from the population, the parameter ν is equal to $n - 1$.

The t distribution is symmetric about 0 and strongly resembles the standard normal density. Figure 7.2.2 gives a plot of the standard normal density versus the t density with $\nu = 2$ degrees of freedom. Note that the normal density has a greater area close to zero and relatively less density in the tails. This is always the case; as ν,

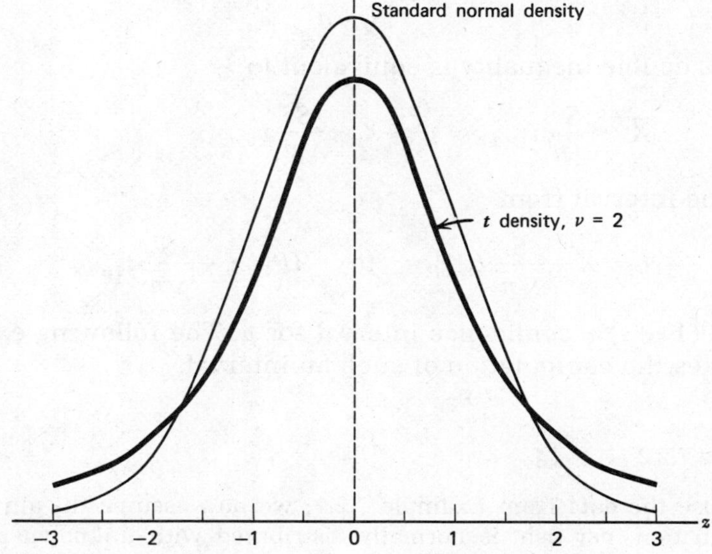

Figure 7.2.2   Standard normal density and t density, $\nu = 2$.

the degrees of freedom for the t distribution increases, the t density actually approaches that of the standard normal, as can be seen in Table A.4.

Table A.4 presents selected percentiles from the t distribution with degrees of freedom ν varying from 1 to 30 and for $\nu = \infty$. The notation used is the same as that for the standard normal and shown in Figure 7.2.1; we use $t_{1-\alpha/2}$ rather than $z_{1-\alpha/2}$, but again, the area to the *right* of $t_{1-\alpha/2}$ is $\alpha/2$ and, because of the symmetry, so is the area to the left of $-t_{1-\alpha/2}$. As can be seen from Table A.4, these percentiles change with the value of ν; thus for $\nu = 2$, $t_{.9} = 1.886$ and $t_{.99} = 6.965$. With $\nu = 15$, these same percentiles are 1.341 and 2.602. The line with $\nu = \infty$, which corresponds to an infinitely large sample size, gives the percentiles for the standard normal density. It was mentioned earlier that the t density approaches the standard normal as its degrees of freedom increase. This phenomenon can be observed by reading down the column for any of the selected percentiles given in Table A.4 and noting that these quantities are getting closer in value to those for the standard normal ($\nu = \infty$) as ν increases.

The percentiles from the t distribution are used in the computation of a confidence interval for μ, with σ unknown, in exactly the same way the standard normal percentiles were used earlier. Thus, given a random sample of size n from a normal population, the probability is $1 - \alpha$ that

$$-t_{1-\alpha/2} \leq \frac{\overline{X} - \mu}{S/\sqrt{n}} \leq t_{1-\alpha/2}$$

and this double inequality is equivalent to

$$\overline{X} - \frac{S}{\sqrt{n}}\, t_{1-\alpha/2} \leq \mu \leq \overline{X} + \frac{S}{\sqrt{n}}\, t_{1-\alpha/2}$$

Then the interval from

$$L = \overline{X} - \frac{S}{\sqrt{n}}\, t_{1-\alpha/2} \qquad \text{to} \qquad U = \overline{X} + \frac{S}{\sqrt{n}}\, t_{1-\alpha/2}$$

is a $100(1 - \alpha)\%$ confidence interval for μ. The following example illustrates the computation of such an interval.

Example 7.2.2

Let us use the data from Example 7.2.1; we now assume the amount of contamination, per fish, is normally distributed with unknown mean μ and unknown standard deviation σ. Since σ is unknown, the confidence interval for μ must be computed with the percentiles from the t distribution, rather than from the standard normal (and σ must be estimated from the sample). Again, a random sample of $n = 9$ fish are selected and their measured contaminations are found to be 28.5, 30.2, 36.8, 29.4, 22.1, 25.0, 29.9, 33.4, and 26.7 (these are the same values used earlier). We find, as before $\Sigma x_i = 262$, and that $\Sigma x_i^2 = 7{,}778.86$; thus

$$\bar{x} = \frac{1}{9}\,(262) = 29.1$$

and
$$s^2 = \frac{1}{8}\left(7778.76 - \frac{(262)^2}{9}\right)$$

$$= \frac{151.65}{8} = 18.96$$

The sample estimate of σ, then, is $s = \sqrt{18.96} = 4.35$, and, from Table A.4, $t_{.975} = 2.306$ with $\nu = 8$ degrees of freedom. Then we find

$$\bar{x} - \frac{(2.306)s}{\sqrt{n}} = 29.1 - \frac{(2.306)(4.35)}{3} = 25.76$$

$$\bar{x} + \frac{(2.306)s}{\sqrt{n}} = 29.1 + \frac{(2.306)(4.35)}{3} = 32.44$$

is the 95% confidence interval for μ. Note that this interval is slightly longer than the one computed earlier; this will generally be true, although it is possible that this interval could be shorter than the other. The interpretation of this confidence interval is exactly the same as before: we are 95% sure or confident that the interval from 25.76 to 32.44 ppm includes the true unknown value of μ, the average contamination of all fish in the lake.

We have seen two different confidence intervals for the mean of a normal distribution, depending on whether or not the standard deviation σ is known. If σ is *known*, it is preferable to use the interval based on the standard normal distribution, since this will quite likely be the shorter of the two. If σ is *not known*, no choice is involved; it is only possible to compute the interval based on the t distribution. If $n \geq 30$, however, no serious error is involved in using the interval based on the standard normal percentiles, using S to estimate σ.

Let us now discuss a confidence interval for σ^2, the variance of a normal population. In Section 7.1 we saw that the recommended estimator for σ^2 is the sample variance $S^2 = \Sigma(X_i - \bar{X})^2/(n - 1)$. It would seem natural, then, that a confidence interval for σ^2 should in some way depend on S^2; that is in fact the case. Given a random sample of size n from a normal population, it can be shown that $(n - 1)S^2/\sigma^2$ has what is called a χ^2 distribution with ν degrees of freedom (χ is the Greek letter chi, pronounced ki with long i); that is, as we know, $(n - 1)S^2/\sigma^2$ will vary from one sample of size n to another. The probability measure that describes its behavior is called a χ^2 distribution; it again, like the t distribution, is independent of the unknown value of σ^2 and has a single parameter ν called its degrees of freedom. For the situation we are considering, $\nu = n - 1$, as with the t density. Unlike the standard normal and t densities used earlier for confidence intervals for μ, the χ^2 density function is *not* symmetric about zero. Figure 7.2.3 gives a picture of the χ^2 density with $\nu = 10$ degrees of freedom. This shape is typical of all χ^2 densities with $\nu \geq 3$. It can be shown that the mean of a χ^2 random variable is ν, its degrees of freedom, and that the variance of χ^2 random variable is 2ν. The χ^2 density has its maximum at $\nu - 2$, so long as $\nu \geq 2$. As ν gets larger, the χ^2 density gets more and more symmetric and, in fact, approaches a normal density.

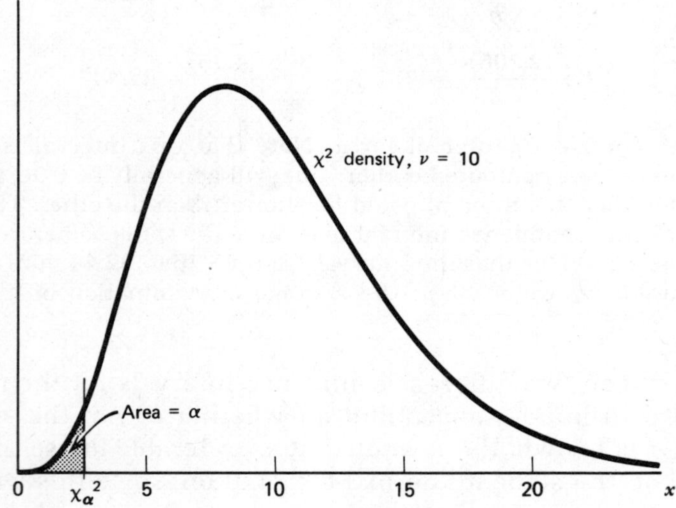

Figure 7.2.3 χ^2 density, $\nu = 10$.

Table A.5 presents selected percentiles for the χ^2 distribution for various degrees of freedom. Because the χ^2 density is positive only for positive values of the variable, it is not symmetric about zero; the point that has 5% of the area to the left of it is *not* the negative of the point that has 5% of the area to the right of it. Different lines in Table A.5 correspond to different numbers of degrees of freedom, and the different columns correspond to different percentiles. Using the same notation introduced earlier for percentiles, we shall let χ_α^2 be the value that has area α, to its left. Notice, then, for $\nu = 2$,

$\chi^2_{.05} = .103$ and $\chi^2_{.90} = 4.61$. With $\nu = 11$, $\chi^2_{.10} = 5.88$ and $\chi^2_{.95} = 19.7$.

Given a random sample of n observations from a normal population, it was mentioned that $(n - 1)S^2/\sigma^2$ has the χ^2 distribution with $\nu = n - 1$; then the probability is $1 - \alpha$ that

$$\chi^2_{\alpha/2} \le \frac{(n-1)S^2}{\sigma^2} \le \chi^2_{1-\alpha/2}$$

This double inequality is equivalent to

$$\frac{(n-1)S^2}{\chi^2_{1-\alpha/2}} \le \sigma^2 \le \frac{(n-1)S^2}{\chi^2_{\alpha/2}}$$

Thus the interval from

$$L = \frac{(n-1)S^2}{\chi^2_{1-\alpha/2}} \qquad \text{to} \qquad U = \frac{(n-1)S^2}{\chi^2_{\alpha/2}}$$

is a $100(1 - \alpha)\%$ confidence interval for σ^2.

Taking square roots we also have that the interval from $S\sqrt{(n-1)}/\chi^2_{1-\alpha/2}$ to $S\sqrt{(n-1)}/\chi^2_{\alpha/2}$ is a $100(1 - \alpha)\%$ confidence interval for σ, the population standard deviation.

Example 7.2.3

Fifteen identical packages of frozen fish were selected and their contents weighed. Each package said, "Net contents 12 ounces." It was found that $\Sigma x_i = 179.8$ ounces and $\Sigma x_i^2 = 2155.76$. Assuming that the weights of packages like this produced by this food packager are normally distributed, we can use these observations to compute a 90% confidence interval for σ^2 (and one for σ). We find

$$\bar{x} = \frac{179.8}{15} = 11.99$$

$$s^2 = \frac{1}{14}\left(2155.76 - \frac{(179.8)^2}{15}\right) = .0398$$

From Table A.5 we find, with $\nu = 14$ degrees of freedom, that $\chi^2_{.05} = 6.57$, $\chi^2_{.95} = 23.7$. Thus, the 90% confidence interval for σ^2 goes from $.0398(14)/23.7 = .024$ to $.0398(14)/6.57 = .085$. The 90% confidence interval for σ is from $\sqrt{.024} = .155$ to $\sqrt{.085} = .292$.

7.2.2   Confidence Intervals for Means of Other Populations

We saw in Section 6.4 that the probability measure for the mean of a sample of size n, selected at random from *any* population, will be well approximated by a normal probability measure. This fact enables us to use the standard normal tables to construct an approximate confidence interval for the population mean μ. For n sufficiently large the probability measure for

$$\frac{\overline{X} - \mu}{S/\sqrt{n}}$$

is approximately standard normal. Then, reasoning as we did earlier, the interval from

$$L = \overline{X} - \frac{S}{\sqrt{n}} z_{1-\alpha/2} \qquad \text{to} \qquad U = \overline{X} + \frac{S}{\sqrt{n}} z_{1-\alpha/2}$$

provides an approximate $100(1 - \alpha)\%$ confidence interval for μ. For $n \geq 30$, the probability that this interval covers μ should be quite close to $1 - \alpha$.

Example 7.2.4

A random sample of 50 households in San Francisco gave the following data for total household income in 1970 (in dollars):

$$\sum_i x_i = 523{,}647 \qquad \sum_i x_i^2 = 5{,}546{,}359{,}900$$

Then we find $\overline{x} = 10{,}473$, $s = 1127$, and the interval from

$$\overline{x} - z_{.975} \frac{s}{\sqrt{n}} = 10{,}473 - (1.96)\frac{1127}{\sqrt{50}} = 10{,}162$$

$$\overline{x} + z_{.975} \frac{s}{\sqrt{n}} = 10{,}473 + (1.96)\frac{1127}{\sqrt{50}} = 10{,}785$$

is a 95% confidence interval for the average income of all households in San Francisco. If we were to repeatedly take samples of size 50 from this population, for each sample compute the observed confidence interval as we just did, we would find that about 95% of the computed intervals did in fact cover the mean μ of the population.

7.2.3   Confidence Intervals for Proportions

If we let Y be the number of successes we will observe in n repeated, independent trials, each with probability of success equal to p, we have seen that Y has a binomial probability measure with parameters n and p. We have also seen that this binomial measure for Y is well approximated by a normal probability measure with $\mu = np$, $\sigma^2 = npq$, if n is large. Thus, for n sufficiently large,

$$\frac{Y - np}{\sqrt{npq}}$$

is approximately a standard normal random variable and the probability is (approximately) $1 - \alpha$ that

$$-z_{1-\alpha/2} \leq \frac{Y - np}{\sqrt{npq}} \leq z_{1-\alpha/2}$$

Algebraically, this inequality is equivalent to

$$\frac{Y}{n} - z_{1-\alpha/2}\sqrt{\frac{pq}{n}} \leq p \leq \frac{Y}{n} + z_{1-\alpha/2}\sqrt{\frac{pq}{n}}$$

We can use this result to derive a confidence interval for p as follows. The product pq is no larger than 1/4 since

$$pq = p(1 - p) = p - p^2 = \frac{1}{4} - \left(p - \frac{1}{2}\right)^2$$

Thus the largest possible value that $\sqrt{pq/n}$ might have is $\sqrt{1/4n} = 1/2\sqrt{n}$ and a conservative confidence interval for p is given by

$$\frac{Y}{n} - \frac{z_{1-\alpha/2}}{2\sqrt{n}} \leq p \leq \frac{Y}{n} + \frac{z_{1-\alpha/2}}{2\sqrt{n}}$$

(The interval is called conservative because we have replaced pq by its largest possible value, leading to the end points of the interval being as far apart as is possible.) A less conservative approach is to replace the product pq by the estimates from the sample data, $\hat{p}\hat{q}$; this will lead to a shorter interval, especially if $\hat{p}$ differs much from 1/2. We shall call this a liberal interval, to distinguish it from the conservative interval above.

Example 7.2.5

To estimate the incidence of a disease among 18-year-olds, a medical research team examined the records of 1000 people, selected at random from the full population of people this age. They found 193 people had the disease, among the 1000 investigated. Let us derive a 95% confidence interval for p, the proportion in the population that have the disease, using first the conservative formula given above and then using the other approach, to see the difference between the two. Given these sample results, the point estimate of p is

$$\hat{p} = \frac{193}{1000} = .193$$

For a 95% confidence interval we have $\alpha = .05$, $1 - \alpha/2 = .975$ and thus we require $z_{.975}$ from Table A.2, which we find to be 1.96. The conservative 95% confidence interval for p, then, has limits

$$\hat{p} - \frac{1}{2\sqrt{n}}\, z_{.975} = .193 - \frac{1}{2\sqrt{1000}}\, 1.96 = .162$$

$$\hat{p} + \frac{1}{2\sqrt{n}}\, z_{.975} = .193 + \frac{1}{2\sqrt{1000}}\, 1.96 = .224$$

If we use $\hat{p}$ to estimate the variance of Y, the 95% confidence interval for p has limits

$$\hat{p} - \sqrt{\frac{\hat{p}\hat{q}}{n}}\, z_{.975} = .193 - \sqrt{\frac{(.193)(.807)}{1000}}\,(1.96) = .169$$

$$\hat{p} + \sqrt{\frac{\hat{p}\hat{q}}{n}}\, z_{.975} = .193 + \sqrt{\frac{(.193)(.807)}{1000}}\,(1.96) = .217$$

Notice that even with $\hat{p}$ as far from .5 as .193, there is not a great deal of difference between the two. As $\hat{p}$ gets closer to zero than this, there is a marked difference in the lengths of the two intervals.

7.2.4   Confidence interval for λ, the Poisson rate parameter

As might be expected, an approximate confidence interval for λ, the parameter of a Poisson process, can be constructed in much the same way, by relying on the fact that sample means tend to be normally distributed. Specifically, if we have a random sample of n observations from a Poisson population with parameter λt, the sample mean $\overline{X}$ will have, for large n, a distribution that is approximately normal with mean equal to λt and variance equal to $\lambda t/n$. Thus $(\overline{X} - \lambda t)/\sqrt{\lambda t/n}$ is approximately a standard normal random variable and will then lie between $-z_{1-\alpha/2}$ and $z_{1-\alpha/2}$ with (approximate) probability $1 - \alpha$. An approximate $100(1 - \alpha)\%$ confidence interval for λt, then, is given by

$$L = \overline{X} - \sqrt{\frac{\overline{X}}{n}}\, z_{1-\alpha/2} \quad \text{and} \quad U = \overline{X} + \sqrt{\frac{\overline{X}}{n}}\, z_{1-\alpha/2}$$

The following example illustrates this computation.

Example 7.2.6

Assume that the arrivals of patients at a University hospital, with broken limbs, occur like events in a Poisson process with parameter λ per day and thus with parameter 7λ per week. In a 10-week period the number of patients arriving with broken limbs were 15, 9, 12, 9, 13, 18, 14, 15, 10, and 7. The point estimate of 7λ, the weekly rate, then is $\bar{x} = 122/10 = 12.2$ and an approximate 90% confidence interval for the weekly rate is given by

$$\bar{x} - \sqrt{\frac{\bar{x}}{n}}\, z_{.95} = 12.2 - \sqrt{\frac{12.2}{10}}\,(1.64) = 10.39$$

$$\bar{x} + \sqrt{\frac{\bar{x}}{n}}\, z_{.95} = 12.2 + \sqrt{\frac{12.2}{10}}\,(1.64) = 14.01$$

If we wanted a 90% confidence interval for λ, the daily rate, notice that $\bar{x}/7 = 1.74$ is the point estimate of λ; this is the observed value of a random variable that is approximately normally distributed with mean λ and variance $\lambda/7(10)$. Thus an approximate 90% confidence interval for λ, the daily rate, for this example is given by

$$\frac{1}{7}\left(\bar{x} - \sqrt{\frac{\bar{x}}{n}}\, z_{.95}\right) = \frac{10.39}{7} = 1.48$$

$$\frac{1}{7}\left(\bar{x} + \sqrt{\frac{\bar{x}}{n}}\, z_{.95}\right) = \frac{14.01}{7} = 2.00$$

7.2.5   One-Sided Confidence Intervals

Both end points of every confidence interval we have discussed are random variables and will vary from one sample to another. This type of interval gives two-sided bounds or limits for the unknown parameter. In many cases it is desired to bound the unknown parameter on only one side, rather than both sides. For example, a manufacturer of light bulbs would be more interested in saying he was 99% sure that the average lifetime of his bulbs is at least a hours (say 1200 hours) rather than saying he is 95% sure the average lies between b and c hours (say between 1210 and 1280 hours). It would seem better for sales to make a statement that limits μ only on one side in a case like this, rather than also limiting the interval from above. *One-sided* confidence intervals provide such an interval.

In constructing the two-sided $100\,(1 - \alpha)\%$ intervals, we split the noncoverage probability of α into two equal parts for the two tails of the distribution and the coverage probability was the middle proportion of $(1 - \alpha)$. To derive a one-sided $100(1 - \alpha)\%$ interval, we simply put all the noncoverage probability of α into one tail of the distribution. This reasoning is illustrated below.

Suppose we have a random sample of size n of a normal random variable with unknown mean μ and known variance σ^2. Then, the probability is $1 - \alpha$ that

$$\frac{\overline{X} - \mu}{\sigma/\sqrt{n}} \leq z_{1-\alpha}$$

and this inequality is equivalent to

$$\overline{X} - \frac{\sigma}{\sqrt{n}} z_{1-\alpha} \leq \mu$$

Thus, the interval from $L = \overline{X} - (\sigma/\sqrt{n}) z_{1-\alpha}$ to ∞ is a $100(1 - \alpha)\%$ one-sided confidence interval for μ. Similarly, the probability is $1 - \alpha$ that

$$-z_{1-\alpha} \leq \frac{\overline{X} - \mu}{\sigma/\sqrt{n}}$$

Inverting this inequality we get

$$\mu \leq \overline{X} + \frac{\sigma}{\sqrt{n}} z_{1-\alpha}$$

and the interval from $-\infty$ to $U = \overline{X} + (\sigma/\sqrt{n}) z_{1-\alpha}$ then is also a $100(1 - \alpha)\%$ one-sided confidence interval for μ. We can derive one-sided confidence intervals for any of the cases treated earlier in the same way.

Example 7.2.7

Assume that a random sample of 15 packages of frozen fish were opened and the net contents of each was weighed, as in Example 7.2.3. Also assume the same results as given in the example, that $\Sigma x_i = 179.8$ and $\Sigma x_i^2 = 2155.76$, thus giving $\overline{x} = 11.99$, $s^2 = .0398$, and $s = \sqrt{.0398} = .2$. Then, since $t_{.95} = 1.761$ with 14 degrees of freedom, we can be 95% sure that the interval from $-\infty$ to

$$\overline{x} + 1.761 \frac{s}{\sqrt{15}} = 11.99 + 1.761 \frac{(0.2)}{\sqrt{15}} = 12.08$$

includes μ; that is, based on these sample results, we are 95% sure that μ is no larger than 12.08 ounces.

Table 7.2 presents a summarization of all the confidence intervals discussed.

Table 7.2 Confidence Interval Summary
Two-Sided $100(1 - \alpha)\%$ Intervals

for Population Mean

Normal population, σ known	$L = \overline{X} - \dfrac{\sigma}{\sqrt{n}}\, z_{1-\alpha/2},$	$U = \overline{X} + \dfrac{\sigma}{\sqrt{n}}\, z_{1-\alpha/2}$
Normal population, σ unknown	$L = \overline{X} - \dfrac{S}{\sqrt{n}}\, t_{1-\alpha/2},$	$U = \overline{X} + \dfrac{S}{\sqrt{n}}\, t_{1-\alpha/2}$
Any population, $n \geq 30$	$L = \overline{X} - \dfrac{S}{\sqrt{n}}\, z_{1-\alpha/2},$	$U = \overline{X} + \dfrac{S}{\sqrt{n}}\, z_{1-\alpha/2}$

for Population Variance

Normal population	$L = \dfrac{(n-1)S^2}{\chi^2_{1-\alpha/2}},$	$U = \dfrac{(n-1)S^2}{\chi^2_{\alpha/2}}$

for Proportion p

Conservative, n large	$L = \hat{p} - \dfrac{1}{2\sqrt{n}}\, z_{1-\alpha/2},$	$U = \hat{p} + \dfrac{1}{2\sqrt{n}}\, z_{1-\alpha/2}$
Liberal, n large	$L = \hat{p} - \sqrt{\dfrac{\hat{p}\hat{q}}{n}}\, z_{1-\alpha/2},$	$U = \hat{p} + \sqrt{\dfrac{\hat{p}\hat{q}}{n}}\, z_{1-\alpha/2}$

for Poisson Parameter λ

n large	$L = \overline{X} - \sqrt{\dfrac{\overline{X}}{n}}\, z_{1-\alpha/2},$	$U = \overline{X} + \sqrt{\dfrac{\overline{X}}{n}}\, z_{1-\alpha/2}$

To construct a one-sided, lower, $100(1 - \alpha)\%$ interval for any parameter listed in Table 7.2, use the value given for L with $\alpha/2$ replaced by α; for an upper interval, use the value given for U with $\alpha/2$ replaced by α.

7.2.6 Sample Sizes

Generally, the bigger the sample size used, the shorter the (two-sided) confidence interval will be for an unknown parameter. This happens because the standard deviations of parameter estimators are inverse functions of n and for the cases we have studied the length of the confidence interval is directly related to the standard deviation of the estimator. For example, given a random sample of size n of a normal random variable with mean μ and *known* variance σ^2, a $100(1 - \alpha)\%$ confidence interval for μ is given by $L = \overline{X} - (\sigma/\sqrt{n}) \, z_{1-\alpha/2}$ and $U = \overline{X} + (\sigma/\sqrt{n}) \, z_{1-\alpha/2}$; the length of this interval is

$$U - L = \frac{2\sigma}{\sqrt{n}} \, z_{1-\alpha/2}$$

a constant times the standard deviation of $\overline{X}$ (which is $\sigma/\sqrt{n}$). Then, if we want the length of this confidence interval to be δ, say, we require

$$\delta = \frac{2\sigma}{\sqrt{n}} \, z_{1-\alpha/2}$$

Solving for n we get

$$n = \frac{4\sigma^2 \, z^2_{1-\alpha/2}}{\delta^2}$$

as the sample size needed. The above solution for n will not necessarily be a whole number. If it is not, we shall always round the solution up to the next larger integer because this will make the length of the resulting interval slightly smaller than δ; if we were to round down, the resulting interval would be longer than δ.

Example 7.2.8

Assume, as in Example 7.2.1, that the amount of mercury contamination per fish in a large lake is a normal random variable with unknown mean μ and $\sigma = 5$. How large a sample must we take so that a 90% confidence interval for μ is 2 ppm long? Here we have $\sigma = 5$, $\alpha = .1$ so $z_{1-\alpha/2} = z_{.95} = 1.64$, $\delta = 2$; the value for n is given by

$$n = \frac{4\sigma^2 \, z^2_{1-\alpha/2}}{\delta^2} = \frac{4(5)^2 \, (1.64)^2}{(2)^2} = 67.24$$

Thus, if we take a sample of 68 fish, the 90% confidence interval for μ is (slightly less than) 2 units long. Recall that this confidence interval for μ consists of $\overline{X}$ plus or minus the same constant and, thus, we could write

$$L = \overline{X} - \frac{\delta}{2} \qquad U = \overline{X} + \frac{\delta}{2}$$

Thus, the probability is $1 - \alpha$ that

$$\overline{X} - \frac{\delta}{2} \leq \mu \leq \overline{X} + \frac{\delta}{2}$$

that is, that

$$|\overline{X} - \mu| \leq \frac{\delta}{2}$$

If we select a random sample of 68 fish from this lake the probability is .9 that $\overline{X}$ differs from μ by no more than $\delta/2 = 1$ ppm.

When we sample from a normal population with σ unknown, the length of the $100(1 - \alpha)\%$ confidence interval for μ is $(2S/\sqrt{n})t_{1-\alpha/2}$, where S is the sample standard deviation. Since S is a random variable and, thus, varies from sample to sample, so will the length of the confidence interval vary from sample to sample. Thus it is not possible in that case to choose a predetermined length for the confidence interval for μ. The probability measure for S could be used to determine the expected or average length of the interval; then n could be chosen to give this expected length a preset value. More advanced textbooks described such a procedure in detail.

The conservative confidence interval for an unknown proportion, p, was given by $L = \hat{p} - (1/2\sqrt{n})z_{1-\alpha/2}$ and $U = \hat{p} + (1/2\sqrt{n})z_{1-\alpha/2}$. The length of this interval is

$$U - L = \frac{1}{\sqrt{n}} z_{1-\alpha/2}$$

Here, again, it is possible to choose the sample size n so that the length is equal to a predetermined value δ. In fact, solving

$$\delta = \frac{1}{\sqrt{n}} z_{1-\alpha/2}$$

gives

$$n = \frac{z_{1-\alpha/2}^2}{\delta^2}$$

Recall that this conservative interval was derived by setting the population variance, pq, equal to its largest possible value of 1/4 (given when $p = q = 1/2$). If, in fact, the true population p differs considerably from 1/2, this value of n will be very conservative (too large) indeed. If we are certain that the true p lies between 0 and p_0, where p_0 is less than 1/2 (or alternatively q lies between 0 and q_0 which is less than 1/2), then the population variance must be smaller than $p_0 q_0 < 1/4$. A less conservative confidence interval for p then is

$$\hat{p} - \sqrt{\frac{p_0 q_0}{n}}\, z_{1-\alpha/2} \leq p \leq \hat{p} + \sqrt{\frac{p_0 q_0}{n}}\, z_{1-\alpha/2}$$

which has length $2\sqrt{(p_0 q_0/n)}\, z_{1-\alpha/2}$. If we want this length to equal δ, we should choose

$$n = \frac{4 p_0 q_0}{\delta^2}\, z^2_{1-\alpha/2} < \frac{z^2_{1-\alpha/2}}{\delta^2}$$

so long as $p_0 q_0 < 1/4$. Both of these concepts are illustrated in the following example.

Example 7.2.9

A political candidate would like to estimate his standing with the elec-torate, one month before the election. He would like to take a sufficiently large sample so that the difference between the sample proportion favoring him and the population proportion favoring him is no more than 1% (.01), with probability .95. How large a sample is required? This is equivalent to saying that the 95% confidence interval for p should have length .02. Using the conservative confidence interval, which is appropriate if p is some-where close to .5, he should take a sample of size

$$n = \frac{z^2_{.975}}{\delta^2} = \frac{(1.96)^2}{(.02)^2} = 9604$$

With a sample of 9604 people the probability is .95 that the sample propor-tion $\hat{p}$ does not differ by more than .01 from the true p. This sample size really is much too large, however, if it can be assumed that p differs more than a little from $1/2$. Suppose the politician is quite sure that at least 80% of the electorate favor him; thus he is willing to assume $p \geq .8 = p_0$. So long as this assumption is valid, a sample of size

$$n = \frac{4 p_0 q_0}{\delta^2}\, z^2_{1-\alpha/2} = \frac{4(.8)(.2)}{(.02)^2}\, (1.96)^2$$

$$= 6146.5 \sim 6147$$

is sufficient to give probability .95 that $|\hat{p} - p| \leq .01$. Note that there is a considerable difference in the number of people required in the sample, because of the different assumption made in the second case.

Exercise 7.2

1. The lifetimes of 15 watt light bulbs are known to be normally distrib-uted with $\sigma = 30$ hours. A sample of 25 bulb lifetimes yielded the fol-lowing statistics:

$$\Sigma x_i = 62{,}050 \qquad \text{and} \qquad \Sigma x_i^2 = 154{,}037{,}500$$

where all lifetimes are in hours. Construct a 95% confidence interval for μ, the mean lifetime of the population of bulb lifetimes.

2. If we assume in question 1 that σ is unknown, construct a 95% confidence interval for μ.

3. Use the data in question 1 to construct a 90% two-sided confidence interval for σ^2 and a 90% two-sided confidence interval for σ.

4. In one afternoon, a university broad jumper made leaps of 24 ft 8 in., 25 ft 2 in., 24 ft 10 in., 26 ft 3 in., 25 ft 7 in., and 25 ft 6 in. Assuming these are observed values of a normal random variable, construct a 90% two-sided confidence interval for the mean distance, μ, that he is able to jump under these conditions.

5. A drugstore finds that its weekly sales of a skin cream occur like events in a Poisson process. Over the past 10 weeks, it has sold a total of 123 bottles of this type of cream. Construct a 90% lower confidence interval for λ, the weekly rate of sales for this item.

6. A large university uses a computer to print the grade reports for all students. During one quarter a total of 58,967 individual grades were processed; a check revealed that 312 of these were in error. Compute an approximate 99% liberal upper confidence interval for p, the probability an error will be made in processing a single grade.

7. The probability an individual seed will germinate, under controlled conditions, is p. Given that 90 out of 100 seeds germinated under these conditions, compute a 95% liberal lower confidence interval for p.

8. Confidence intervals can also be constructed for functions of a parameter, given an interval for the parameter itself. Assume in problem 7 that three seeds will be planted together in the same location (called a hill). Then the number that will germinate in a hill is a random variable Y with a binomial probability measure, with parameters $n = 3$ and p, under the assumption that the three seeds in the same hill will germinate independently of one another. The probability that at least one seed in the hill will germinate then is

$$P(Y \geq 1) = 1 - P(Y = 0) = 1 - (1 - p)^3$$

a function of p, the probability a single seed will germinate. The inequality

$$L \leq p$$

can be manipulated as follows:

$$1 - p \leq 1 - L$$

$$(1 - p)^3 \leq (1 - L)^3$$

$$1 - (1 - L)^3 \leq 1 - (1 - p)^3 = P(Y \geq 1)$$

to give a lower inequality for $P(Y \geq 1)$, the function of p. Since one inequality is true if and only if the other is, the probabilities attached to them are equal. Thus the final inequality gives a lower confidence interval for $P(Y \geq 1)$. Using the results of question 7, evaluate a 95% lower confidence interval for the probability that at least one seed will germinate in a hill of three seeds.

9. Verify the algebra used in deriving the confidence intervals for the mean of a population. That is, show that

$$\frac{\overline{X} - \mu}{\sigma/\sqrt{n}} \leq z$$

is equivalent to $\overline{X} - z(\sigma/\sqrt{n}) \leq \mu$ and that

$$-z \leq \frac{\overline{X} - \mu}{\sigma/\sqrt{n}}$$

is equivalent to $\mu \leq \overline{X} + z(\sigma/\sqrt{n})$.

10. Verify the algebra used in deriving a confidence interval for the variance of a population; that is, show that

$$\frac{(n-1)S^2}{\sigma^2} \leq \chi^2$$

is equivalent to $(n-1)S^2/\chi^2 \leq \sigma^2$ and that

$$\chi^2 \leq \frac{(n-1)S^2}{\sigma^2}$$

is equivalent to $\sigma^2 \leq (n-1)S^2/\chi^2$.

11. What is the area under the standard normal curve between $z_{1-\alpha+\epsilon}$ and z_ϵ, where $0 \leq \epsilon \leq 1/2 - \alpha$? Why did we choose $\epsilon = \alpha/2$ for the confidence intervals derived in the text?

12. The quality control section of a manufacturing firm is in charge of deriving 90% confidence limits for the proportion of defectives in each incoming lot of material received. Given 100 arriving lots, approximate the probability that at least 95 of the computed intervals do cover the corresponding p. Approximate the probability that exactly 90 of them cover their corresponding value for p.

13. Use the data in question 1, assuming $\sigma = 30$, to construct a 90% lower confidence limit for the probability a bulb of this type will burn at least 2475 hours. (Read the discussion given in question 8.)

14. A toothpaste manufacturer believes that 10% of all purchasers prefer his brand. Assuming his belief is correct, how large a sample should he take to have a 90% chance that the sample proportion does not differ from the true proportion by more than .01? How large a sample would be required if in fact p is close to .5?

15. In Section 7.2.1 we saw two different confidence intervals for the mean μ of a normal population, depending on whether σ was known or unknown. The statement was made there that the interval with σ known will generally be shorter than the interval with σ unknown. What must be true for the interval with σ known to be shorter than the interval with σ unknown?

16. A random sample of 100 college freshmen was selected; each was weighed and it was found that

$$\sum_{i=1}^{100} x_i = 15{,}075 \qquad \sum_{i=1}^{100} x_i^2 = 2{,}313{,}362.41$$

    where the weight of the ith student was x_i. Construct a 95% confidence interval for μ, the mean of the population of weights of all college freshmen.

17. An inveterate bettor at the horse races went to the track 196 times in one year, and made bets on the horses each time he went. The total amount he won over these 196 times was $-\$35$. (That is, he ended up \$35 behind.) Assume that these 196 trips represent a random sample from some population of wins and losses (wins are positive, losses negative). Given that the sums of squares of his wins and losses over these 196 days equaled 2,190,100, compute a 90% confidence interval for μ, the mean of the population of wins and losses.

18. A reading ability test was given to a random sample of 100 third grade students. Letting x_i represent the score achieved by student i, it was found that

$$\sum_{i=1}^{100} x_i = 12{,}516 \qquad \sum_{i=1}^{100} x_i^2 = 1{,}606{,}102$$

    Construct a 95% confidence interval for the mean score of all third grade students on this exam.

19. One hundred fifty adults volunteered to follow the same diet-exercise regimen for one month. The amount of weight lost, over this period, was observed for each and was denoted by x_i for the ith person. Given that

$$\sum_{i=1}^{150} x_i = 1387 \qquad \sum x_i^2 = 15{,}209$$

    compute a 99% confidence interval for the weight an adult could expect to lose over a one-month period, following this diet-exercise regimen.

20. A thumbtack was tossed in the air 125 times and landed point up 87 of these times. Construct a (conservative) 95% confidence interval for the probability this type of tack will land point up when tossed in the air.

21. Assuming the population of distances sampled from in question 4 is normal in form, use the data presented there to construct a 95% confidence interval for the population standard deviation.

22. How many times must a fair coin be flipped to have probability .99 that the sample proportion of heads will be within .005 of the true probability of heads?

23. Seventy-three voters out of 100 contacted from a large electorate said that they were in favor of a proposed school bond issue. Assuming these 100 people are a random sample of the full electorate, construct a (liberal) 90% two-sided confidence interval for the proportion of voters in the full electorate that favor the school bond issue.

24. Use the data given in problem 23 to compute a 90% lower confidence interval for the proportion of voters in the electorate that favor the school bond issue. If an election were held, do you believe the school bonds would pass?

25. An item constructed on an assembly line is either defective or not defective, when it is completed. A random sample of 200 items selected from the line contained 10 defectives. Construct a liberal 95% lower confidence interval for p, the proportion of defectives produced by this line.

26. For the data given in question 25, construct a liberal 95% upper confidence interval for p. This is the interval that the owner of the assembly line would generally use; it enables him to say he is essentially sure the proportion of defectives is no greater than a certain amount.

27. A successful real estate firm made 292 sales of private residences during a 365-day period. Assuming their sales occur like events in a Poisson process, construct an approximate two-sided 99% confidence interval for λ, the daily rate of sales.

28. For the data given in problem 27, construct a 99% two-sided confidence interval for the probability the firm will make at least one sale on any given day. (Read the discussion in problem 8.)

29. Over a one-week, 40-hour period, a large business office made and received a total of 4000 telephone calls. Assuming the instants at which the calls started behaved like events in a Poisson process, compute a 90% upper confidence limit for λ, the total number of calls per hour.

30. January 1, 1965 to February 9, 1971, inclusive, is an interval of 2231 days. During this time, at least one earthquake of magnitude 4 or more on the Richter scale was recorded somewhere in the world on 143 separate days. Assuming there is a constant probability p of one or more earthquakes of magnitude 4 or more occurring in a 24-hour period, and that occurrences are independent from day to day, compute a two-sided, liberal, 95% confidence interval for the probability of one or more earthquakes occurring on any given day.

31. Ten thousand feet of extruded nylon filament was examined for defects and a total of 50 were found. Assuming these defects are located along the length of the filament like events in a Poisson process, construct a 95% confidence interval for λ, the number of defects per 100 feet.

32. A cigarette manufacturer believes that about 25% of all smokers prefer one of his brands. How large a sample of smokers should he select so that the difference between the sample proportion and the true proportion is no more than .02, with probability .99?

33. The answer to question 32 refers to the number of *smokers* who should be sampled. Not all people are cigarette smokers and, since cigarette smokers are not all restricted to one geographic area, or labeled in some other way, they first must be located. Assume that 10% of the adult population are smokers; then the probability an adult selected at random would be a smoker *and* prefer a brand of the manufacturer mentioned in 32 is $(.1)(.25) = .025$, if 25% of all smokers prefer one of his brands. How large a sample of adults should he select if he wants the sample proportion of the people that smoke and prefer his brand to differ from the true proportion by no more than .002, with probability .99?

34. A television program rating firm takes a random sample of 2000 households to determine their rankings of the programs on the air at any given time. If 10% of all the households in the United States are watching a given program, what is the probability that the sample proportion will be within .01 of .10?

35. For the case described in question 34, what is the probability the sample proportion will be within .01 of the true proportion, given that true proportion is .05?

7.3 SUMMARY

Methods of estimation fall into two classes: point estimators for parameters and interval estimators for parameters. The (point) estimator for a parameter is the formula used and is itself a random variable with a probability measure describing its behavior over repeated samples. The observed value of an estimator, given a specific sample from the population, is called the estimate of the parameter. If the expected value of an estimator is equal to the parameter being estimated, the estimator is unbiased. Among several unbiased point estimators we prefer the one with the smallest variance; small variance of the unbiased estimator implies that, from one sample to another, it varies relatively little about the true parameter value. Thus, it will in general equal a value within any given distance of the true parameter value in a higher proportion of samples than will another estimator with larger variance. Table 7.1 lists estimators for the distribution parameters we have studied in this book.

A $100(1 - \alpha)\%$ confidence interval for an unknown parameter is specified by two random variables, U and L; the probability that the two random variables bracket the unknown parameter value is $1 - \alpha$. Thus, over repeated samples, the fraction $1 - \alpha$ gives the proportion of observed sample intervals that include the unknown parameter value. Taking only one sample and computing the observed interval for that sample, we are $100(1 - \alpha)\%$ confident that we have bracketed the unknown parameter value. One-sided confidence intervals are at times of interest; rather than bracketing the parameter and bounding it from two sides, a one-sided confidence interval bounds a parameter on only one side. Table 7.2 summarizes the confidence intervals for the distributional parameters we have studied. The length of a two-sided confidence interval is a decreasing function of n, the sample size. In certain cases it is possible to specify the length of the confidence interval in advance and then solve for the sample size n appropriate for that length.

Exercise 7.3

1. A fair coin is to be flipped n times. How large should n be so that the observed proportion of heads will differ from 1/2 by no more than .01, with probability .99?

2. Automobiles arrive at a city parking lot similar to events in a Poisson process with parameter λ per hour. The total number of cars to arrive in 50, nonoverlapping 15-minute intervals was 1497.
   (a) Estimate λ.
   (b) Compute the standard error of your estimate in (a).

3. Estimate (approximately) the probability that at least 110 cars will arrive in one hour at the parking lot mentioned in question 2.

4. Of 2000 randomly selected households telephoned between 7 and 8 P.M. the same evening, 327 had a television set on.
   (a) Compute a 90% liberal two-sided confidence interval for p, the proportion of all households with a television set on during this period.
   (b) If 100 of the sets turned on were all tuned to Network A, compute a liberal , 90% two-sided confidence interval for the proportion of households that were watching television and watching Network A during this period.

5. During one spring a college broad jumper made 100 jumps in intercollegiate competition. The average distance he jumped was $\bar{x} = 24.2$ feet and the standard deviation of the distance was $s = 1.1$ feet; assume these 100 distances are the observed values of a random sample of size $n = 100$ selected from a normal population.
   (a) Estimate the probability this athlete will jump at least 25 feet on any given try.
   (b) Estimate the probability that he jumps less than 23 feet on any given attempt.

6. A random sample of 1000 1972 income tax forms were examined in detail; 679 of those examined were found to have one or more errors or inconsistencies. Compute a 95% liberal lower confidence interval for the proportion of all 1972 income tax forms with one or more errors.

7. The average number of children per family in a random sample of 100 families selected from the same country was 3.2; the standard deviation of the number of children per family was .5. Compute a 95% confidence interval for the mean number of children per family, for all families living in this country.

8. A statistician working for the research department of a large industry computed 110 90% confidence intervals in 1972. What is the (approximate) probability that at least 100 of these intervals covered the parameter being estimated?

CHAPTER 8
STATISTICAL INFERENCE: TESTS OF HYPOTHESES

As was mentioned at the start of Chapter 7, there are two major areas of classical statistical techniques, estimation, which was introduced in Chapter 7 and tests of hypotheses, which we shall study in this chapter. Essentially all areas of science have grown, and continue to grow, by employing the scientific method, an objective procedure for gathering data and evaluating their worth. Part of this scientific method is concerned with putting forth hypotheses and in choosing objective rules for deciding whether the hypotheses are true or not. The methods of this chapter are appropriate to this type of decision-making procedure.

As was mentioned earlier, a hypothesis, for our purposes, will be a statement about a probability measure. The statements "The probability is .7 that Joe will make a shot from the free throw line," "The expected yield per acre using this hybrid corn is 90 bushels," and "The expected length of life of this light bulb is 750 hours" are all hypotheses; each is making a statement about a probability measure for a population. Any time such a statement or hypothesis is made it may be true or it may be false. A statistical test of a hypothesis is a rule for deciding, on the basis of a random sample from the population, whether to accept or to reject the hypothesis.

The methodology of statistical tests of hypotheses is very similar to our judical system in its treatment of criminal cases. In a United States court the defendent is presumed innocent until he is proven guilty. The hypothesis then, is that the defendant is innocent. A jury is selected, evidence is presented from both sides, and the jury must then decide whether the defendant is innocent (accept the hypothesis) or guilty (reject the hypothesis).

Similarly, in a statistical test of a hypothesis we first must state what the hypothesis is, then gather the evidence (take a sample from the population) and, on the basis of the evidence (the sample), we must decide to accept or reject the hypothesis. For example, suppose we wanted to test the hypothesis that the probability is .7 that Joe will make a shot from the free throw line. The pertinent evidence to be gathered to test this hypothesis would be the numbers of shots he makes and misses in shooting from the free throw line. Suppose we decide to have him attempt $n = 10$ shots from the free throw line. Then if he makes 7 shots we would undoubtedly accept the hypothesis, based on this sample, since that is exactly the number we would expect him to make if the hypothesis were true; this sample result is perfectly consistent with the hypothesis so we should not reject it if we observe 7 baskets in 10 shots. If, on the other hand, he misses all 10 shots, or if he makes only 1 or 2 out of the 10, we would reject the hypothesis, since it is quite unlikely he would do that poorly if the hypothesis were true; these sample results are not consistent with the hypothesis. What about the other possibilities that could be observed from the results of 10 shots, namely, that he makes 3, 4, 5, 6, 8, 9, or 10 baskets? Since a decision must be made, no matter what is observed in the sample, we must be prepared to choose one of the two actions, accept or reject, for each of these possibilities as well. In this chapter we shall discuss the methodology applicable to making this decision.

8.1   TESTS AND ERRORS

In a statistical test of a hypothesis, as well as our judicial system, there are two distinct types of errors that can be committed: we may reject the hypothesis when it is true (called a type I error) or we may accept it when it is false (called a type II error). Table 8.1 illustrates these two types of error.

Table 8.1   Two Possible Types of Error

	Hypothesis Is:	
We Decide To:	True	False
Accept hypothesis	No error	Type II error
Reject hypothesis	Type I error	No error

No matter what we observe in the sample we select, we must choose one or the other of the two possible decisions, accept the hypothesis or reject the hypothesis. Notice also that if we decide to accept the hypothesis, the only possible error we could commit is a type II error; if we decide to reject the hypothesis, the only possible error we may make is a type I. That is, it is not possible for us to simultaneously commit both errors in making our decision. We *may* commit one or the other, but not both at the same time.

This requirement that we must make a decision, no matter what the outcome observed in the sample, can be described as follows. Generally, in advance of selecting the sample, we can list the range of possibilities that we may observe. Since we must choose either *accept* or *reject*, for each of these possibilities, we must partition this range into two parts; the outcomes for which we will accept the hypothesis versus the outcomes for which we reject the hypothesis. For example, in the case of 10 shots by Joe from the free throw line, we can categorize what will be observed by the random variable Y, the number of shots he makes, with range $R_Y = \{0,1,2,3,\ldots,$ $10\}$. We must decide, then, which elements of R_Y will lead us to accept the hypothesis versus which elements will lead us to reject the hypothesis. For example, one rule we could use would be

$$\text{Accept if} \quad Y = 5, 6, 7, 8, 9$$

$$\text{Reject if} \quad Y = 0, 1, 2, 3, 4, 10$$

A different possible rule (different partition of R_Y) is

$$\text{Accept if} \quad Y = 3, 4, 5, 6, 7, 8, 9, 10$$

$$\text{Reject if} \quad Y = 0, 1, 2$$

There are a large number of different possible partitions that could be chosen. What would lead us to prefer one instead of another?

As has been mentioned, there are two different errors that can be committed, called type I and type II. It would be ideal if we could choose the rule or partition that simultaneously minimizes the probability of occurrence of both these errors. Unfortunately such an ideal rule is impossible. For any given sample size n it can be shown that as the probability of a type I error is decreased, the probability of a type II error is necessarily increased, and vice versa. Thus it is not possible to simultaneously minimize the two probabilities of error.

If we let α represent the probability of a type I error and let β represent the probability of a type II error, the above discussion says we cannot simultaneously minimize α and β. What the standard procedures for testing hypotheses do, then, is to consider all possible tests (partitions of R_Y) which fix the value of α, say at .05; the best test, then, is defined to be the one in this class (with $\alpha = .05$) that gives the smallest possible value for β. This type of philosophy underlies the choice of tests that we shall discuss in this book. The following example evaluates α for the two different partitions of R_Y discussed earlier.

Example 8.1.1

Assume Joe is going to attempt 10 shots from the free throw line and we want to test the hypothesis that his probability of making a shot is $p = .7$. We earlier mentioned that the random variable to be observed then is Y, the number of baskets he makes. The range of Y is $R_Y = \{0, 1, 2, \ldots, 10\}$ and, assuming that his shots are independent with common (unknown) probability p that he makes a basket, the probability measure for Y then is binomial with parameters $n = 10$ and p. Consider the two tests (partitions of R_Y) presented earlier:

$$\text{Test 1: Accept } p = .7 \quad \text{if} \quad Y = 5, 6, 7, 8, 9$$

$$\text{Reject } p = .7 \quad \text{if} \quad Y = 0, 1, 2, 3, 4, 10$$

$$\text{Test 2: Accept } p = .7 \quad \text{if} \quad Y = 3, 4, 5, 6, 7, 8, 9, 10$$

$$\text{Reject } p = .7 \quad \text{if} \quad Y = 0, 1, 2$$

Let us evaluate $\alpha = P(\text{type I error})$ for both of these tests. First, we recall that a type I error means we reject the hypothesis when in fact it is true. Thus, for test 1 we want to evaluate

$$P(Y = 0 \text{ or } 1 \text{ or } 2 \text{ or } 3 \text{ or } 4 \text{ or } 10)$$

under the assumption the hypothesis is true, that is, that $p = .7$. We find from Table A.1 with $n = 10$, $p = .7$, that

$$\alpha = P(Y = 0 \text{ or } 1 \text{ or } 2 \text{ or } 3 \text{ or } 4 \text{ or } 10) = 0 + .0001 + .0014 + .0090 +$$
$$.0368 + .0282$$

$$= .0755$$

(To evaluate this probability we have to actually use $p = .3$ and transform the values of interest by $10 - Y = 10, 9, 8, 7, 6, 0$; see the earlier discussion in Section 5.1.) Thus, if we use test 1 there is a 7.5% chance we will commit a type I error; in 7.5% of all samples of a binomial random variable with $n = 10$, p $= .7$ we would observe $Y = 0$ or 1 or 2 or 3 or 4 or 10.

For test 2 we again want to evaluate the probability that we would reject the hypothesis, under the assumption that $p = .7$. Thus we require

$$\alpha = P(Y = 0 \text{ or } 1 \text{ or } 2) = 0 + .0001 + .0014 = .0015$$

Test 2 has a considerably smaller value for α. It also has a higher value for β. We could anticipate a smaller value of α for test 2 because with test 1 we reject all the test 2 rejection values (0, 1, 2) plus several more (3, 4, 10).

In the above example we did not evaluate β, the probability of type II error. Recall that we commit a type II error if we accept the hypothesis when it is false. Thus, for test 1 above we would want to evaluate

$$\beta = P(Y = 5 \text{ or } 6 \text{ or } 7 \text{ or } 8 \text{ or } 9)$$

under the assumption that $p \neq .7$. But the value for this probability depends on p, the probability that Joe will make a basket from the free throw line. Assuming that $p \neq .7$ does not give a unique value for β; we get a different value for β as we assume different values for p. This is generally true in testing hypotheses; if the hypothesis to be tested is not true there will be a range of alternative values to be considered. β varies with the different alternatives of interest.

Example 8.1.2

Let us evaluate β for the two tests given in Example 8.1.1 under the assumption that $p = .5$ and also under the assumption that $p = .9$. (These are two particular alternatives to $p = .7$.) Assuming $p = .5$, for test 1 we have

$$\beta = P(Y = 5 \text{ or } 6 \text{ or } 7 \text{ or } 8 \text{ or } 9) = .2461 + .2051 + .1172 + .0439 + .0098$$

$$= .6221$$

while for test 2 we find

$$\beta = P(Y = 3 \text{ or } 4 \text{ or } 5 \text{ or } 6 \text{ or } 7 \text{ or } 8 \text{ or } 9 \text{ or } 10)$$

$$= .1172 + .2051 + .2461 + .2051 + .1172 + .0439 + .0098 + .0010$$

$$= .9454$$

Assuming $p = .9$, we find for test 1

$$\beta = P(Y = 5 \text{ or } 6 \text{ or } 7 \text{ or } 8 \text{ or } 9) = .0015 + .0112 + .0574 + .1937 + .3874$$

$$= .6512$$

and for test 2 we have

$$\beta = P(Y = 3 \text{ or } 4 \text{ or } 5 \text{ or } 6 \text{ or } 7 \text{ or } 8 \text{ or } 9 \text{ or } 10)$$

$$= 0 + .0001 + .0015 + .0112 + .0574 + .1937 + .3874 + .3487$$

$$= 1.0000$$

Notice we do have a larger value for β, for both these alternatives, if we use test 2 rather than test 1.

We shall in succeeding sections see the types of tests that are used for hypotheses about the distributional parameters we have studied. The hypothesis to be tested is called the *null hypothesis* and is denoted by H_0. As already mentioned, the standard philosophy used in choosing a test of H_0 is to consider all tests with α, the probability of type I error, fixed; within this group of tests we will use the one with the smallest β, if possible. Recall that β is the probability we accept H_0 when it is false. If H_0 is false, then some other statement or hypothesis must be true, which gives the parameter values to be considered in computing β. Thus, hypotheses always occur in pairs; this second hypothesis, of interest in computing probabilities of type II error, is called the *alternative* hypothesis and is denoted by H_1.

In the above two examples the null hypothesis is that $p = .7$, to be written $H_0 : p = .7$. Which partition of R_Y is to be used will depend on the alternative hypothesis H_1. Thus, we shall find one partition to be best if we want to test

$$H_0 : p = .7 \qquad \text{versus} \qquad H_1 : p \neq .7$$

and other partitions will be used to test

$$H_0 : p \leq .7 \quad \text{versus} \quad H_1 : p > .7$$

or to test

$$H_0 : p \geq .7 \quad \text{versus} \quad H_1 : p < .7$$

The equality case will always occur in H_0, never in H_1. Table 8.2 defines a number of terms that are commonly used in statistical tests of hypotheses.

Table 8.2   Terms Used in Tests of Hypotheses

Term	Definition
Null hypothesis	Hypothesis to be tested.
Alternative hypothesis	Hypothesis to be accepted if H_0 is rejected.
P(type I error)	P(Reject H_0, assuming H_0 is true).
P(type II error)	P(Accept H_0, assuming H_1 is true).
Test statistic	Numerical quantity whose observed value is used to decide whether to accept or reject H_0.
Critical region	Values of test statistic for which H_0 is rejected.
Rejection region	Same as critical region.
Acceptance region	Values of test statistic for which H_0 is accepted.
Level of significance	Value of α.

In both of the preceding examples the test statistic, whose range is partitioned (into the critical region versus the acceptance region) was Y, the number of shots made. In any statistical test of a hypothesis the following four steps are taken.

1. Define H_0 and H_1.
2. Decide on α, the sample size n and the appropriate test statistic.
3. Choose the critical region (or equivalently choose the acceptance region).
4. Select the sample, observe the value of the test statistic, and either accept or reject H_0.

In any practical applications, there are of course implications coming from step 4 in the applied context. Hypotheses are tested, for example, to see if a theory is supported by actual observed data, to see if a claim made by a manufacturer is correct, and to decide which is better of two or more possible ways of doing something. Thus the acceptance or rejection of H_0 will be followed in practical cases by the implications of the subject matter area. In this book, it is not possible because of limitations of space and time to go into detail on these aspects of the full problem. The importance of these further steps should not be minimized, although we are not able to discuss them here.

Example 8.1.3

Test tubes have been broken by students in a university chemistry laboratory, over the past academic year, in accordance with a Poisson process at the rate of four per week. With the start of the new academic year, chemistry department officials want to test early in the year whether it appears the rate of breakage is greater this year. They will collect five weeks worth of data and then want to make their decision whether to accept $H_0 : \lambda \leq 4$ or to accept the alternative $H_1 : \lambda > 4$, where λ is the rate of breakage per week; they are willing to have one chance in 10 of rejecting H_0 when it is true ($\alpha = .1$). (If they reject H_0, it would appear that the rate of breakage is greater this year than last and some special measures may be called for.) Letting X_1, X_2, X_3, X_4, X_5 be the number of broken tubes that will be observed during the five weeks, a reasonable test statistic would be $Y = \sum_{i=1}^{5} X_i$, the total number of tubes broken during the period. (Equivalently, we could use $\overline{X} = Y/5$, the average number of broken tubes per week as the test statistic). As we know, if H_0 is true, then each of the X_i's is a Poisson random variable with parameter equal to $\lambda = 4$ (at most) and, thus, Y, their sum, is again a Poisson random variable with parameter $5\lambda = 20$ (at most). Then, to choose the critical region, with $\alpha = .1$, we must ask ourselves, would small values or large values for Y be inconsistent with H_0. Clearly, since the Poisson parameter for the probability measure for Y is its mean value, large values of Y would be inconsistent with H_0. Thus, we would like to find c such that

$$P(Y \geq c, \text{ given } 5\lambda = 20) = .1$$

and then we will reject H_0 if we observe a value of Y as large or larger than c. Our Poisson table does not extend to a distribution with its parameter equal to 20, but we recall from Section 6.4 that the Poisson probability measure is well approximated by the normal as long as its parameter is at least 10; thus we can use the normal approximation to the Poisson to find the (approximate) value of c. The approximating normal measure has $\mu = 20$, $\sigma = \sqrt{20} = 4.47$, and

$$P(Y \geq c) \doteq P\left(Z \geq \frac{c - \frac{1}{2} - 20}{4.47}\right)$$

$$= 1 - N_z\left(\frac{c - 20.5}{4.47}\right)$$

Thus, we require

$$\frac{c - 20.5}{4.47} = z_{.9} = 1.28$$

and we have $c = 20.5 + 4.47\,(1.28) = 26.22 = 27$. The chemistry department officials should reject H_0 if they find 27 or more broken tubes in the first five weeks and they then have $\alpha = .1$.

Suppose tubes are being broken at the rate of $\lambda = 6$ per week; what is the probability H_0 will be accepted? (Equivalently, what is the value for β if $\lambda = 6$?) The test being used accepts H_0 if and only if $Y \leq 26$; thus we must evaluate

$$P(Y \leq 26, \text{ given } 5\lambda = 30)$$

to answer this question. Under the assumption that $5\lambda = 30$, the exact probability measure for Y is approximated by a normal measure with $\mu = 30$ and $\sigma = \sqrt{30} = 5.48$. We find

$$P(Y \leq 26) = N_z\left(\frac{26 + \frac{1}{2} - 30}{5.48}\right)$$

$$= N_z\,(-.64)$$

$$= .2611$$

In using this test there is about a 26% chance that they will accept H_0 if the tubes are being broken at the rate of six per week.

Example 8.1.4

Assume that it is known that the yield to be obtained, per acre, using a particular hybrid corn is a normal random variable with unknown μ and $\sigma = 3$. To test

$$H_0 : \mu \geq 85 \qquad \text{versus} \qquad H_1 : \mu < 85$$

with $\alpha = .05$, $n = 10$, what should be used as the critical region? Letting $X_1, X_2, \ldots, X_{10}$ be the yields to be observed from the 10 plots, a reasonable test statistic appears to be

$$Y = \sum_{i=1}^{10} X_i$$

(or, equivalently, $\overline{X} = Y/10$). If H_0 is true, then the probability measure for Y is normal with mean $10\mu = 850$ (at least) and standard deviation $3\sqrt{10} = 9.48$; small values of Y would be inconsistent with H_0 in this case. Thus to have $\alpha = .05$ we would want to find c such that

$$P(Y \leq c) = .05$$

under the assumption Y is normal, mean 850, and standard deviation 9.48. We find with these assumptions

$$P(Y \leq c) = N_z \left(\frac{c - 850}{9.48} \right) = .05$$

so

$$\frac{c - 850}{9.48} = z_{.05} = -1.64$$

and

$$c = 850 - (9.48)(1.64)$$

$$= 834.45$$

If the sum of the yields from the 10 plots is no bigger than 834.45, then H_0 should be rejected.

In using a statistical test of a hypothesis, we are *not* proving or disproving statements; thus we say we either accept or reject a given hypothesis on the basis of a sample. Indeed, it is logically impossible to prove that the value of a parameter must be a certain number, based only on the results of a sample from a population (so long as the sample size is smaller than the population size).

Exercise 8.1

1. For the situation discussed in Example 8.1.1, assume we decide to reject $H_0 : p = .7$ if Y equals 0, 2, or 4. Evaluate α for this test.

2.   Evaluate β for the test given in problem 1 if $p = .5$.

3.   For the situation discussed in Example 8.1.3, what should be used as the critical region if we want $\alpha = .01$? [That is, what is the value of c if we require $P(Y \geq c) = .01$, given Y is Poisson with parameter 20.]

4.   Evalute β for the test given in question 3, under the assumption that $\lambda = 6$.

5.   For the situation described in Example 8.1.4, what is the value for β if
(a) $\mu = 80$?
(b) $\mu = 75$?

6.   It is assumed that the population of lifetimes of 40-watt light bulbs has a normal probability measure with unknown μ and $\sigma = 50$. A random sample of $n = 30$ bulbs are to be tested (left on until they burn out); it is desired to test

$$H_0 : \mu \geq 700 \quad \text{versus} \quad H_1 : \mu < 700.$$

(a)   What would you use as a test statistic for this case? (That is, what would you look for in the 30 failure times to try to decide whether to accept or reject H_0?
(b)   What form of critical region would you use (what values of the test statistic should lead you to reject H_0)?

7.   It is desired to test

$$H_0 : p = .5 \quad \text{versus} \quad H_1 : p \neq .5$$

where p is the probability of getting a head when a new 50¢ piece is flipped one time. The coin will be flipped $n = 15$ times and H_0 will be rejected if the number of heads we observe is less than four or more than 11.
(a)   Evaluate α for this test.
(b)   What is the value for β if $p = .4$?
(c)   What is the value for β if $p = .6$?

8.   It is assumed that fatal automobile accidents occur in Monterey County, California, like events in a Poisson process with parameter λ (measurements made in months). To test $H_0 : \lambda \leq 5$ versus $H_1 : \lambda > 5$ it is decided to accept H_0 if the total number of fatal accidents in eight months is no larger than 47.
(a)   What is the value for α in using this test?
(b)   If, in fact, $\lambda = 8$ (per month) what is the value of β?

9.   Scores made by high school seniors on an ability test are assumed to be normal with unknown μ and $\sigma = 100$. To test $H_0 : \mu = 500$ versus $H_1 : \mu \neq 500$ a random sample of 25 seniors will take the test; if $|\overline{X} - 500| < 40$, H_0 will be accepted, where $\overline{X}$ is the average of the 25 scores made by the seniors who take the exam.
(a)   Evaluate α for this test.
(b)   Evaluate β, assuming $\mu = 550$.

10. For the case described in problem 9, show that the test

$$\text{accept } H_0 \quad \text{if} \quad \overline{X} < 533.8$$

has (essentially) the same value for α as does the test given in problem 9. Which of these would you prefer for testing

$$H_0 : \mu = 500 \quad \text{versus} \quad H_1 : \mu \neq 500$$

8.2 SOME STANDARD TESTS

As we saw in Section 8.1, there are two different errors that can be committed in statistical tests of hypotheses; it is not possible to choose a test that simultaneously minimizes the probabilities of occurrence of both errors. The philosophy adopted, then, is to examine tests that have the same value of α, and among these to choose the one that has the smallest possible value for β. As was also mentioned in Section 8.1, the alternative hypothesis will in general include many possible values for the unknown parameter, and β may very well be different for these different possible values. In these cases it may be that there is no test with the smallest β for *all* the possibilities specified by H_1; in such cases compromise is called for in choosing the test. We shall in the following subsections examine the commonly chosen tests for some standard cases.

8.2.1 Tests about Means

As with confidence intervals, the unrealistic case of assuming a normal population with unknown mean μ and *known* variance s^2 gives a simple illustration to outline the basic facts and reasoning used in testing hypotheses about a population mean. Suppose we have a random sample of size n from this normal population and want to test a null hypothesis about the value of μ versus some alternative hypothesis about the value of μ. The sample mean, $\overline{X}$, is the estimator we would use for μ in this case; thus we should expect $\overline{X}$ to serve well as the test statistic for this hypothesis. That is, it would seem reasonable to base our decision whether to accept or reject H_0 on the value we observe for $\overline{X}$ in the sample; if $\overline{X}$ equals a value "consistent with" H_0, we should accept H_0 and if it equals a value "consistent with" H_1, we should reject H_0 (accept H_1). What, then, might we mean by the value of $\overline{X}$ being consistent with one hypothesis or the other?

Table 8.3 presents the critical regions for various possible hypotheses of interest about μ. In each case $\overline{X}$ is the test statistic employed.

Table 8.3 Critical Regions of Size α for Normal, σ Known

H_0	H_1	Reject H_0 if
$\mu \le \mu_0$	$\mu > \mu_0$	$\overline{X} > \mu_0 + \dfrac{\sigma}{\sqrt{n}}\, z_{1-\alpha}$
$\mu \ge \mu_0$	$\mu < \mu_0$	$\overline{X} < \mu_0 - \dfrac{\sigma}{\sqrt{n}}\, z_{1-\alpha}$
$\mu = \mu_0$	$\mu \ne \mu_0$	$\overline{X} > \mu_0 + \dfrac{\sigma}{\sqrt{n}}\, z_{1-\alpha/2}$
		or $\overline{X} < \mu_0 - \dfrac{\sigma}{\sqrt{n}}\, z_{1-\alpha/2}$

Figure 82.1 graphically presents the critical region for the test given in line 1 in Table 8.3. For line 1, we want to test $H_0 : \mu \le \mu_0$ versus $H_1 : \mu > \mu_0$. If in fact $\mu = \mu_0$, then we know that the probabil-

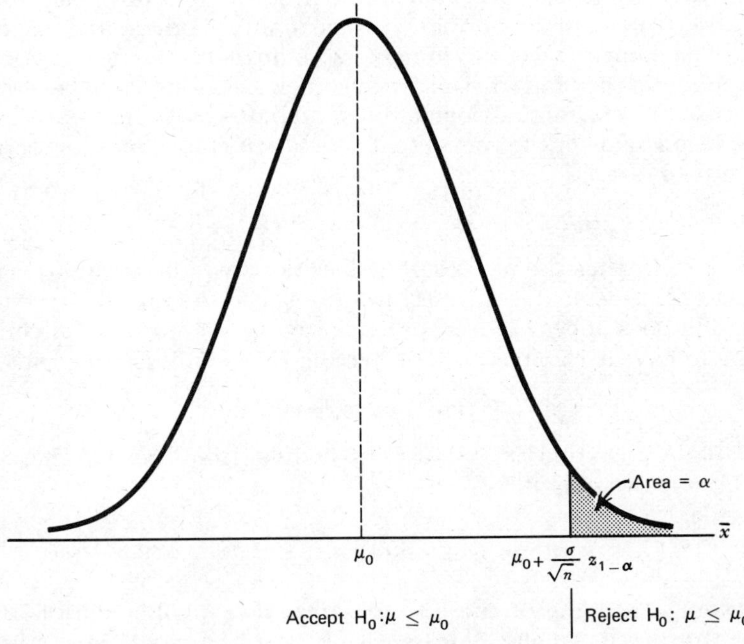

Figure 8.2.1 Testing $H_0 : \mu \le \mu_0$ versus $H_1 : \mu > \mu_0$.

ity measure for $\overline{X}$ is normal with mean μ_0 and variance σ^2/n. Thus, as is pictured in Figure 8.2.1, we would find $\overline{X}$ bigger than $\mu_0 + (\sigma/\sqrt{n})\, z_{1-\alpha}$ with probability α; that is, if we repeatedly take samples of size n from the normal population, in $100\alpha\%$ of the samples selected we would find $\overline{X} > \mu_0 + (\sigma/\sqrt{n})\, z_{1-\alpha}$. The probability we would reject H_0, then, if $\mu = \mu_0$, is equal to α; if the value of μ is smaller than μ_0, we would reject H_0 with probability less than α, with this test, because then the true density of $\overline{X}$ is centered at some point to the left of μ_0 and the area under this shifted density is smaller than if it is centered at μ_0. On the other hand, if $\mu > \mu_0$ (some value specified by H_1), the probability density for $\overline{X}$ is shifted to the right and the probability we would reject H_0 becomes greater and greater. It can be shown that this rule [reject H_0 if $\overline{X} > \mu_0 + (\sigma/\sqrt{n})\, z_{1-\alpha}$] has the smallest value for β for *every* value of μ specified by $H_1 : \mu > \mu_0$, among all possible tests with $P(\text{type I error}) = \alpha$. The following example illustrates this test.

Example 8.2.1

The research division of a light bulb manufacturer has developed a new filament coating that they feel will make light bulbs last longer. The currently used process produces light bulbs with normally distributed lifetimes, $\mu = 800$, $\sigma = 100$. It is assumed that lifetimes of bulbs made with the new coating will again be normal with mean μ (unknown) and again with $\sigma = 100$. The company decides to make 25 bulbs with the new coating and to burn them till they fail (these lifetimes then are assumed to be a random sample from the conceptual population of all bulbs made this way). Letting μ be the mean value of the conceptual population of lifetimes, the company then wants to test

$$H_0 : \mu \leq 800 \qquad \text{versus} \qquad H_1 : > 800$$

Accepting H_0 implies the new coating does not give a better bulb (in terms of average lifetimes); rejecting H_0 would lead to the conclusion that the new coating does appear to lead to longer average lifetime. The company is willing to have a 1/20 chance of rejecting H_0 when it is true; thus,

$$P(\text{type I error}) = .05$$

From Table A.2 we find $z_{1-\alpha} = z_{.95} = 1.64$; thus, from line 1, Table 8.3, H_0 should be rejected only if

$$\overline{X} > \mu_0 + \frac{\sigma}{\sqrt{n}}\, z_{1-\alpha} = 800 + \frac{100}{5} \cdot 1.64 = 832.8$$

If the average lifetime of the 25 bulbs tested is smaller than 832.8, H_0 should be accepted. Among all tests with $P(\text{type I error}) = .05$, no other test

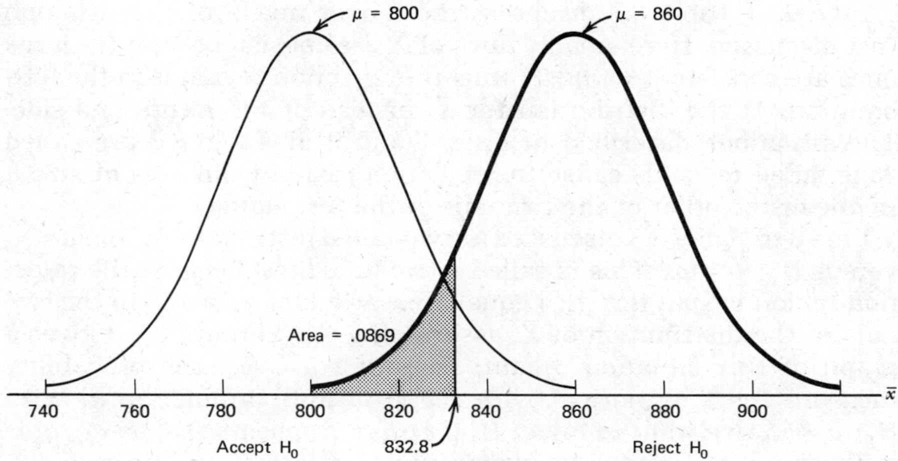

Figure 8.2.2 $P(\text{type II error}) = .0869$ if $\mu = 860$.

has a smaller value of β, no matter which value μ equals above 800. With this test it is quite straightforward to evaluate β for any desired value of μ. For example, if we assume $\mu = 830$

$$\beta = P(\text{we accept } H_0 \text{ with } \mu = 830)$$

$$= P(\overline{X} \le 832.8, \text{ given } \mu = 830)$$

$$= N_z \left(\frac{832.8 - 830}{20} \right)$$

$$= N_z (.14)$$

$$= .5557$$

If we assume $\mu = 860$,

$$\beta = P(\text{we accept } H_0 \text{ with } \mu = 860)$$

$$= P(\overline{X} \le 832.8, \text{ given } \mu = 860)$$

$$= N_z \left(\frac{832.8 - 860}{20} \right)$$

$$= N_z (-1.36)$$

$$= .0869$$

No other test with $\alpha = .05$ gives smaller values of β, for these values of μ (or any others). Figure 8.2.2 illustrates this computation of β, assuming $\mu = 860$.

Line 2 of Table 8.3 discusses the mirror image of the situation just discussed. Here small values of $\overline{X}$ are consistent with H_1, large ones are consistent with H_0; thus the rejection region is in the left-hand tail of the distribution for $\overline{X}$, instead of the right-hand side. The situations described in lines 1 and 2 of Table 8.3 are called "one-tailed tests" because the rejection region is an area of size α in one or the other of the two tails of the test statistic.

Line 3 of Table 8.3 discussed a "two-tailed test," namely, $H_0: \mu = \mu_0$ versus $H_1: \mu \neq \mu_0$. This is called a two-tailed test because the rejection region is split into two equal areas, each of area $\alpha/2$, in the two tails of the distribution of $\overline{X}$, assuming $\mu = \mu_0$. Figure 8.2.3 gives a graph of this situation. Again, assuming $\mu = \mu_0$, the probability measure for $\overline{X}$ is normal with mean μ_0 and variance σ^2/n. With $H_1: \mu \neq \mu_0$, we want to reject H_0 if *either* it appears that $\mu > \mu_0$ or if it appears $\mu < \mu_0$; thus we would like to reject H_0 in this case if $\overline{X}$ differs too much from μ_0 in either direction and split the total probability of type I error into two equal parts, as shown. The following example illustrates this test.

Example 8.2.2

The amount of beer that an automatic bottle-filling machine puts into a

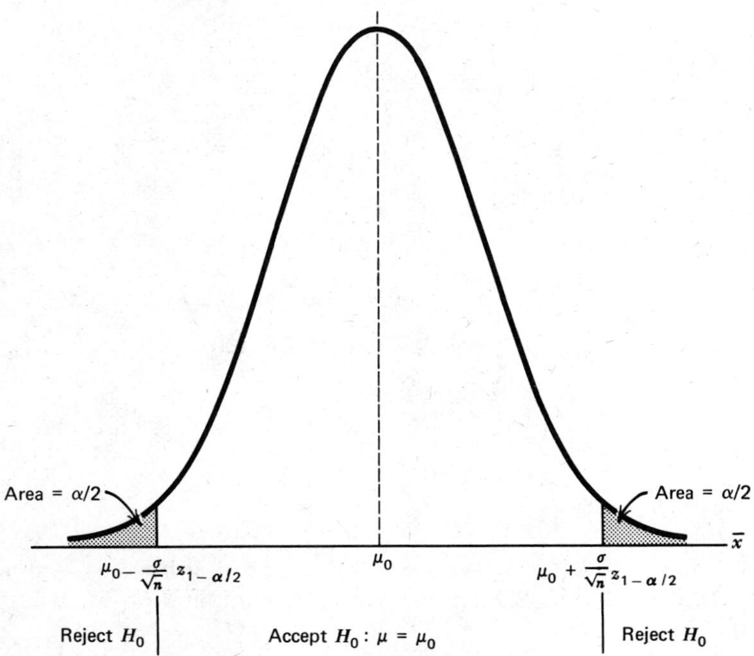

Figure 8.2.3 Testing $H_0: \mu = \mu_0$ versus $H_1: \mu \neq \mu_0$.

12-ounce bottle is a normal random variable with mean μ and $\sigma = .1$ ounce. Controls on the machine must be periodically set to try to keep μ, the average content, close to 12 ounces. If μ is too far from 12, then too many bottles will be overfilled or underfilled. To determine if the controls are set correctly, the contents of nine bottles are determined. Thus, it is desired to test

$$H_0 : \mu = 12 \qquad \text{versus} \qquad H_1 : \mu \neq 12$$

a two-tailed test given in Table 8.3, with $\mu_0 = 12$; it is decided to use $\alpha = .1$. Then, from Table A.2, $z_{1-\alpha/2} = z_{.95} = 1.64$ and H_0 should be rejected if

$$\overline{X} > \mu_0 + \frac{\sigma}{\sqrt{n}} z_{1-\alpha/2} = 12 + \frac{.1}{3} (1.64) = 12.05$$

or if

$$\overline{X} < \mu_0 - \frac{\sigma}{\sqrt{n}} z_{1-\alpha/2} = 12 - \frac{.1}{3} (1.64) = 11.95$$

Thus, if the average contents of the nine bottles examined is 11.92 ounces, for example, H_0 is rejected (and presumably the controls on the machine are reset).

If we have a random sample of size n of a normal random variable with both μ and σ unknown, the percentiles of the t distribution with $n-1$ degrees of freedom and S, instead of σ, are used in defining the critical regions for testing hypotheses about μ. Table 8.4 gives the critical regions for the different possible hypotheses about μ, with σ unknown.

Table 8.4   Critical Regions of Size α for Normal, σ Unknown

H_0	H_1	Reject H_0 if
$\mu \leq \mu_0$	$\mu > \mu_0$	$\overline{X} > \mu_0 + \dfrac{S}{\sqrt{n}} t_{1-\alpha}$
$\mu \geq \mu_0$	$\mu < \mu_0$	$\overline{X} < \mu_0 - \dfrac{S}{\sqrt{n}} t_{1-\alpha}$
$\mu = \mu_0$	$\mu \neq \mu_0$	$\overline{X} > \mu_0 + \dfrac{S}{\sqrt{n}} t_{1-\alpha/2}$
		or $\overline{X} < \mu_0 - \dfrac{S}{\sqrt{n}} t_{1-\alpha/2}$

Notice that these are exactly the same procedures given in Table 8.3, the only change being that the percentiles of the t distribution, with $n - 1$ degrees of freedom, replace the standard normal percentiles and S, the sample standard deviation, replaces σ.

Example 8.2.3

An automobile manufacturer claims that one of its car models will typically get at least 22 miles per gallon, even with full emission control equipment in operation. A consumer action group obtains eight of these cars, runs each over the same 200-mile course under the same conditions and wants to test

$$H_0: \mu \geq 22 \qquad \text{versus} \qquad H_1: \mu < 22$$

with $\alpha = .05$. They are willing to assume that the population of miles per gallon that cars of this type will achieve under these conditions is normal with unknown mean μ and unknown variance σ^2. The eight readings obtained were

$$19.8, \quad 21.7, \quad 22.3, \quad 20.5, \quad 20.3, \quad 21.1, \quad 22.0, \quad 20.1$$

Should they accept H_0? According to Table 8.4, we must compute $\bar{x}$ and s, from these sample values; we will accept H_0 if $\bar{x} > 22 - (s/\sqrt{8}) \, t_{1-\alpha}$, where $t_{.95} = 1.895$ with $n - 1 = 7$ degrees of freedom. From the above eight numbers it is found that

$$\Sigma x_i = 167.8 \qquad \Sigma x_i^2 = 3525.78$$

so

$$\bar{x} = \frac{167.8}{8} = 20.97$$

$$s^2 = \frac{1}{7}\left(3525.78 - \left(\frac{167.8}{8}\right)^2\right) = 0.882$$

from which $s = \sqrt{.882} = .939$. Then,

$$22 - \frac{.939}{2.828}\,(1.895) = 21.37$$

Since $\bar{x}$ does not exceed 21.37, the group would reject H_0 and conclude that the manufacturers' claim is incorrect, based on these sample results. This would undoubtedly lead to some further action on their part. The chance they are wrong in rejecting H_0 is $\alpha = .05$.

Because of the central limit theorem, the same sort of procedure is appropriate in testing hypotheses about the mean μ of any population, regardless of whether it is normal or not, for sufficiently large sample sizes. The rule $n \geq 30$ should prove adequate for most sorts of populations in applying these tests. Table 8.5 presents the critical regions for these approximate tests.

Table 8.5   Approximate Critical Regions of Size α Tests about μ, any Population, $n \geq 30$

H_0	H_1	Reject H_0 if
$\mu \leq \mu_0$	$\mu > \mu_0$	$\overline{X} > \mu_0 + \dfrac{S}{\sqrt{n}} z_{1-\alpha}$
$\mu \geq \mu_0$	$\mu < \mu_0$	$\overline{X} < \mu_0 - \dfrac{S}{\sqrt{n}} z_{1-\alpha}$
$\mu = \mu_0$	$\mu_0 \neq \mu_0$	$\overline{X} > \mu_0 + \dfrac{S}{\sqrt{n}} z_{1-\alpha/2}$
		or $\overline{X} < \mu - \dfrac{S}{\sqrt{n}} z_{1-\alpha/2}$

Example 8.2.4

The defense in a criminal case in a California county recently charged that the jury rolls, from which jury members are selected, were not representative of the population at large in the county. Specifically, the defense contended that the average salary of wage earners selected as jury members was greater than the average wage earner salary for the county. The most recent census found that the average salary in the county was $8500. The sum of the salaries of the last 100 wage earners selected for jury duty at this court was 1,289,047 and the sum of their squares was 22,449,226,900. The defense attorney used these results to test, (with $\alpha = .1$),

$$H_0: \mu \leq 8500 \qquad \text{versus} \qquad H_1: \mu > 8500$$

where μ is the mean of the population of salaries of wage earners selected for jury duty. Note that

$$\overline{x} = 12,890.47$$

$$s^2 = \frac{1}{99}\left(22,449,226,900 - \frac{(1,289,047)^2}{100}\right)$$

$$= 58,917,231$$

Therefore, $s = \sqrt{58917231} = 7676$. Then from Table A.2, $z_{.9} = 1.28$ and

$$8500 + \frac{7676}{10}(1.28) = 9482.5$$

Since $\overline{x} > 9482.5$, H_0 is rejected.

It has been mentioned that as the probability of a type I error, α, is decreased, then necessarily the probability of a type II error, β, is increased. This statement assumes that the sample size, n, is held fixed. It is also true that if we consider a fixed rejection rule, such as reject H_0 if we find $\overline{X} < 21.37$, then *both* α and β will decrease as the sample size n is increased. As we take larger and larger sample sizes both α and β will get smaller.

If we are sampling from a normal population whose variance σ^2 is known, and want to test $H_o: \mu \le \mu_0$ versus $H_1: \mu > \mu_0$, we can fix α at any desired value and also fix β (for some specific $\mu_1 > \mu_0$) by choosing the sample size n as follows. Table 8.3 says we should reject H_0 if,

$$\overline{X} > \mu_0 + \frac{\sigma}{\sqrt{n}} z_{1-\alpha} = c$$

We should accept H_0 if $\overline{X} \le c$. If in fact $\mu = \mu_1$, the probability we accept H_0, then, is

$$\beta = P(\overline{X} \le c) = N_z\left(\frac{c - \mu_1}{\sigma/\sqrt{n}}\right)$$

since $\overline{X}$ then is normal with mean μ_1 and σ^2/n. But since $N_Z(z_\beta) = \beta$, we have

$$z_\beta = \frac{c - \mu_1}{\sigma/\sqrt{n}} = \frac{u_0 - \mu_1}{\sigma/\sqrt{n}} + z_{1-\alpha}$$

If we solve this equation for n, we get

$$n = \frac{(z_\beta - z_{1-\alpha})^2 \, \sigma^2}{(\mu_0 - \mu_1)^2}$$

If we use this value for n (or round up to the next largest integer if the solution is not an integer) as our sample size, the probability of a type I error is α and the probability of a type II error (assuming $\mu = \mu_1$) is β. If, for example, $\mu_0 = 1$, $\mu_1 = 2$, $\sigma^2 = 1$, and we want $\alpha = .01$, $\beta = .05$, then $z_{.99} = 2.33$, $z_{.05} = -1.64$ and the value for n is

$$n = \frac{(-1.64 - 2.33)^2 \, (1)}{(1 - 2)^2} = 15.76$$

If we select a sample of size 16 from this population and reject H_0 if we find

$$\overline{X} > \mu_0 + \frac{\sigma}{\sqrt{n}} z_{1-\alpha} = 1 + \frac{1}{4}(2.33) = 1.58$$

we have $\alpha = .01$, $\beta = .05$ (for the alternative that $\mu = 2$).

Example 8.2.5

Assume as in Example 7.2.1 that the mercury contamination, per fish, in a large lake is normal with $\sigma = 5$. If we want to test $H_0: \mu \leq 30$ versus $H_1: \mu > 30$ with $\alpha = .05$, and also want to have only one chance in 100 of accepting H_0 if μ is as large as 32, then we should use a sample of size

$$n = \frac{(z_\beta - z_{1-\alpha})^2 \, \sigma^2}{(\mu_0 - \mu_1)^2} = \frac{(-2.33 - 1.64)^2}{(30 - 32)^2} (5)^2$$

$$= 98.5 \approx 99$$

If we want $\alpha = .05$ and also want one chance in 20 of accepting H_0 if $\mu = 31$, we need a sample of size

$$n = \frac{(-1.64 - 1.64)^2}{(30 - 31)^2} (5)^2 = 269$$

In testing hypotheses about the mean of a normal population, with σ^2 unknown, again both α and β are decreasing functions of n, but the determination of n for desired values of α and β is not so straightforward. With the use of specialized tables it is possible to find, for this case as well, the value of n that will give desired values for α and β, as a function of $\Delta = (\mu_1 - \mu_0)/\sigma$. (The interested reader is invited to consult the book *Elementary Applied Statistics*, by Marvin Lenter, Tarrytown-on-Hudson, N.Y.: Bogden and Quigley, 1972, for a discussion of this technique.)

8.2.2   Tests about Variances

To test hypotheses about the variance or standard deviation of a normal random variable, we would probably expect to use the sample variance S^2 (or S) as the test statistic, since these quantities are the point estimators for σ^2 and σ. We might also anticipate that the percentiles of the χ^2 distribution with $n - 1$ degrees of freedom will be involved in defining the critical region for tests about σ^2 since we saw in Section 7.2 that $(n - 1)S^2/\sigma^2$ has a χ^2 distribution. Table 8.6 presents the critical regions for the various possible tests about σ^2.

Table 8.6   Tests about σ^2, Normal Population

H_0	H_1	Reject H_0 if
$\sigma^2 \leq \sigma_0^2$	$\sigma^2 > \sigma_0^2$	$S^2 > \chi^2_{1-\alpha} \, \sigma_0^2/(n-1)$
$\sigma^2 \geq \sigma_0^2$	$\sigma^2 < \sigma_0^2$	$S^2 < \chi^2_{\alpha} \, \sigma_0^2/(n-1)$
$\sigma^2 = \sigma_0^2$	$\sigma^2 \neq \sigma_0^2$	$S^2 > \chi^2_{1-\alpha/2} \, \sigma_0^2/(n-1)$
		or $S^2 < \chi^2_{\alpha/2} \, \sigma_0^2/(n-1)$

These hypotheses can equally well be stated in terms of σ rather than σ^2; the three possible H_0's are, respectively, $\sigma \leq \sigma_0$, $\sigma \geq \sigma_0$, and $\sigma = \sigma_0$. The critical regions remain the same (or equivalently may be expressed in terms of the square roots of the quantities involved).

Example 8.2.6

The manufacturer of a precision scale claims that repeated weighings of the same object will give rise to a population of measurements that has a normal probability measure with μ equal to the true weight of the object and σ no larger than .02 mg. The same object is weighed $n = 10$ times, and it is found that $\bar{x} = 1.2$ mg and $s = .03$ mg. Based on these sample results, should we accept the manufacturer's claim, with $\alpha = .05$? We want to test

$$H_0: \sigma \leq .02 \qquad \text{versus} \qquad H_1 \quad \sigma > .02 \qquad \text{with} \qquad \alpha = .05$$

From Table 8.6 we see that we shall reject H_0 if $S^2 > \chi^2_{.95}(.02)^2/9$; from Table A.5 in the appendix, with nine degrees of freedom, we find $\chi^2_{.95} = 16.9$ so we should reject H_0 if $S^2 > (16.9)(.0004)/9 = .000751$. From the sample data, $s^2 = (.03)^2 = .0009 > .000751$, so H_0 is rejected. The probability we have incorrectly rejected H_0 is .05.

8.2.3 Tests about Proportions

If n independent, repeated Bernoulli trials are performed, p is the probability for success for each, and Y is the total number of successes observed, then Y has a binomial probability measure with parameters n and p. If p is unknown, we saw in Section 7.1 that Y/n is the point estimator for p. Thus, if we want to test a hypothesis about p, the natural test statistic to use is Y/n. For large values of n we saw in Section 6.4 that the binomial probability measure is well approximated by the normal with parameters $\mu = np$ and $\sigma^2 = npq$. Thus, for n large we would expect the critical regions for tests about p to involve the normal percentiles. Table 8.7 gives the critical regions for the various possible tests about p with n large.

Table 8.7 Tests about p, n Large
(Y = Total Number of Successes in n Trials)

H_0	H_1	Reject H_0 if
$p \leq p_0$	$p > p_0$	$\dfrac{Y}{n} > p_0 + z_{1-\alpha}\sqrt{p_0 q_0/n}$
$p \geq p_0$	$p < p_0$	$\dfrac{Y}{n} < p_0 - z_{1-\alpha}\sqrt{p_0 q_0/n}$
$p = p_0$	$p \neq p_0$	$\dfrac{Y}{n} > p_0 + z_{1-\alpha/2}\sqrt{p_0 q_0/n}$
		or $\dfrac{Y}{n} < p_0 - z_{1-\alpha/2}\sqrt{p_0 q_0/n}$

Example 8.2.7

A Congressman feels that 60% of the voters in his district support his stand on a bill pending in Congress. To test the hypothesis that .6 is the correct proportion, he will select a random sample of $n = 300$ voters in his district and use their responses to test $H_0: p = .6$ versus $H_1: p \neq .6$ with $\alpha = .1$. Then, from Table 8.7 we see that he should reject H_0 if the sample proportion of people supporting his stand (Y/n) exceeds

$$p_0 + z_{1-\alpha/2}\sqrt{p_0 q_0/n} = .6 + 1.64\sqrt{(.6)(.4)/300} = .646$$

or if this sample proportion is smaller than

$$p_0 - z_{1-\alpha/2}\sqrt{p_0 q_0/n} = .6 - 1.64\sqrt{(.6)(.4)/300} = .554$$

As with tests about the mean of a normal random variable, with σ known, it is possible with tests about p to select a sample size n such that the probability of type I error is α, given $p = p_0$, and at the same time the probability of a type II error, given $p = p_1$ (some particular value from the alternative H_1), is β. Suppose we want to test $H_0: p \leq p_0$ versus $H_1: p > p_0$; we want to have a probability of type I error equal to α and, if $p = p_1$ (some value bigger than p_0), we want our probability of type II error to equal β. How large should n be?

From Table 8.7 we see that we should accept H_0 if

$$\frac{Y}{n} \leq p_0 + z_{1-\alpha}\sqrt{p_0 q_0/n}$$

Since, if $p = p_1$, the probability measure for Y/n is approximately normal with mean p_1 and variance $p_1 q_1/n$.

$$\beta = P(\text{Accept } H_0) = P\left(\frac{Y}{n} \leq p_0 + z_{1-\alpha}\sqrt{p_0 q_0/n}\right)$$

$$\doteq N_z\left(\frac{p_0 + z_{1-\alpha}\sqrt{p_0 q_0/n} - p_1}{\sqrt{p_1 q_1/n}}\right)$$

Thus

$$z_\beta = \frac{p_0 - p_1 + z_{1-\alpha}\sqrt{p_0 q_0/n}}{\sqrt{p_1 q_1/n}}$$

If we solve this equation for n, we get

$$n = \frac{(z_\beta\sqrt{p_1 q_1} - z_{1-\alpha}\sqrt{p_0 q_0})^2}{(p_0 - p_1)^2}$$

Again, we should round up to the next larger integer if this solution is not a whole number. If we select a sample of this size, our probability of type I error is α and, given $p = p_1$, our probability of type II error is β.

Example 8.2.8

Assume that a manufacturer receives a small item, needed to assemble his product, in lots of size 10,000. The supplier claims that no more than 5% of the items supplied are defective. We want to test the hypothesis his claim is true, with $\alpha = .05$, and if in fact 10% of the items are defective, with $\beta = .05$. How large a sample size, n, do we need? Thus we want to test

$$H_0: p \leq .05 \qquad H_1: p > .05$$

with $\alpha = .05$ and with $\beta = .05$ if $p = .10$. Then, $z_{.95} = 1.64$, $z_{.05} = -1.64$, $p_0 = .05$, and $p_1 = .10$ and we should choose a sample of size

$$n = \frac{(z_\beta \sqrt{p_1 q_1} - z_{1-\alpha} \sqrt{p_0 q_0})^2}{(p_0 - p_1)^2}$$

$$= \frac{[-1.64\sqrt{(.1)(.9)} - 1.64\sqrt{(.05)(.95)}]^2}{(.05 - .10)^2}$$

$$= 297.8 = 298$$

Thus, if we select a sample of $n = 298$ items from the incoming lot and reject $H_0: p \leq .05$ if $Y/n > .05 + 1.64\sqrt{(.05)(.95)/298} = .071$, we have $\alpha = .05$ and, also, $\beta = .05$ if $p = .1$.

Exercise 8.2

1. Assume that the distribution of lengths of northern pike in Lake Washington is normal in form and that $\sigma = 4$. A sample of 16 northern pikes is to be randomly selected from the lake and, if $25 \leq \bar{x} \leq 29$, the hypothesis that $\mu = 27$ will be accepted. What is the probability of a type I error in using this rule?

2. As in question 1, assume $\sigma = 4$ and a sample of 16 fish is to be selected; what should the acceptance region be if we are to have $\alpha = .01$? $\alpha = .001$?

3. With the acceptance region given in question 1, what is the probability of a type II error if in fact $\mu = 24$? $\mu = 28$? $\mu = 30$? $\mu = 32$?

4. Four hundred pieces of a fish line manufacturer's 30-pound test line were tested; the average breaking strength of the samples was 29 pounds and the sample standard deviation was five pounds. Does it appear likely the manufacturer's claim is justified, with $\alpha = .01$?

5. The research division of a light bulb manufacturer has devised a new method of coating the filament; would you accept the claim that the population mean lifetime for bulbs made this way is 1500 hours if the sample mean of 100 bulbs tested was 1490 hours and the sample standard deviation was 100 hours? Use $\alpha = .05$ and a two-tailed test.

6. Assume that the heights of 21-year-old males are normally distributed with $\sigma = 3$ inches. Given a random sample of size 10 yielded a sample mean height of 71 inches, would you accept the hypothesis that the population average height is 5 feet 9 inches, versus the alternative it is not, with $\alpha = .01$? What is the probability of type II error with your test, given that the average height is 6 feet?

7. The Fair Trades Bureau randomly selects 25 packages of hamburger meat, packaged by a supermarket, each saying the net contents are one pound. They know that the weights of such packages are normally distributed with $\sigma = .5$ ounce. If the average weight of the 25 packages they select is less than 15.8 ounces, they will reject the hypothesis that the average weight of packages put out by this market is one pound.
    (a) What is the probability of type I error with this test?
    (b) If the mean weight of the market's packages is 15.9 ounces, what is the probability of type II error with this procedure?

8. Assume that the time needed for a 30-year-old male to move his foot from the floorboard to the brake pedal of an automobile is a normal random variable. In 10 independent tests of this reaction time, assume that the average time required was .5 seconds and the standard deviation was .1 second. Would you accept the hypothesis that his mean reaction time is .45 seconds, with $\alpha = .05$ and a two-tailed test?

9. In sampling from a normal population with known standard deviation σ, show that the two-tailed test of $\mu = \mu_0$ discussed in the text, with $\alpha = P(\text{type I error})$, is equivalent to the following: Select the sample and compute a two-sided $100(1 - \alpha)\%$ confidence interval for μ. If μ_0 falls inside the confidence interval, accept the hypothesis that $\mu = \mu_0$; if it does not, reject the hypothesis.

10. Draw a parallel between one-sided tests and one-sided confidence intervals, like that discussed in problem 9.

11. Assume you are given a random sample of 16 observations from a normal distribution and want to test the hypothesis that the mean is equal to 10, with $\alpha = .05$, versus the alternative that the mean is not equal to 10.
    (a) Assuming the population variance is known to be equal to 1, what would you use as an acceptance region for this test?
    (b) For the test described above, what is the probability of a type II error, assuming the mean is equal to 9?
    (c) If the value of the population variance is not known, what would you use as an acceptance region?

12. Use the data given in Exercise 7.2, problem 19, to test the hypothesis that the mean weight loss in following that diet-exercise regimen is no more than seven pounds, versus the alternative that the mean loss is greater than seven pounds, with $\alpha = .05$.

13. A supplier of electric typewriters claims that his machine will operate in normal business usage for at least 12 months with no breakdowns. A university purchased 20 of these machines. They are willing to assume that the length of service for this machine, until a breakdown occurs, is a normal random variable. For the university's 20 machines they find the sum of the times until the first breakdown was 221 months and the sum of the squares of these numbers is 2471. With $\alpha = .05$, should the university believe the manufacturer's claim?

14. Use the data given in problem 16, of Exercise 7.2, to test the hypothesis that the mean weight of college freshmen is 160 pounds versus the (one-sided) alternative that it is less than 160 pounds, with $\alpha = .05$.

15. Use the data given in problem 1, Exercise 7.2, to test the hypothesis that the mean lifetime of those bulbs is 2100 hours versus the two-sided) alternative that it is not, with $\alpha = .10$. What is the probability of type II error if in fact $\mu = 2200$ hours, with this test?

16. For a normal population with $\sigma = 2$, how large a sample size is needed to test $H_0: \mu \geq 5$, $H_1: \mu < 5$ with $\alpha = .05$ and $\beta = .1$ if $\mu = 3$? If $\mu = 4$?

17. To test hypotheses about the parameter, λ, of a Poisson process, again the central limit theorem can be used, for large n. If $X_1, X_2, \ldots, X_n$ are independent Poisson random variables, each with parameter λ, then for sufficiently large sample sizes $\overline{X} = \dfrac{1}{n} \sum X_i$ has approximately a normal probability measure with $\mu = \lambda$, $\sigma^2 = \lambda/n$. For sufficiently large n what would you use as a critical region for testing

    (a) $H_0: \lambda \leq \lambda_0$ versus $H_1: \lambda > \lambda_0$?

    (b) $H_0: \lambda \geq \lambda_0$ versus $H_1: \lambda < \lambda_0$?

    (c) $H_0: \lambda = \lambda_0$ versus $H_1: \lambda \neq \lambda_0$?

18. In a large city it is assumed that suicides occur like events in a Poisson process with parameter λ (per month). Given that over a 12-month period 130 suicides were recorded in this city, would you accept

$$H_0: \lambda \leq 10 \quad \text{versus} \quad H_1: \lambda > 10$$

    with $\alpha = .05$?

19. Assume that the lifetimes of bicycle tires are normal with unknown μ and $\sigma = 100$ miles. How large a sample is needed to test $H_0: \mu \leq 1000$ versus $H_1: \mu > 1000$ with $\alpha = .1$ and $\beta = .1$ if $\mu = 1100$?

20. In designing educational and psychological tests, it is of importance that the variance of responses is neither too small nor too large. If the test is too easy, and everyone gets all the questions right, then $\sigma^2 = 0$; similarly if the test is too hard and everyone gets all the questions wrong, again $\sigma^2 = 0$. Neither of these is desirable. On the other hand, if σ^2 gets too large the scores are extremely dispersed about the mean value and become more difficult to interpret (and an assumed normal distribution for the scores may become less tenable). Suppose an "ideal" test is one with $\sigma = 100$; we design a test, give it to 30 students and find the standard deviation of the scores to be $s = 80$. Using a two-tailed test and $\alpha = .1$, would you accept the hypothesis that $\sigma = 100$?

21. Modern computers use various methods to generate observed values of a standard normal random variable; these observed values are then used in subsequent calculations of interest. Assume a new procedure has been suggested and we suspect that possibly the variance of the numbers generated by the procedure is smaller than 1. Twenty-five values are generated, and it is found that their standard deviation is .9. Would you accept $H_0: \sigma \geq 1$ versus $H_1: \sigma < 1$, with $\alpha = .05$?

22. Show that the following procedure is equivalent to the two-tailed test of $\sigma^2 = \sigma_0^2$ discussed in this section: Select a sample of size n and compute a two-sided $100(1 - \alpha)\%$ confidence interval for σ^2. If the interval includes σ_0^2, then accept the hypothesis $\sigma^2 = \sigma_0^2$; otherwise reject it.

23. Use the data given in problem 1, Exercise 7.2, to test the hypothesis that the standard deviation of those 15-watt light bulbs is 30 hours, versus the alternative it is not, with $\alpha = .10$.

24. Use the data given in problem 4, Exercise 7.2, to test the hypothesis that the standard deviation of the distances this broad jumper can leap is four inches, with $\alpha = .10$, versus the alternative it is not four inches.

25. The interior sole length of men's shoes, size $9\frac{1}{2}$, should be $10\frac{3}{4}$ inches. Ten shoes of this size were randomly selected from one day's production and the length was measured for each (to the nearest .05 inch). The sample standard deviation of the 10 measurements was found to be .05. Using this sample data, would you accept the hypothesis that $\sigma \leq .03$ versus the alternative $\sigma > .03$, with $\alpha = .1$?

26. A television set manufacturer guarantees to repair any set of his that fails within its first year of service. The manufacturer makes this guarantee because he believes at least 90% of his sets will not fail until after a year of service. Of 200 such sets sold by a dealer in one year, 27 required service within their first year of use. Based on this sample, do you believe the manufacturer is correct in assuming at least 90% of his sets will not fail in their first year of service? Use $\alpha = .05$.

27. A supplier of feed for a commercial beef finishing lot says that 80% of all cattle finished with his product will be ready for market in six weeks. Of 500 cattle fed with this product, 450 were ready for market in six weeks. Do you think the feed supplier is too modest in his claims? Use $\alpha = .01$.

28. It was found that 52 of 200 randomly selected university students were suffering from the same communicable disease. Using $\alpha = .1$, would you believe the university administration's claim that they did not have an epidemic of this disease on campus? (If 15% or more of the students have any disease, it is called an epidemic.)

29. The supplier of small parts claims that the lots he provides contain no more than 10% defectives. The buyer will select a random sample of size 100 from each lot and test the hypothesis that $p = .1$ versus the alternative that $p > .1$ with $\alpha = .05$. What is the largest number of defects that could be found in the sample such that the lot will still be accepted?

30. A college football buff claims he can predict the winner of any game with a probability of at least .75. On one fall weekend he was correct on 26 games out of 40. Would you believe his claim with $\alpha = .05$?

31. Bulk wine is sold in gallon bottles, each containing a "Sentry Guard" paper seal on the top to keep out the air. Let p be the proportion of these seals that are really airtight at the time the wine is purchased. Assume that 16 out of 50 gallons examined had seals that were not in fact airtight. With $\alpha = .05$, what is the largest possible value for p_0 such that we would accept the hypothesis $p \geq p_0$ versus the alternative $p < p_0$, given this set of sample data?

32. Genetic theory predicts that the proportion of gray offspring to be produced by a mating of two flies should be equal to 1/3. Given a mating which produced 58 offspring, 21 of which were gray, test the hypothesis that the theory is correct, with $\alpha = .05$.

33. Using the data given in problem 6, Exercise 7.2, would you accept the hypothesis that the university computer center makes a mistake in transcribing grades no more than 1/2 of 1% of the time, versus the alternative that the proportion of errors is greater than 1/2 of 1%? Use $\alpha = .1$.

34. Use the data given in problem 20, Exercise 7.2, to test the hypothesis that $p = 2/3$ versus the alternative $p \neq 2/3$, where p is the probability a thumbtack of this sort will land with its point up. Use $\alpha = .05$.

35. Use the data given in problem 23, Exercise 7.2 to test

$$H_0: p \leq .6 \quad \text{versus} \quad H_1: p > .6$$

where p is the proportion of voters in the electorate that favor the school bond issue. Use $\alpha = .05$.

36.   Use the data given in problem 25, Exercise 7.2, to test

$$H_0 : p \leq .03 \qquad \text{versus} \qquad H_1 : p > .03$$

where p is the probability the assembly line turns out a defective item. Use $\alpha = .01$.

37.   (See question 17 above.) Use the data given in question 27, Exercise 7.2, to test the hypothesis that the rate of sales made by the real estate firm is $\lambda \leq .7$ per day versus the alternative $\lambda > .7$, with $\alpha = .05$.

38.   (See question 17 above.) Use the data in problem 29, Exercise 7.2, to test the hypothesis that $\lambda \leq 90$ per hour, versus the alternative $\lambda > 90$, where λ is the number of telephone calls made and received by the firm per hour. Use $\alpha = .01$.

39.   What is β, assuming $p = .2$, for the test described in question 31 above?

40.   For the test described in question 32 above, evaluate β if $p = .6$.

41.   For the test described in question 30 above, evaluate β if $p = .6$.

42.   For question 33 above, evaluate β if $p = .01$.

43.   Assume p is the probability of success for a Bernoulli trial. How many independent trials should be performed if we want to test $H_0 : p \leq .5$ versus $H_1 : p > .5$, with $\alpha = .05$ and $\beta = .05$, given $p = .51$?

44.   Let p be the proportion of United States adults who smoke cigarettes regularly. How large a sample of adults do we need to test $H_0 : p \leq .4$ versus $H_1 : p > .4$ with $\alpha = .01$ and $\beta = .01$, given $p = .48$?

8.3   SOME χ^2 TESTS

In Section 7.2 we discussed the χ^2 density and have seen its use in computing confidence intervals for σ^2, in sampling from a normal population, as well as in testing hypotheses about σ^2. This distribution is also useful in testing several other types of hypotheses. We shall examine two of these situations in this chapter.

8.3.1   Goodness of Fit Tests

Suppose we are given a set of sample results, a random sample selected from a population, and wanted to test that the probability measure for the population is of some specified type, say, normal. How could we make such a test? The methodology for this situation is described in this subsection. These tests are called goodness of fit tests becuase they measure how closely an assumed population measure "fits" the results obtained in a sample. As with the tests previously discussed, we shall accept the hypothesis that the distribution is of the specified form if the sample values are "consistent" with this specified population measure, and shall reject the hypothesis if the sample results are not consistent with the specified measure. They way in which this consistency is measured is presented below.

Suppose we are given a set of sample data that has been summarized into classes (see Chapter 2). Thus, if class 1 goes from 0 to 10, for example, we have recorded the frequency of occurrence of sample values in this interval; if class 2 consists of all the values greater than 10 and no larger than 20, we have also recorded the frequency of occurrence of sample values in this class, and so on, for each class used. Table 8.8 reproduces the distribution of scores of 120 students on a statistics exam, presented earlier in question 1, Exercise 2.2; this table has a mean of 80.08 and a standard deviation of 9.33. How could we test that the population of test scores, from which these numbers were sampled, is normal in form?

Table 8.8 Scores of 120 Students on a Statistics Exam

Class (Upper End Point Included)	Frequency f_i
60–65	6
65–70	13
70–75	17
75–80	25
80–85	23
85–90	16
90–95	12
95–100	8
Total	120

If in fact the population is normal, then the histogram summarizing the 120 scores should be pretty well approximated by a normal curve. Figure 8.3.1 presents the histogram of the scores and a superimposed normal density with $\mu = 80.08, \sigma = 9.33$ (the same parameter values that the histogram has). Notice that the density does pretty well approximate the histogram and, graphically, this sample would appear consistent with a normal measure for the population. An

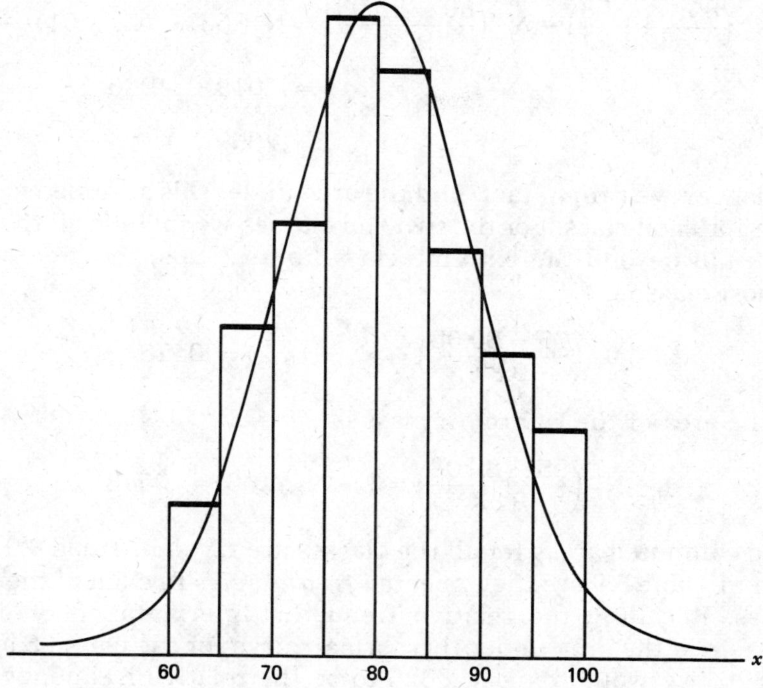

Figure 8.3.1 Statistics exam scores and normal density with the same mean and variance.

objective test of the hypothesis that these scores were selected from a population with a normal probability measure can be made using a statistic whose distribution is well approximated by a χ^2 density, for large sample sizes. This statistic is discussed below.

As was mentioned earlier, it can be verified that the mean of the frequency Table 8.8 is 80.08 and the standard deviation is 9.33. If we assume a normal distribution with these values for μ and σ, respectively, we can compute the area under the assumed normal distribution over each of the classes summarized in the table. For example, the area under this normal curve between 70 and 75 is

$$N_z\left(\frac{75-80.08}{9.33}\right) - N_z\left(\frac{70\ \ 80.08}{9.33}\right) = N_z(-.54) - N_z(-1.08)$$

$$= .2946 - .1401$$

$$= .1545$$

the area between 80 and 85 is

$$N_z \left(\frac{85 - 80.08}{9.33} \right) - N_z \left(\frac{80 - 80.08}{9.33} \right) = N_z(.53) - N_z(-.01)$$

$$= .7019 - .4960$$

$$= .2059$$

In this way we can, in fact, find the area under this assumed normal curve for each class. For the two end classes we include all the area in the tail beyond the extreme class limits. Thus, the area of the leftmost class is

$$N_z \left(\frac{65 - 80.08}{9.33} \right) = N_z(-1.62) = .0526$$

and the area of the rightmost class is

$$1 - N_z \left(\frac{95 - 80.08}{9.33} \right) = 1 - N_z(1.60) = .0548$$

These computed areas for all the classes are given in Table 8.9. Also given in Table 8.9 are the *expected frequencies* in each of the eight classes. Recalling the relative frequency interpretation of probability, since the area under the normal curve for the class 65 or less is .0526, we would expect .0526 to be the relative frequency with which we would observe scores in this interval; thus the number out of 120 that we would expect to fall in this interval is $(.0526)(120)$ $= 6.3$. These values are given in the column labeled expected frequency; each is simply the product of the normal probability for the class times 120, the sample size. The final column reproduces the actual frequencies observed for the classes. The χ^2 statistic for test-

Table 8.9   Normal Areas, Expected Frequencies and Actual Frequencies for Statistics Exam Scores

Class (Upper End Point Included)	Area under Normal Density	Expected Frequency e_i	Observed Frequency f_i
65 or less	.0526	6.3	6
65–70	.0875	10.5	13
70–75	.1545	18.5	17
75–80	.2014	24.2	25
80–85	.2059	24.7	23
85–90	.1535	18.4	16
90–95	.0898	10.8	12
More than 95	.0548	6.6	8
Total	1.0000	120.0	120

ing the hypothesis that this sample came from a normal population is based on a comparison of these expected frequencies and actual frequencies. The statistic is

$$w = \sum_{i=1}^{k} \frac{(f_i - e_i)^2}{e_i}$$

where k is the number of classes used (eight for the case being discussed). For large values of n, the sample size, the probability measure for w is well approximated by a χ^2 density if the population measure really is of the form used in computing the expected frequencies. The degrees of freedom for the χ^2 density is equal to the number of classes used less the number of parameters estimated from the data to compute the expected frequencies, less 1. If k is the number of classes and r is the number of parameters estimated from the data to compute the expected frequencies, then the degrees of freedom for this test is $k - r - 1$. For the statistics exam scores, $k = 8$, $r = 2$ (μ and σ) so the degrees of freedom for w is $k - r - 1 = 8 - 2 - 1 = 5$.

Notice that the observed value of w will be small if there is good ageeement between the observed and expected frequencies, and the value of w will be large if there is not good agreement between these two frequencies. Since we want to accept the hypothesis that the exam scores came from a normal population if the sample results are consistent with this assumption, and reject it if they are not consistent, we will reject the hypothesis if the value for w is too large (indicating big differences between the e_i's and f_i's). w will exceed $\chi^2_{1-\sigma}$, with probability α, if the hypothesis is true; thus, if we reject H_0 only if $w > \chi^2_{1-\alpha}$, our probability of type I error is equal to α.

For the statistics exam data given in Table 8.9 we find

$$w = \frac{(6 - 6.3)^2}{6.3} + \frac{(13 - 10.5)^2}{10.5} + \ldots + \frac{(8 - 6.6)^2}{6.6} = 1.62$$

With five degrees of freedom we find $\chi^2_{.95} = 11.1$ and, thus, with $\alpha = .05$ we would accept the hypothesis that this sample was selected from a normal population

Example 8.3.1

A single die was rolled 402 times; the frequencies of occurrence of the six faces are given in the following table:

Face number	1	2	3	4	5	6
Frequency	60	73	68	59	63	79

With this data we can test the hypothesis that the die is fair with $\alpha = .10$. If it is fair the probability of each face occurring must be 1/6; thus the expected number of rolls giving each face should be $402/6 = 67$. Then

$$w = \sum_{i=1}^{6} \frac{(f_i - e_i)^2}{e_i}$$

$$= \frac{(60 - 67)^2}{67} + \frac{(73 - 67)^2}{67} + \ldots + \frac{(79 - 67)^2}{67} = 4.63$$

w is the observed value of an approximate χ^2 random variable with $k - r - 1 = 6 - 0 - 1 = 5$ degrees of freedom. Since $\chi^2_{.9} = 9.24$, we would accept the hypothesis that the die is fair, based upon this sample.

This type of test can be employed to test the hypothesis that a sample was selected from any specific type of distribution. All that is necessary is that the probability law be available to compute the expected numbers for each class. If parameters of the distribution must be estimated, before these expected numbers can be computed, then one degree of freedom is "lost" for each one. The approximate χ^2 distribution is a good approximation to the exact distribution of w, as long as the expected number for each class is at least 5. If in the original summarization this is not satisfied, then the number of classes can be reduced, by combining adjacent classes, until it is satisfied.

Example 8.3.2

The life histories of 1000 people, all living through the same span of time, gave rise to the ages at death presented in the following table. By using the class marks (midpoints) and assigning 95 as the midpoint for the last class, it can be verified that the mean of the table is 68.35 and the standard deviation is 18.24. Let us test the hypothesis that these ages were selected at random from a normal population with $\alpha = .05$. To do so we must first compute the expected frequencies for each class, using the estimates for μ and σ. Thus, the probability an observation from a normal distribution with $\mu = 68.35$, $\sigma = 18.24$ will be less than 10 is

$$P(X < 10) = P\left(Z < \frac{10 - 68.35}{18.24}\right)$$

$$= N_z(-3.20) = 0$$

The probability an observation would fall between 10 and 20 is

$$P(10 \leq X < 20) = P\left(\frac{10 - 68.35}{18.24} \leq Z < \frac{20 - 68.35}{18.24}\right)$$

$$= N_z(-2.65) - N_z(-3.20)$$

$$= .0040, \text{ etc.}$$

Proceeding in this way we get the normal probabilities displayed in the following table; multiplying each by 1000 gives the expected frequencies in the classes.

Notice that the first two expected frequencies are less than five. To arrive at classes, each with $e_i \geq 5$, we combine the first three classes, giving

Age at Death	Frequency f_i	Normal Probability	Expected Number, e_i
$0 \leq x < 10$	19	0	0
$10 \leq x < 20$	14	.0040	4.0
$20 \leq x < 30$	18	.0139	13.9
$30 \leq x < 40$	24	.0427	42.7
$40 \leq x < 50$	48	.0956	95.6
$50 \leq x < 60$	106	.1666	166.6
$60 \leq x < 70$	211	.2131	213.1
$70 \leq x < 80$	297	.2030	203.0
$80 \leq x < 90$	216	.1441	144.1
$90 \leq x$	47	.1170	117.0
Total	1000		

$f_1 = 51$ (the sum of the first three observed frequencies) and $e_1 = 17.9$ (the sum of the first three expected frequencies). We then have a total of $k = 8$ classes each with $e_i \geq 5$. The degrees of freedom for the approximately χ^2 density is $8 - 2 - 1 = 5$ and we find from Table A.5 that $\chi^2_{.95} = 11.1$. The observed value for w is

$$w = \sum_{i=1}^{8} \frac{(f_i - e_i)^2}{e_i}$$

$$= \frac{(51 - 17.9)^2}{17.9} + \frac{(24 - 42.7)^2}{42.7} + \ldots + \frac{(47 - 117)^2}{117}$$

$$= 258.1$$

Thus we resoundingly reject the hypothesis of normality for the population of ages at death. A comparison of the e_i's and f_i's shows great disparity between them for all classes with the single exception of $60 \leq x < 70$. Because of this lack of consistency we would undoubtedly reject H_0 even if we did not compute the value for w. It would not be possible to use techniques that assume a normal population for sample results of this kind, because this sample does not very closely resemble what we would expect in a sample from a normal distribution.

8.3.2 Contingency Tables

A contingency table summarizes the number of observations in a sample that lie in various classes according to two (or more) different variables of classification. For example, suppose we had available the medical records of 1000 deceased adults and that the record for each indicates two things:

1. Did the person die of a heart attack?
2. Was the person a cigarette smoker?

Here, then, we have a sample of size 1000, and *each* sample observation can be classified according to two variables: cause of death and smoking habits. These observations, then, could be summarized in a 2×2 table and we might end up with results like the following:

	Died of		
	Heart Attack	Other Causes	Total
Nonsmokers	102	506	608
Smokers	99	293	392
Total	201	799	1000

Given the results of such a sample we would be interested in whether or not the two variables of classification are independent in the population. By the two variables being independent in the population we mean the probability measure of either variable is the same, regardless of the value of the other variable. In this example, then, independence of the two variables would mean the proportion of adults who die of heart attacks is the same for both smokers and nonsmokers. The statistic used to test this hypothesis of independence again has a probability measure that is well approximated by a χ^2 density, for large sample sizes.

Some additional examples are given below. Assume a sample of n people is selected, and each person is classified as having blue eyes, brown eyes, or other eye color, as well as being classified as having brown hair, black hair, blond hair or other. Are these two variables independent in the population? A sample of n registered adults is selected, and each is classified according to the party he is registered with (Republican, Democratic, other). Each adult is also classified according to the candidate he voted for in the last presidential election. Are these two variables independent in the population? A random sample of n people is selected from the United States. Each person is classified according to the section of the country in which he resides and also according to his religious beliefs. Are these two variables independent in the population?

The tabular summarization used for the numbers observed in the various classes for the two variables of classification is called a *contingency table*. Table 8.10 shows the type of notation we shall adopt for the observations in the sample. The first entry in the first row, x_{11}, is the number of sample observations that fell in the first

Table 8.10   Notation for Sample Observations

		Variable 2 Class					
		1	2	3	$\cdots$	c	Total
	1	x_{11}	x_{12}	x_{13}	$\cdots$	x_{1c}	$x_{1\cdot}$
	2	x_{21}	x_{22}	x_{23}	$\cdots$	x_{2c}	$x_{2\cdot}$
	3	x_{31}	x_{32}	x_{33}	$\cdots$	x_{3c}	$x_{3\cdot}$
Variable	·	·	·	·		·	·
1	·	·	·	·		·	·
	·	·	·	·		·	·
	r	x_{r1}	x_{r2}	x_{r3}	$\cdots$	x_{rc}	$r_{r\cdot}$
Total		$x_{\cdot 1}$	$x_{\cdot 2}$	$x_{\cdot 3}$	$\cdots$	$x_{\cdot c}$	$n = x_{\cdot\cdot}$

class according to variable 1, and also fell in the first class according to variable 2; that is, the first subscript identifies the class for the first variable of classification and the second subscript identifies the class for the second variable of classification. Thus x_{23} is the number of sample observations that fell in class 2 for variable 1 *and* fell in class 3 for variable 2. x_{32} is the number of sample observations in class 3 for variable 1 and class 2 for variable 2. Notice that we certainly would not necessarily have $x_{23} = x_{32}$ and in general $x_{ij} \neq x_{ji}$. A subscript is replaced by a dot to denote summation over that subscript. Thus

$$x_{1\cdot} = \sum_{j=1}^{c} x_{1j}$$

so $x_{1\cdot}$ is the grand total of all sample observations that fell in class 1 of variable 1 (since it has added together all those observations across the first row of Table 8.10). Similarly

$$x_{\cdot 3} = \sum_{i=1}^{r} x_{i3}$$

represents the total of all sample observations that fell in class 3 of variable 2 (it is the total for column 3). The grand total of all observations then is

$$x.. = \sum_{i=1}^{r} x_i. = \sum_{j=1}^{c} x._{j} = n$$

the total sample size.

The test of independence in a contingency table proceeds in much the same way as the χ^2 test already described in this section. From the sample we can observe the x_{ij}'s, the entries in the cells in Table 8.10. We then must compute the *expected* number for each class, e_{ij}, under the assumption that the two variables of classification are independent. The test statistic then is

$$v = \sum_{i} \sum_{j} \frac{(x_{ij} - e_{ij})^2}{e_{ij}}$$

If in fact the hypothesis is true, v has approximately a χ^2 distribution with $(r - 1)(c - 1)$ degrees of freedom. Note that the form of this test statistic is identical with the one discussed earlier in this section. For each cell we form the difference between what is observed and what is expected, square it, divide by the number expected and then sum over all the cells in the table. If $v \geq \chi^2_{1-\alpha}$ for $(r - 1)(c - 1)$ degrees of freedom, we reject the hypothesis of independence; the (approximate) probability of a type I error is α, with this rule. As with the previous test, the distribution of v gets closer to that of a χ^2 random variable the larger that n is. As long as each expected number is at least 5, the true distribution of v is well approximated by that of a χ^2 random variable with $(r-1)(c-1)$ degrees of freedom.

How, then, do we compute the e_{ij}'s, the expected numbers for the various cells under the assumption that the two variables of classification are independent? It will be recalled that two variables are independent if the probability measure for the first one remains unchanged, no matter what the value of the second variable. This implies that if the variables are independent, the probability a sample observation will fall in cell ij is given by the *product* of the probabilities that it will fall in row i and column j. We would estimate the probability an observation will fall in row i by $x_i./n$ (the total number of observations in row i over the sample size) and would estimate the probability an observation will fall in column j by $x._{j}/n$; thus, assuming independence, the estimated probability an individual will be classified in cell ij is $x_i. x._{j}/n^2$ and the expected number in cell ij, in a sample of size n, is

$$e_{ij} = n\frac{x_i. \, x._{j}}{n^2} = \frac{x_i. \, x._{j}}{n}$$

The test statistic can then be computed using these expected numbers. The following examples illustrate this procedure.

Example 8.3.3

Let us test the independence of cause of death (heart attack or other) and smoking habits using the data presented earlier, using $\alpha = .05$. The following table reproduces the observed values quoted earlier and the expected numbers for the classes.

| | Died of | | |
	Heart Attack	Other	Total
Nonsmokers	$x_{11} = 102$ $e_{11} = (608)(201)/1000$ $\quad = 122.21$	$x_{12} = 506$ $e_{12} = (608)(799)/1000$ $\quad = 485.79$	$x_{1.} = 608$
Smokers	$x_{21} = 99$ $e_{21} = (392)(201)/1000$ $\quad = 78.79$	$x_{22} = 293$ $e_{22} = (392)(799)/1000$ $\quad = 313.21$	$x_{2.} = 392$
	$x_{.1} = 201$	$x_{.2} = 799$	1000

For this data we have $r = 2$ rows and $c = 2$ columns; thus the approximate probability measure for v is χ^2 with $(2-1)(2-1) = 1$ degree of freedom; we find $\chi^2_{.95} = 3.84$. The observed value for v is

$$v = \sum_i \sum_j \frac{(x_{ij} - e_{ij})^2}{e_{ij}}$$

$$= \frac{(102 - 122.21)^2}{122.21} + \frac{(506 - 485.79)^2}{485.79} + \frac{(99 - 78.79)^2}{78.79} + \frac{(293 - 313.21)^2}{313.21}$$

$$= 10.67$$

Since $10.67 > 3.84$ we would reject the hypothesis that these two variables are independent, with $\alpha = .05$. Thus, based on this sample, it would appear that the probability of an adult dying of a heart attack is not the same for both smokers and nonsmokers.

Example 8.3.4

At a recent graduation a large university granted a total of 1494 baccalaureate degrees from the colleges of sciences, arts and humanities, and engineering. This university counts an A as 4 points, a B as 3 points, etc., and requires every graduate to have at least a 2.0 average to graduate. The numbers of different students graduating from the different colleges, and their graduating grades, are summarized in the following table:

		College			
		Science	Arts	Engineering	Total
Graduating	3.34–4.00	98	218	35	351
Average	2.68–3.33	78	317	107	502
	2.00–2.67	207	229	205	641
	Total	383	764	347	1494

Let us use this data to test the hypothesis that the two variables of classification (college and graduating average) are independent with $\alpha = .05$; independence of the two would imply that the distributions of graduating averages are the same for the three colleges. To test the hypothesis we must first compute the expected number for each cell. These are displayed in the following table (the numbers in parentheses are the observed numbers in each class, already given in the preceding table).

Expected Numbers

		College		
		Science	Arts	Engineering
Graduating	3.34–4.00	89.98 (98)	179.50 (218)	81.52 (35)
Average	2.68–3.33	128.69 (78)	256.71 (317)	116.60 (107)
	2.00–2.67	164.33 (207)	327.79 (229)	148.88 (205)

The expected number, e_{11}, of graduates in the College of Science with an average between 3.34 and 4.00 is

$$\frac{x_1 . x_{.1}}{n} = \frac{(351)(383)}{1494} = 89.98$$

as shown; similarly, the expected number in cell 23 (grade point between 2.68 and 3.33, college of engineering) is

$$\frac{x_2 . x_{.3}}{n} = \frac{(502)(347)}{1494} = 116.60$$

as shown. To test the hypothesis of independence, then, we compute

$$v = \sum_i \sum_j \frac{(x_{ij} - e_{ij})^2}{e_{ij}} = \frac{(98 - 89.98)^2}{89.98} + \frac{(218 - 179.50)^2}{179.50} +$$

$$\cdots + \frac{(205 - 148.88)^2}{148.88} = 132.44$$

With $(3-1)(3-1) = 2 \cdot 2 = 4$ degrees of freedom, we find $\chi^2_{.95} = 9.49$, so we would resoundingly reject the hypothesis of independence. On the basis of this sample it would appear that the distributions of grades of graduating students are not the same for the three colleges. This could be explained by several different factors, including differences in grading philosophies of the faculties, differences in students' abilities, and differences in courses required.

Notice that the philosophy behind this test of independence in a contingency table is identical with the philosophy used in the goodness of fit test. From the sample we get the observed numbers, f_i or x_{ij}. Then, assuming the hypothesis is true, we compute the expected numbers, e_i or e_{ij}; if the observed and expected counts are similar ("small" value of the χ^2 statistic) we will accept the hypothesis. If there are sufficiently large discrepancies between what is observed and what is expected, we will reject the hypothesis.

Exercise 8.3

1. Genetic theory predicts that three types of offspring, AA, Aa, aa, should be observed in the proportions 1/4, 1/2, 1/4, respectively, from a mating of two flies. Given that one mating produced 178 offspring of which 50, 88, 40, respectively, were of types AA, Aa, aa, test the hypothesis that the theory is consistent with this sample result, with $\alpha = .05$.

2. Over a 72-hour three-day weekend in the United States a total of 400 fatal car crashes occurred. The following table summarizes the distribution of these accidents over the 72 one-hour periods.

Number of Accidents	Number of One-Hour Periods
2 or less	7
3	9
4	12
5	12
6	10
7	8
8	7
9 or more	7

   Test the hypothesis that this data comes from observations of a Poisson process with $\lambda = 5.5$/hour, using $\alpha = .01$.

3. In each of 100 games a professional baseball player was at bat exactly three times. The numbers of hits he got, per game, are summarized below, for the 100 games.

Number of Hits	Number of Games
0	35
1	42
2	19
3	4

(a) Based on these 100 games, what would you estimate his probability of getting a hit to be?

(b) What would you estimate as his probability of getting at least one hit in three times at bat?

(c) With $\alpha = .05$, test the hypothesis that this distribution is a random sample of a binomial random variable with $n = 3$, $p = .3$.

4. The following table summarizes the distance a university broad jumper jumped, in 150 attempts.

Distance (Upper End Point Included)	Number of Times
20 to 22 feet	9
22 to 23 feet	12
23 to 24 feet	14
24 to 25 feet	32
25 to 26 feet	47
26 to 27 feet	36

(a) Verify that this frequency table has a mean of 24.83 feet and a standard deviation of 1.53 feet.

(b) Test the hypothesis that this data is a random sample from a normal distribution with $\alpha = .10$.

5. In one calendar year, 1687 bachelor's degrees were granted by a large university. The grade point distribution of those graduating is presented in the following table. ($A = 4.0$, $B = 3.0$, etc.)

Grade Point (Upper End Point Included)	Number
2.0 –2.25	364
2.25–2.50	422
2.50–2.75	383
2.75–3.00	272
3.00–3.25	150
3.25–3.50	66
3.50–4.00	30

Test the hypothesis that the grade point averages of graduating seniors are normally distributed, with $\alpha = .10$.

6. At the time of the drawing for the 1973 draft numbers it was expected that anyone getting number 126 or higher would not be drafted. The following table summarizes, for each month, the number of birthdates receiving one of the integers between 1 and 125.

January	February	March	April	May	June
10	10	12	12	11	13

July	August	September	October	November	December
10	15	6	10	5	11

With $\alpha = .05$, would you accept the hypothesis that these 125 numbers were randomly allocated to the 365 birthdates? (*Hint.* If the assignment is really random, the probability of a number being assigned to January is 31/365 and to February is 28/365, for example.)

7. The contents of 116 "half-gallon" containers of milk were measured; the distribution of actual contents (to the nearest .05 liquid ounce) were as follows:

Contents	Number
63.5, 63.55	8
63.6, 63.65	12
63.7, 63.75	20
63.8, 63.85	32
63.9, 63.95	22
64.0, 64.05	16
64.1, 64.15	6

Test the hypothesis that the actual amount of milk, per container, is a normal random variable, with $\alpha = .1$.

8. Four coins were flipped 100 times. The following table summarizes the number of heads that occurred, per flip.

Number of Heads	Number of Flips
0	12
1	27
2	33
3	22
4	6

Use this data to test the hypothesis that each of the four coins is fair $\alpha = .05$. (*Hint*. If all four are fair and independent, the number of heads per flip should be a binomial random variable with $n = 4$, $p = 1/2$.)

9. The records of a new car dealer show the following number of times that the purchasers of 200 new cars, with one-year warranties, returned to him with warranty claims during their first year of ownership.

Number of Claims	Number of Owners
0	22
1	56
2	52
3	38
4	20
5 or more	12

With $\alpha = .1$, test the hypothesis that the number of times a new car owner will return with a warranty claim, within the first year of ownership, is a Poisson random variable with $\lambda = 2$.

10. The first 32 United States Presidents to die had death dates between 1799 and 1963, inclusive. The following table summarizes the quarters of the year in which their deaths occurred. Test the hypothesis that the death dates of U.S. Presidents are uniformly distributed over the four quarters of the year with $\alpha = .05$. (This would imply that a President's death date is equally likely to fall in each of the four quarters.)

	Quarter			
	1	2	3	4
Number of Deaths	9	8	10	5

11. It is well established that the average age of death for United States citizens has been considerably lengthened over the last 175 years. The following table summarizes the age at death of the first 16 U.S. Presidents to die; it also summarizes the age at death of the second 16 U.S. Presidents to die.

	Age at Death	
	Age ≤ 70	Age > 70
First 16	6	10
Second 16	13	3

With $\alpha = .05$, test the hypothesis that these two variables of classification are independent. Note that the implication here is that the later Presidents tend to die at a younger age, exactly the opposite of the general trend in the population over this same span of years. Can you think of any explanations for this phenomenon?

12. From 1936 to 1971, inclusive, the Preakness was run 36 times (once per year). It is a $1\frac{3}{16}$-mile race for three-year-old race horses. The following table summarizes the winning times for this race, from 1936 to 1953, versus the winning times for the years from 1954 to 1971. Test the hypothesis that these two variables of classification are independent, with $\alpha = .01$.

Winning Time

	Time $\leq$ 1:57	Time $>$ 1:57
1936–1953	3	15
1954–1971	13	5

13. A 1971 survey of 2000 people, each 25 years of age or more, yielded the following information about educational attainment and race. Test the hypothesis that these two variables are independent, with $\alpha = .01$.

	Had Not Completed High School	Had Completed High School	Had Completed College
Caucasian	728	866	220
Non-Caucasian	106	66	14

8.4 SUMMARY

A hypothesis is a statement about a population probability measure; the statement may be true or false, but, lacking perfect knowledge of the population, we do not know which of these is the case. To perform a statistical test of a hypothesis, we select a random sample from the population and, on the basis of the sample results, decide to accept or reject the hypothesis. Two distinct errors may be committed when using a statistical test of a hypothesis. A type I error occurs if the hypothesis is rejected when it is true; a type II error occurs if the hypothesis is accepted when it is false. The standard statistical tests are designed so that the probability of a type I error, α, is easily controlled. Among all tests with the same value of α the test with the smallest probability of type II error is preferred.

Tables 8.3 and 8.4 give the tests for hypotheses about the mean of a normal population. Table 8.5 gives approximate tests for hypotheses about the mean of any population, for large sample sizes. Table 8.6 presents tests for hypotheses about the variance, or standard deviation, of a normal population and Table 8.7 gives tests for hypotheses about a population proportion p.

A statistic whose probability measure is well approximated by a χ^2 density can be used to test that a population distribution has any given form; the χ^2 approximation gets better the larger the sample size. Similarly, a statistic whose probability measure is also well approximated by a χ^2 density can be used to test the hypothesis that two methods of classification in the population are independent. Both of these χ^2 statistics make comparisons of observed frequencies with expected frequencies, assuming the hypothesis is true.

Exercise 8.4

1. A random sample of $n = 15$ mature Valencia oranges were selected from the same tree and weighed. The estimated standard deviation of the weights was $s = 4.2$ ounces. Assuming the population of weights of oranges produced by this tree is normal, would you accept

    $$H_0: \sigma^2 \leq 12 \qquad \text{versus} \qquad H_1: \sigma^2 > 12$$

    with $\alpha = .1$?

2. A random sample of $n = 50$ freshman grade point averages (at the end of the year) gave $\bar{x} = 2.63$ and $s = .42$. Based on these sample results, would you accept

    $$H_0: \mu = 2.5 \qquad \text{versus} \qquad H_1: \mu \neq 2.5$$

    with $\alpha = .05$, where μ is the average end of year grade point average for all freshmen at this university?

3. A random sample of 500 full-time employed adults from the same state yielded the information given below.

Annual Salary level x (thousands of dollars)

	$x \leq 10$	$10 < x \leq 20$	$x > 20$	Total
Males	128	96	95	319
Females	108	59	14	181

Test the hypothesis that these two variables of classification are independent with $\alpha = .01$.

4. A random number generator produces sequences of whole numbers $(0, 1, 2, \ldots, 9)$ presumably in a random, haphazard way. If it really is producing these numbers randomly, the probability would be 1/10 that each position in the sequence has a 0 in it or a 1 or a 2, etc. Furthermore, then, if it is working correctly, about 1/10 of the whole numbers in any sequence should be 0's, 1/10 should be 1's, 1/10 should be 2's, etc. A sequence of 1000 whole numbers produced by a particular generator resulted in the following counts:

					Number of					
0	1	2	3	4	5	6	7	8	9	Total
110	95	119	79	100	98	115	103	84	97	1000

Test the hypothesis that the generator is working correctly, with $\alpha = .05$.

5. Of 300 constituents interviewed in a Congressman's district, 137 said they favored his stand on the sizes of welfare payments that should be made to unwed mothers. Based on these results, would you accept the hypothesis that a majority (at least .5) of his constituents favor his stand on this issue, with $\alpha = .05$?

6. If in fact $p = .45$, what is the probability of a type II error with the test you used in question 5 above?

7. During a nine-month academic year, a college hospital treated 112 cases involving broken bones. Assuming this type of case arrives at the hospital like events in a Poisson process with parameter λ (per month), would you accept $H_0: \lambda \geq 15$ versus $H_1: \lambda < 15$, with $\alpha = .05$? (Refer to problem 17 of Exercise 8.2.)

8. If $\lambda = 12$, what is the probability of type II error for the test you used in problem 7 above?

9. The results of 400 "serious" automobile accidents yielded the following data:

	Driver Was		
Driver was	Wearing Seat Belt	Not Wearing Seat Belt	Total
Killed	10	87	97
Not Killed	157	146	303
Total	167	233	400

Test the hypothesis that these two variables of classification are independent with $\alpha = .01$.

CHAPTER 9
COMPARATIVE EXPERIMENTS

Much research is concerned with *comparative experiments,* ones in which it is desired to compare two theories or two procedures. For example, an automobile tire manufacturing firm may have two distinct formulas for making the rubber used in its tread. Under some set of "standard" driving conditions, they can conceive of the population of miles driven until wearout of tires made with formula 1 as well as the population of miles driven until wearout for tires made with formula 2. All other things being equal, such as cost of manufacture and distribution, it then would be of interest to them to be able to compare these two conceptual populations in some way. Which conceptual population has the greater mean? Which has the greater proportion of tires with lifetimes exceeding 30,000 miles, or some other arbitrary distance? A comparative experiment is used to answer these questions. Samples are selected from each of the two populations and, using the observed values from the two, inferences about the differences in the populations are made.

As a second example, suppose two different vaccines are available for the prevention of flu. Then, again, under standard conditions it would be of great interest to determine which of the two is the more effective; that is, which of the two appears to have the higher probability of preventing the incidence of flu or which of the two results in the larger proportion of the population not getting the flu. This again would ideally call for the use of a comparative experiment. Two sets of individuals are used, which are as similar as possible. Those individuals in the first group would be innoculated with vaccine A, those in the second group with vaccine B. Then the number of incidences of flu in these two sample groups are used to make inferences about the proportions of the whole population that get the flu, when using vaccine A versus vaccine B.

9.1   ESTIMATION

In comparative experiments, two samples are available, one from each of the two populations to be compared. To distinguish between the two sets of sample values, we shall use X as the population random variable for one population and Y for the other. Thus, the sample values for the first population will be denoted $X_1, X_2, \ldots, X_m$ and and those for the second $Y_1, Y_2, \ldots, Y_n$. (Note that two different sample sizes, m and n, are used for the most general case.)

Let us first discuss estimation of the difference in means of the two populations. From Chapter 7, we recall that we would use

$$\overline{X} = \frac{1}{m} \sum_{i=1}^{m} X_i$$

to estimate μ_X and

$$\overline{Y} = \frac{1}{n} \sum_{j=1}^{n} Y_j$$

to estimate μ_Y. The estimator for the difference $\mu_X - \mu_Y$ is $\overline{X} - \overline{Y}$, the difference in the two sample means. The variance of this estimator can be evaluated using Theorem 6.1; in fact, from the results discussed there,

$$\sigma_{\overline{X}-\overline{Y}}^2 = \sigma_{\overline{X}}^2 + \sigma_{\overline{Y}}^2$$

so long as $\overline{X}$ and $\overline{Y}$ are independent (again, the probability measure for $\overline{Y}$ is unaffected by the particular value $\overline{X}$ is equal to). When the two samples are drawn from completely separate physical entities, this assumption of independence is generally made. Since $\sigma_{\overline{X}}^2 = \sigma_X^2/m$, $\sigma_{\overline{Y}}^2 = \sigma_Y^2/n$ we have, then,

$$\sigma_{\overline{X}-Y}^2 = \frac{\sigma_X^2}{m} + \frac{\sigma_Y^2}{n}$$

is the variance of the estimator of the difference in the two means. If the two population variances are equal ($\sigma_X^2 = \sigma_Y^2 = \sigma^2$), then

$$\sigma_{\overline{X}-\overline{Y}}^2 = \sigma^2 \left(\frac{1}{m} + \frac{1}{n} \right)$$

To estimate the population variances we would use S_X^2 to estimate σ_X^2, S_Y^2 to estimate σ_Y^2 (S_X^2 is the sample variance of the values from the X population; S_Y^2 is defined analogously). Thus the estimated variance of $\overline{X} - \overline{Y}$ is

$$S_{\overline{X}-\overline{Y}}^2 = \frac{S_X^2}{m} + \frac{S_Y^2}{n}$$

in the case that no assumption is made about σ_X^2 and σ_Y^2. If it is assumed that $\sigma_X^2 = \sigma_Y^2 = \sigma^2$, then both samples contain information about the common variance σ^2 and the information from the two samples must be combined to estimate σ^2. The estimator for this common value of σ^2 is

$$S^2 = \frac{(m-1)S_X^2 + (n-1)S_Y^2}{(m-1)+(n-1)}$$

$$= a\,S_X^2 + (1-a)S_Y^2$$

where $a = (m-1)/(m+n-2)$. Notice that S^2 is a weighted average of S_X^2 and S_Y^2, the two individual sample variances. S^2 is called the *pooled estimator* for σ^2.

Example 9.1.1

A tire manufacturer has available two different synthetic rubbers that could be used in manufacturing automobile tires. He wants to estimate the difference in expected tire lifetime using these two rubbers. Thus he has one conceptual population of X values, tire lifetimes for tires made using synthetic rubber 1, and a second conceptual population of Y values, consisting of tire lifetimes for tires made using synthetic rubber 2. He wants to estimate $\mu_X - \mu_Y$. To do this he tests $m = 10$ tires made using synthetic rubber 1 under standard conditions and also tests $n = 12$ tires made with synthetic rubber 2. For each tire tested, the number of miles traveled until wearout is measured. He finds, for synthetic rubber 1, $\Sigma x_i = 315$, $\Sigma x_i^2 = 9960$, where the x_i's are measured in thousands of miles. For synthetic rubber 2 he finds $\Sigma y_i = 408$, $\Sigma y_i^2 = 13{,}909$, where again the measurements are made in thousands of miles. With these values we find $\overline{x} = 315/10 = 31.5$, $\overline{y} = 408/12 = 34.0$,

$$s_X^2 = [9960 - 10(31.5)^2]/9 = 4.17$$
$$s_Y^2 = [13{,}909 - 12(34)^2]/11 = 3.36$$

Thus his estimate of the difference in expected lifetimes, using the two synthetics, is

$$\overline{x} - \overline{y} = 31.5 - 34.0 = -2.5$$

Based on these sample results he would estimate that, on the average, a tire made with synthetic rubber 2 will go 2500 miles further than a tire made with synthetic rubber 1. It is also of interest, as in Chapter 7, to estimate the standard deviation of this quantity, to get an idea of how variable $\overline{x} - \overline{y}$ may be from one repetition to another of the experiment. If we make no assumptions about the relative magnitudes of s_X^2 and s_Y^2, the estimated standard deviation of $\overline{X} - \overline{Y}$ is

$$s_{\bar{X}-\bar{Y}} = \sqrt{s_{\bar{X}}^2 + s_{\bar{Y}}^2} = \sqrt{\frac{s_X^2}{10} + \frac{s_Y^2}{12}}$$

$$= \sqrt{.417 + .280} = .835$$

If we assume that the variances of the two conceptual populations are equal ($\sigma_X^2 = \sigma_Y^2$), the estimated standard deviation of $\bar{X} - \bar{Y}$ is

$$s_{\bar{X}-\bar{Y}} = \sqrt{s^2 \left(\frac{1}{10} + \frac{1}{12}\right)}$$

$$= \sqrt{\frac{9s_X^2 + 11s_Y^2}{20} \left(\frac{1}{10} + \frac{1}{12}\right)}$$

$$= \sqrt{.683} = .827$$

Note that there is very little difference in these two estimates for this data. This is because the values of s_X^2 and s_Y^2 are quite close.

If the populations sampled from are both normal, the two population variances are equal, and the samples are independent, it can be shown that

$$\frac{\bar{X} - \bar{Y} - (\mu_X - \mu_Y)}{S\sqrt{1/m + 1/n}}$$

has the t distribution with $m + n - 2$ degrees of freedom; S is the square root of the pooled estimator of the common variance defined earlier. This fact can be used to derive a $100(1 - \alpha)\%$ confidence interval for $\mu_X - \mu_Y$ in the same way the t distribution was used in Chapter 7 to derive a confidence interval for the mean of a normal random variable. The resulting two-sided interval has end points

$$L = \bar{X} - \bar{Y} - t_{1-\alpha/2} \, S\sqrt{1/m + 1/n}$$

$$U = \bar{X} - \bar{Y} + t_{1-\alpha/2} \, S\sqrt{1/m + 1/n}$$

where the t percentiles are chosen from the t distribution with $m + n - 2$ degrees of freedom.

Example 9.1.2

Let us assume the two populations of tire lifetimes, discussed in Example 9.1.1, are both normal and that $\sigma_X^2 = \sigma_Y^2$. With the data given in that example let us compute a 90% confidence interval for $\mu_X - \mu_Y$. To do so we use the t distribution with $m + n - 2 = 20$ degrees of freedom; we find $t_{.95} = 1.725$, and from the previous calculations $\bar{x} - \bar{y} = -2.5$, $s\sqrt{1/10 + 1/12} = .827$. Thus a 90% two-sided confidence interval for $\mu_X - \mu_Y$ goes from

$$-2.5 - (1.725)(.827) = -3.93$$

to

$$-2.5 + (1.725)(.827) = -1.07$$

In order for the confidence interval for $\mu_X - \mu_Y$ just discussed to be exact, it is necessary that the two populations be normal, that the samples be independent, and that the population variances be equal. (Problems 10 and 11 of Exercise 9.3 discuss a test of equal variances.) If the populations are normal with unequal variances, there is no exact procedure to get a confidence interval for $\mu_X - \mu_Y$. However, the procedure described in problem 5 of Exercise 9.1 below gives an approximate interval that should be sufficient for most applications. If one or the other (or both) of the two populations are not normal in form, an approximate confidence interval for $\mu_X - \mu_Y$ is described in problem 6 below, for large samples. If the two samples are not independent, then special procedures are called for; we do not have the space here to delve into them. However, see problem 9 below for one special case of related samples and the resulting computations for a confidence interval about $\mu_X - \mu_Y$.

Now let us examine comparative experiments for proportions. Assume we have population 1 with proportion p_1 of successes and population 2 with proportion p_2 of successes. Let X be the number of successes observed in a random sample of size m from population 1 and let Y be the number of successes in a random sample of size n from population 2. Then, as we saw in Chapter 7, the unbiased estimator of p_1 is X/m and the unbiased estimator of p_2 is Y/n; the difference, $X/m - Y/n$, is an unbiased estimator of $p_1 - p_2$. The variance of this estimator is $p_1 q_1/m + p_2 q_2/n$, which we may estimate using

$$\frac{X}{m}\left(1 - \frac{X}{m}\right)\frac{1}{m} + \frac{Y}{n}\left(1 - \frac{Y}{n}\right)\frac{1}{n} = \frac{X(m-X)}{m^3} + \frac{Y(n-Y)}{n^3}$$

Example 9.1.3

Assume that 200 volunteers have agreed to participate in an experiment to compare the efficacies of two flu vaccines. They are split into two groups of 100 each, so that group 1 is as similar as possible, overall, to group 2 in terms of locations of residence, health habits, possible exposure to the flu, etc. All the individuals in group 1 are innoculated with vaccine 1, and those in group 2 with vaccine 2. At the conclusion of the flu season it was found that 32 individuals in group 1 had contracted the flu while 13 of the individuals in group 2 had contracted the flu. The point estimate of $p_1 - p_2$, then, is $\hat{p}_1 - \hat{p}_2 = .32 - .13 = .19$. Based on these sample results we would estimate that 19% less of the population would get flu if all were innoculated with vaccine 2 versus vaccine 1. The standard error of this estimate is

$$\sqrt{\frac{\hat{p}_1(1-\hat{p}_1)}{100} + \frac{\hat{p}_2(1-\hat{p}_2)}{100}} = .058$$

For sufficiently large sample sizes m and n we know, from Theorem 6.1 and Theorem 6.5 (the central limit theorem), that the probability measure of $X/m - Y/n$ is well approximated by a normal density with mean $p_1 - p_2$ and variance $p_1 q_1/m + p_2 q_2/n$; thus

$$\frac{X/m - Y/n - (p_1 - p_2)}{\sqrt{p_1 q_1/m + p_2 q_2/n}}$$

is approximately a standard normal random variable. Just as in Section 7.2, then, a conservative $100(1 - \alpha)\%$ confidence interval for $p_1 - p_2$ is given by replacing both $p_1 q_1$ and $p_2 q_2$ by 1/4 (the largest possible value for each), leading to the (two-sided) confidence limits

$$L = \frac{X}{m} - \frac{Y}{n} - \tfrac{1}{2} z_{1-\alpha/2} \sqrt{\frac{1}{m} + \frac{1}{n}}$$

$$U = \frac{X}{m} - \frac{Y}{n} + \tfrac{1}{2} z_{1-\alpha/2} \sqrt{\frac{1}{m} + \frac{1}{n}}$$

The liberal $100(1 - \alpha)\%$ two-sided confidence limits for $p_1 - p_2$ are

$$L = \frac{X}{m} - \frac{Y}{n} - z_{1-\alpha/2} \sqrt{\frac{X(m - X)}{m^3} + \frac{Y(n - Y)}{n^3}}$$

$$U = \frac{X}{m} - \frac{Y}{n} + z_{1-\alpha/2} \sqrt{\frac{X(m - X)}{m^3} + \frac{Y(n - Y)}{n^3}}$$

derived by replacing p_1 by X/m and p_2 by Y/n, rather than 1/2.

Example 9.1.4

Let us use the flu vaccine data from Example 9.1.3 to compute both the liberal and conservative 95% confidence intervals for the difference in the proportions of people that will get the flu, using vaccine 1 versus using vaccine 2. From Table A.2 we find $z_{.975} = 1.96$ and thus the liberal 95% confidence limits for $p_1 - p_2$ go from $.19 - 1.96(.058) = .08$ to $.19 + 1.96(.058) = .30$. The conservative 95% limits extend from $.19 - (1/2)(1.96) \times \sqrt{1/100 + 1/100} = .05$ to $.19 + (1/2)(1.96)\sqrt{1/100 + 1/100} = .33$.

Exercise 9.1

1. To compare the mileages he could get with two different gasolines, Joe made eight runs over the same measured 30-mile course, four times

with Brand A and four times with Brand B. x's are used to represent the amounts of gasoline of Brand A used in their four runs and y's are used for Brand B. It was found that $\Sigma x_i = 60.2$, $\Sigma x_i^2 = 924.76$, $\Sigma y_i = 55.4$, and $\Sigma y_i^2 = 779.29$. Estimate the difference in expected amounts of gasolines needed to travel 30 miles with these two brands and construct a 90% confidence interval for the difference, assuming equal population variances.

2. An agricultural experiment station planted 12 plots with hybrid corn A and 12 plots with hybrid corn B; the plots were as similar as possible and as far as possible subjected to the same conditions during the growing season. The average yield per plot from hybrid A was 55.7 and the sample standard deviation of the observed yields was 5.2; the average yield per plot for hybrid B was 58.2 and the sample standard deviation was 4.9. What is the point estimate of the difference in expected yields for plots of this size? Construct a 95% confidence interval for the expected difference in yield, assuming equal population variances.

3. Forty-one out of 320 people interviewed at random in San Francisco were smokers and 32 out of 250 interviewed at random in Los Angeles were smokers. Construct an approximate liberal 99% confidence interval for the expected difference in the proportions of smokers in these two cities.

4. In a freshman math class with 500 students, 48 were given an A. Of 600 students in a freshman history class, 75 were given an A. Compute a conservative 99% confidence interval for the expected difference in proportions of students to get A's in these two subjects.

5. (*Unequal variances*). Suppose we have two normal populations but are unwilling to assume they have equal variances. Given a sample of m from population 1 and a sample of n from population 2, define

$$s^2 = \frac{\Sigma(x_i - \bar{x})^2}{m(m-1)} + \frac{\Sigma(y_i - \bar{y})^2}{n(n-1)}$$

Then the interval from

$$\bar{x} - \bar{y} - s\, t_{1-\alpha/2} \qquad \text{to} \qquad \bar{x} - \bar{y} + s\, t_{1-\alpha/2}$$

where we use the t distribution with degrees of freedom equal to the *smaller* of $m-1$, $n-1$, is an approximate $100(1-\alpha)\%$ confidence interval for $\mu_X - \mu_Y$. Use the data given in Example 9.1.1 to place a confidence interval on $\mu_X - \mu_Y$, with this approximate method and compare it with the one derived in that example.

6. (*Large samples*). Given large samples from each of two populations, which need not be normal in form, an approximate $100(1-\alpha)\%$ confidence interval for the difference in their means is given by the interval from $\bar{x} - \bar{y} - s\, z_{1-\alpha/2}$ to $\bar{x} - \bar{y} + s\, z_{1-\alpha/2}$, where s^2 is as defined in problem 5. Assume a random sample of 100 family incomes of house-

holds in Iowa had a sample mean of 6.5 and sample standard deviation of 2.1 (thousands of dollars). A random sample of 150 family incomes of households in Alabama had a sample mean of 3.7 and a sample standard deviation of 1.8. Construct a 99% confidence interval for the difference in mean family incomes in Iowa and Alabama.

7. Assume that the heights of United States adult citizens are normally distributed, as are the heights of French adult citizens; further assume that the variances of the two populations are equal. Using x_i's to represent the sample heights of $m = 8$ American adults and y_i's to represent the sample heights of $n = 6$ French adults, assume it was found that

$$\Sigma x_i = 562 \qquad \Sigma x_i^2 = 39{,}543$$

$$\Sigma y_i = 410 \qquad \Sigma y_i^2 = 28{,}048$$

where all measurements are made in inches. Construct a 95% confidence interval for the difference in mean heights of United States and French adults.

8. Assume that the rainfall in Monterey, California, in January of any year is a normal random variable; the rainfall in February is also a normal random variable. Assume the two population variances are equal. Using x_i's to represent the observed rainfall in January over $m = 10$ years, and y_i's to represent the rainfall in February over the same $n = 10$ years, suppose

$$\Sigma x_i = 48.7 \qquad \Sigma x_i^2 = 247.20$$

$$\Sigma y_i = 44.9 \qquad \Sigma y_i^2 = 211.40$$

where again the measurements are made in inches.
   (a) What is the point estimate of $\mu_X - \mu_Y$, the difference in expected rainfall in the two months?
   (b) Construct a 95% confidence interval for the difference in rainfall in these two months.

9. (*Paired t distribution*). In the confidence interval for $\mu_X - \mu_Y$ discussed in the text, it was assumed that the samples from the two normal populations were independent. This really is not a reasonable assumption for cases like the rainfall data presented in problem 8. If it were raining on January 31 in any given year, it would seem likely that the same storm may remain active on February 1, thus making contributions to the total rainfall in both months. Conversely, if it is bright and sunny on January 31, it is likely to be the same on February 1 and both monthly totals are affected by the lack of a storm over this period. Thus it might not be reasonable to assume that the total rainfall observed in January is independent of the February total for the same year. In such a case the confidence interval for $\mu_X - \mu_Y$ discussed in the text, is not appropriate. Note that the observations made are naturally paired together, being the two monthly totals for the same year. We could then form the difference of the January total

less the February total, $X_i - Y_i$, for the same year and then work with these differences to get a confidence interval for $\mu_X - \mu_Y$. If X and Y are each normal random variables, then their difference $X - Y$ is again a normal random variable. Also if we define

$$D_i = X_i - Y_i$$

that is, each pair is used to define one difference, then the interval from

$$\overline{D} - t_{1-\alpha/2} \, S_D/\sqrt{n} \qquad \text{to} \qquad \overline{D} + t_{1-\alpha/2} \, S_D/\sqrt{n}$$

forms a $100(1 - \alpha)\%$ confidence interval for $\mu_X - \mu_Y$, the difference in the population means. S_D is the sample standard deviation of the differences (the D_i's) defined in the usual way, and the constant $t_{1-\alpha/2}$ is chosen from the t distribution with $n - 1$ degrees of freedom, where n is the *number of pairs* in the sample. (Thus we have n X's and n Y's in the original sample.) For the same case discussed in problem 8, $n = 10$ pairs of rainfall totals for January and February, assume that

$$\Sigma d_i = \Sigma(x_i - y_i) = 3.8$$

$$\Sigma d_i^2 = \Sigma(x_i - y_i)^2 = 16.69$$

Use the procedure just discussed to construct a 95% confidence interval for $\mu_X - \mu_Y$.

10. Fifteen college students volunteered to participate in an experiment, designed to study the degradation of reaction time as a consequence of their consuming three bottles of beer in one hour. Each student's reaction time was measured when he first arrived; he then consumed three bottles of beer over a one-hour period and his reaction time was measured a second time. Let D_i be the difference in reaction time (after minus before) for the ith student and assume it was found that

$$\Sigma d_i = 3.75 \qquad \Sigma d_i^2 = 30$$

Construct a 90% confidence interval for the expected difference in reaction time after consumption of three bottles of beer at this rate.

11. Eight professional golfers each drove two times from the same tee, one time with a golf ball of type 1 and the second time with a golf ball of type 2 (half the golfers hit type 1 first and the other half hit type 2 first). Let D_i be the difference in the lengths of the two drives by the ith golfer (type 1 distance less type 2 distance for each). Assume it was found that

$$\Sigma d_i = -192 \qquad \Sigma d_i^2 = 74{,}608$$

where the measurements were made in yards. Construct a 95% confidence interval for the expected difference a drive would travel, using these two types of balls.

12. A presidential candidate is interested in the proportion of voters that

will vote for him in California, versus the proportion that will vote for him in New York. Of 150 voters interviewed in California 58 said they would vote for him; of 200 voters interviewed in New York, 98 said they would vote for him. Given this data construct a 95% liberal confidence interval for the difference, $p_1 - p_2$, in proportions of voters in the two states that will vote for him.

13. The California highway patrol periodically sets up inspection stations along the highways of the state. They then stop some of the cars passing by and inspect their equipment (e.g., lights, brakes, and horn) to see if it is functioning properly. Of 100 cars that were at least two years old, 18 failed the inspection while of 100 cars that were less than two years old, 11 failed the inspection. Construct a liberal 95% confidence interval for the difference in the proportions of cars in these two age groups that could pass the inspection.

14. On two different major earth faults the following statistics were gathered:

| | | Days Between Quakes | |
Fault	Number of Quakes	Mean	Variance
1	100	14	225
2	87	16	600

It would appear that the variances of the two distributions that these are samples from are not equal, and they are not normal in form. Use the technique of problem 6 to construct a 95% confidence interval for the difference in average days between quakes for the two faults.

15. Two different lakes were stocked with the same species of fish. After two years, samples of fish were removed from both lakes. The length of each fish was measured, with the following results:

Lake	Number of Fish	Sample Mean	Sample Variance
1	8	23	9
2	12	27.5	25

Use the technique of problem 5 to compute a 90% confidence interval for the difference in mean lengths of fish in the two lakes.

16. If $m = n$ show that the pooled variance estimator $S^2 = \frac{1}{2}(S_X^2 + S_Y^2)$, the average of the two sample variances.

17. What would you use as a $100(1 - \alpha)\%$ two-sided confidence interval for $\mu_Y - \mu_X$, assuming normal populations and equal variances?

9.2 TESTS OF HYPOTHESES

In many comparative experiments it is desired to make a decision about whether or not two population means are equal, rather than to estimate the size of the difference between the two. The methodology for making decisions about statements concerning probability measures is provided by statistical tests of hypotheses, as was discussed in Chapter 8.

The statistics introduced in the last section, for deriving confidence intervals, also prove useful in testing hypotheses in comparative experiments. We shall use the same notation introduced there, assuming we have a random sample of size m from one population (with X as the population random variable) and a random sample of size n from a second population (with Y as the population random variable). Then, if we assume both populations are normal, with equal variances ($\sigma_X^2 = \sigma_Y^2 = \sigma^2$), and the samples are independent, the random variable

$$\frac{\overline{X} - \overline{Y} - (\mu_X - \mu_Y)}{S\sqrt{1/m + 1/n}}$$

has the t distribution with $m + n - 2$ degrees of freedom (S^2 is the pooled estimator of σ^2, defined in Section 9.1). Under the assumption that $\mu_X = \mu_Y$, note then that this random variable becomes

$$\frac{\overline{X} - \overline{Y}}{S\sqrt{1/m + 1/n}}$$

This statistic is useful in testing hypotheses about the values of μ_X and μ_Y. For example, to test $H_0: \mu_X = \mu_Y$ versus $H_1: \mu_X \neq \mu_Y$, we would reject H_0 if

$$\frac{\overline{X} - \overline{Y}}{S\sqrt{1/m + 1/n}}$$

exceeds $t_{1-\alpha/2}$ in absolute value; using this rule we would have probability α of committing a type I error. Table 9.1 summarizes the possible tests about μ_X and μ_Y in comparative experiments, for independent samples from normal populations with equal variances. A comparison of these rules with those given earlier in Table 8.3 and 8.4 shows they are exactly the same types of procedures as the ones used earlier for tests of hypotheses about the mean of a normal population; the difference $\overline{X} - \overline{Y}$ replaces the sample mean, μ_0 is 0 and S is the pooled estimator of the variance.

Table 9.1 Tests of Size α about μ_X, μ_Y, Samples from Normal Populations, Equal Variances

H_0	H_1	Reject H_0 if
$\mu_X \leq \mu_Y$	$\mu_X > \mu_Y$	$\overline{X} - \overline{Y} > t_{1-\alpha}\, S\sqrt{1/m + 1/n}$
$\mu_X \geq \mu_Y$	$\mu_X < \mu_Y$	$\overline{X} - \overline{Y} < -t_{1-\alpha}\, S\sqrt{1/m + 1/n}$
$\mu_X = \mu_Y$	$\mu_X \neq \mu_Y$	$\overline{X} - \overline{Y} > t_{1-\alpha/2}\, S\sqrt{1/ m + 1/n}$
		or $\overline{X} - \overline{Y} < -t_{1-\alpha/2}\, S\sqrt{1/m + 1/n}$

Example 9.2.1

The state department of fish and game is in charge of monitoring the health of fish in the lakes in the state. To estimate the mercury content in the fish in Clear Lake the department will annually select a sample of fish from the lake and measure the amount of mercury in each. Assume that the mercury contents of all fish in the lake form a normal population for each year that a sample is taken. We further assume that the variance of this normal population is σ^2, for any year. The mean or average content may or may not change from year to year. Let X be the population random variable for last year's population and let Y be the population random variable for this year's contents. Last year a sample of $m = 10$ fish were examined, and this year a sample of $n = 13$ fish were examined. The department wants to test

$$H_0: \mu_X \geq \mu_Y \qquad \text{versus} \qquad H_1: \mu_X < \mu_Y$$

with $\alpha = .05$. Note that if $\mu_X \geq \mu_Y$, then the average content this year is no larger than last year. Referring to Table 9.1, we should reject H_0 if we find $\overline{X} - \overline{Y} < -1.721\, S\sqrt{1/10 + 1/13}$, where $t_{.95} = 1.721$ is chosen from Table A.4 with $10 + 13 - 2 = 21$ degrees of freedom. The two samples yield the following results:

$$\Sigma x_i = 50.0 \qquad \Sigma x_i^2 = 273.04$$

$$\Sigma y_j = 84.5 \qquad \Sigma y_j^2 = 592.57$$

From these quantities we find

$$\overline{x} = 5.0 \qquad \overline{y} = 6.5 \qquad s = 1.78$$

Thus $\overline{x} - \overline{y} = -1.5$

and

$$-1.721 s\sqrt{1/10 + 1/13} = -1.721(1.78)(.421) = -1.29$$

Therefore we reject H_0. The probability we are wrong in rejecting H_0 is $\alpha = .05$. It is quite unlikely to get this large an increase in the sample mean content if in fact this year's population mean content is no larger than last year's.

As has been mentioned, the tests listed in Table 9.1 assume that the two populations are normal, the population variances are equal, and the samples are independent. If the population variances are unequal, then there is in general no exact test for the hypotheses given in Table 9.1. Problem 5 in Exercise 9.2 describes approximate tests of the hypotheses that should be quite adequate for most applications. Given related samples (of the right kind) from two normal populations, exact tests are available, as discussed in problem 12 below. If one or both of the populations are not normal, problem 6 below gives approximate tests that are quite good for large sample sizes.

Tests about two population proportions can be made, for large sample sizes, again using the fact that the sample proportions have probability measures that are well approximated by normal densities. Again we shall use the notation introduced in Section 9.1,

Table 9.2   Tests of Size α about p_1 and p_2

Conservative Tests

H_0	H_1	Reject H_0 if
$p_1 \leq p_2$	$p_1 > p_2$	$X/m - Y/n > \frac{1}{2}z_{1-\alpha}\sqrt{1/m + 1/n}$
$p_1 \geq p_2$	$p_1 < p_2$	$X/m - Y/n < -\frac{1}{2}z_{1-\alpha}\sqrt{1/m + 1/n}$
$p_1 = p_2$	$p_1 \neq p_2$	$X/m - Y/n > \frac{1}{2}z_{1-\alpha/2}\sqrt{1/m + 1/n}$
		or $X/m - Y/n < -\frac{1}{2}z_{1-\alpha/2}\sqrt{1/m + 1/n}$

Liberal Tests
$$\hat{p} = (X+Y)/(m+n) \qquad \hat{q} = 1 - \hat{p}$$

H_0	H_1	Reject H_0 if
$p_1 \leq p_2$	$p_1 > p_2$	$X/m - Y/n > z_{1-\alpha}\sqrt{\hat{p}\hat{q}(1/m + 1/n)}$
$p_1 \geq p_2$	$p_1 < p_2$	$X/m - Y/n < -z_{1-\alpha}\sqrt{\hat{p}\hat{q}(1/m + 1/n)}$
$p_1 = p_2$	$p_1 \neq p_2$	$X/m - Y/n > z_{1-\alpha/2}\sqrt{\hat{p}\hat{q}(1/m + 1/n)}$
		or $X/m - Y/n < -z_{1-\alpha/2}\sqrt{\hat{p}\hat{q}(1/m + 1/n)}$

and assume we are given population 1 that contains proportion p_1 of successes; population 2 contains proportion p_2 of successes. A random sample of size m is selected from population 1 and X is the total number of successes in the sample; Y is the total number of successes in an independent sample of size n selected from population 2. Then, for m and n sufficiently large

$$\frac{X/m - Y/n - (p_1 - p_2)}{\sqrt{p_1 q_1/m + p_2 q_2/n}}$$

has a probability measure that is well approximated by the standard normal density. Notice that if $p_1 = p_2 = p$, say, then

$$\frac{X/m - Y/n}{\sqrt{pq(1/m + 1/n)}}$$

is approximately standard normal. This is the basis of both the conservative and liberal tests of hypotheses listed in Table 9.2. To get the conservative tests the factor pq is replaced by its largest possible value, which is $1/4$. The liberal tests replace this factor by $\hat{p}\hat{q}$, where $\hat{p} = (X+Y)/(m + n)$ is the sample estimate of the common probability of a success in the two populations.

Example 9.2.2

In a random sample of 100 seniors selected from the University of California system in 1960, 11 had a cumulative grade point average of 3 or more (4.0 = A). In a random sample of 150 seniors selected from the University of California system in 1970, 32 had a grade point average of 3 or more. Does this data indicate a change in the proportion of students in this university with average grades of B or better over these 10 years? We shall use this data to test

$$H_0 : p_1 = p_2 \quad \text{versus} \quad H_1 : p_1 \neq p_2$$

with $\alpha = .1$, using both the conservative test and the liberal test to illustrate both; p_1 is the proportion of seniors with a grade point average of B or better in 1960 and p_2 is the same proportion in 1970. We find then, from Table 2, $z_{.95} = 1.64$, and $x/m = 11/100 = .110$. $y/n = 32/150 = .213$, $x/m - y/n = -.113$, and $\hat{p} = (x + y)/(m + n) = 43/250 = .172$; thus $z_{.95}\sqrt{\hat{p}\hat{q}(1/m + 1/n)} = 1.64\sqrt{(.172)(.828)(.0167)} = .080$, and with the liberal test we would reject H_0. Also

$$\tfrac{1}{2}z_{.95}\sqrt{1/m + 1/n} = (\tfrac{1}{2})(1.64)\sqrt{.0167} = .106$$

so we would also reject H_0 with the conservative test. Based on these samples, it appears the proportions of seniors with grade points of 3.0 or more are not the same in the two years.

Exercise 9.2

1. Two different wholesalers will sell "50-pound" boxes of bananas to a grocer for the same price. The grocer weighs five boxes supplied by wholesaler 1 and finds their average weight to be 49.8 pounds and the standard deviation of their weights to be 2.7 pounds; he weighs three boxes supplied by wholesaler 2 and finds their average weight to be 50.3 pounds and their standard deviation to be three pounds. Assuming these to be random samples from two normal populations with equal variances, would you accept the hypothesis that the average weights of banana boxes supplied by the two wholesalers are equal, with $\alpha = .1$? (Use a two-sided test.)

2. Of 100 people interviewed when leaving a Las Vegas casino, 62 said they had come out ahead in gambling, while of 120 people interviewed when leaving a Reno casino, 75 said they had come out ahead. Using $\alpha = .1$, test the hypothesis that the proportions of winners (or at least of those that say they are winners) are the same for the patrons of these two casinos.

3. The nicotine content of 10 cigarettes of Brand A was measured; the sample statistics were $\Sigma x_i = 223.6$, $\Sigma x_i^2 = 5040.20$. Similarly the nicotine content of 10 cigarettes of Brand B was measured, giving $\Sigma y_i = 197.4$, $\Sigma y_i^2 = 3928.20$. With $\alpha = .05$, test the hypothesis that the mean nicotine content is the same for both brands. (Use a two-tailed test.)

4. Thirty-five out of 100 students interviewed at a large university were cigarette smokers, whereas 10 out of 50 interviewed at a small college were cigarette smokers. With $\alpha = .1$, test the hypothesis that the proportion of students that smoke at the two schools is the same, versus the alternative that the large university proportion is greater.

5. To test the hypothesis that the means of two normal populations are equal, when it does not seem reasonable to assume they have equal variances, the statistic used to derive a confidence interval for $\mu_X - \mu_Y$ in problem 5, Exercise 9.1 may be used. That is, let

$$S^2 = \frac{\Sigma(X_i - \overline{X})^2}{m(m-1)} + \frac{\Sigma(Y_i - \overline{Y})^2}{n(n-1)}$$

then, if $\mu_X = \mu_Y$,

$$\frac{\overline{X} - \overline{Y}}{S}$$

has an approximate t distribution with degrees of freedom equal to the smaller of $m - 1$, $n - 1$. This statistic may be used to test the hypothesis that $\mu_X = \mu_Y$. Suppose it is desired to compare two different highway paints for their lasting ability on asphalt highways. One strip of paint of each type is painted across a highway at each of five locations, having roughly the same traffic density and speed. The number of days until the paint has worn a specified amount is the variable used in comparing the two. Assume that for the five strips of type 1 the average number of days until they wore down the given amount was 192 and the standard deviation was 22 days. For the five strips of type 2, the average number of days was 174 and the standard deviation was 10 days. With $\alpha = .05$, test the hypothesis that the means of the two populations of days of wear are equal,

(a)   Using the approximate procedure just described where it is not assumed the two variances are equal.

(b)   Using the procedure that assumes the two variances are equal.

6.   To test the equality of means of two populations, either of which may be nonnormal we can again rely on the fact that the sample mean has an approximate normal distribution for large samples. Thus, let $\overline{X}$ be the mean of the first sample, and $S_X{}^2$ the variance; $\overline{Y}$ and $S_Y{}^2$ are the mean and variance of the second sample. If the two sample sizes, m and n, are sufficiently large, then

$$\frac{\overline{X} - \overline{Y}}{\sqrt{\dfrac{S_X{}^2}{m} + \dfrac{S_Y{}^2}{n}}}$$

is approximately a standard normal random variable, if the two population means are equal. Describe how you would use this fact to test the hypothesis that $\mu_X = \mu_Y$, if the alternative hypothesis is

(a) $\mu_X \neq \mu_Y$

(b) $\mu_X < \mu_Y$

(c) $\mu_X > \mu_Y$

7.   Assume we have independent random samples of size m and n, respectively, from two normal populations with equal variances. How would you test $H_0: \mu_X - \mu_Y = 3$ versus $H_1: \mu_X - \mu_Y \neq 3$?

8.   Samples of two different dog foods were selected and analyzed to determine their protein content. For dog food A, $m = 12$ packages were selected, and it was found that

$$\sum_{i=1}^{12} x_i = 87 \qquad \sum_{i=1}^{12} x_i{}^2 = 674.75$$

similarly, $n = 10$ packages of dog food B were selected and it was found that

$$\sum_{i=1}^{10} y_i = 86 \qquad \sum_{i=1}^{10} y_i{}^2 = 748.60$$

It was suspected that the protein contents for each are normally distributed, but that the variances are unequal. Would you accept the hypothesis that the mean protein contents are the same for the two brands, with $\alpha = .05$? Use the procedure discussed in problem 5 above.

9. Two different plant nutrients were compared for their effect on the growth of one-year-old camellias; 24 camellia plants were used. Twelve were selected at random to be treated with nutrient 1 and the remaining 12 were treated with nutrient 2. A measure, which depended both on the amount of vertical growth and increase in circumference, was computed after six months' usage of the nutrients. Assume that the measure used is normally distributed, and that the variances in the measure are the same for both nutrients. For plants treated with nutrient 1, the mean value of the measure was 22 and the standard deviation of these measures was 3; for plants treated with nutrient 2, the mean value of the measure was 15 and the standard deviation was 4. Test the hypothesis that the means of the measurements are the same for the two nutrients, with $\alpha = .01$.

10. Joe flipped a new 50-cent piece 25 times and observed 10 heads. Hugh then flipped the same coin 25 times and observed 15 heads. With $\alpha = .1$, test the hypothesis that the probability of observing a head with this coin is the same, irrespective of whether Joe or Hugh is doing the flipping.

11. Use the data presented in question 7, Exercise 9.1, to test the hypothesis that the mean heights of United States and French adults are the same, with $\alpha = .05$.

12. (*Paired t-test*). The distribution discussed in problem 9, Exercise 9.1, can also be used to test the hypothesis that two population means are equal, when samples of pairs, one element from each population, are available, and it does not seem reasonable to assume the two are independent. Specifically, if again we let D_i be the difference between the two elements of the same pair, it can be shown that

$$\frac{\overline{D}\sqrt{n}}{S_D}$$

has the t distribution with $n - 1$ degrees of freedom, if the two population means are equal; $\overline{D}$ is the average of the differences, S_D^2 is the sample variance of the differences, and n is the number of *pairs* in the sample. With the two-sided alternative $\mu_X \neq \mu_Y$, we would reject the hypothesis of equal population means if

$$|\overline{D}| > \frac{t_{1-\alpha/2}\, S_D}{\sqrt{n}}$$

where $t_{1-\alpha/2}$ is from the t distribution with $n - 1$ degrees of freedom. Apply this test to the rainfall data presented in problem 9, Exercise 9.1, with $\alpha = .1$.

13. Use the data presented in problem 11, Exercise 9.1, to test the hypothesis that the mean distances traveled by the two types of golf balls are equal, versus the two-sided alternative that they are not, with $\alpha = .05$.

14. Use the data presented in problem 12, Exercise 9.1, to test the hypothesis that equal proportions of voters in New York and California favor the candidate, versus the one-sided alternative that a greater proportion of voters in New York favor him, with $\alpha = .05$.

9.3 THE ONE-WAY ANALYSIS OF VARIANCE

We have seen some techniques of inference appropriate for comparative experiments involving samples from two populations. What can be done if it is desired to compare more than two populations? A tire manufacturer may very well have more than two different synthetic rubbers he could choose from; more than two different flu vaccines may be available for possible usage in preventing flu. In making comparisons over years (mercury content of fish, grade point averages), we certainly may want to make inferences involving more than two years. In this section we shall briefly study techniques of inference appropriate to studies involving three or more populations.

The techniques of estimation already discussed in the case of one or two populations can be straightforwardly applied to studies involving three or more populations. For example, assume we have available independent samples from three populations; let us use n_1, n_2, and n_3, respectively, to denote the sample sizes from the three populations. We are assuming, then, that we have available a total of $n_1 + n_2 + n_3$ numbers, if the samples are all combined. To differentiate the sample values from population 1 versus those from populations 2 and 3, we shall use a double subscripting system and formally denote our sample values by X_{ij}, $i = 1, 2, 3, j = 1, 2, \ldots, n_i$. The first subscript denotes the population the sample value came from and the second subscript the several sample values from the sample population. Then, the means for the three samples are

$$\overline{X}_1 = \frac{1}{n_1} \sum_j X_{1j}$$

$$\overline{X}_2 = \frac{1}{n_2} \sum_j X_{2j}$$

$$\overline{X}_3 = \frac{1}{n_3} \sum_j X_{3j}$$

These quantities would be used to estimate their respective population means. Functions of these quantities would be used to estimate the corresponding functions of the population means. For example, to estimate the difference between the means of populations 1 and 2, $\mu_1 - \mu_2$, we would use $\overline{X}_1 - \overline{X}_2$; and to estimate the difference between the means of populations 2 and 3, $\mu_2 - \mu_3$, we would use $\overline{X}_2 - \overline{X}_3$. The standard errors of such estimates can be obtained using an extension of Theorem 6.1. Assuming σ_1^2, σ_2^2, σ_3^2 represent the three population variances, then

$$\sigma_{\overline{X}_1}^2 = \frac{\sigma_1^2}{n_1} \qquad \sigma_{\overline{X}_2}^2 = \frac{\sigma_2^2}{n_2} \qquad \sigma_{\overline{X}_3}^2 = \frac{\sigma_3^2}{n_3}$$

as before. Furthermore, if c_1, c_2, and c_3 are any known constants and the samples are independent, then the variance of $c_1\overline{X}_1 + c_2\overline{X}_2 + c_3\overline{X}_3$ is

$$V(c_1\overline{X}_1 + c_2\overline{X}_2 + c_3\overline{X}_3) = c_1^2\sigma_{\overline{X}_1}^2 + c_2^2\sigma_{\overline{X}_2}^2 + c_3^2\sigma_{\overline{X}_3}^2$$
$$= c_1^2\sigma_1^2/n_1 + c_2^2\sigma_2^2/n_2 + c_3^2\sigma_3^2/n_3$$

Thus, if we choose $c_1 = 1$, $c_2 = -1$, and $c_3 = 0$

$$V(c_1\overline{X}_1 + c_2\overline{X}_2 + c_3\overline{X}_3) = V(\overline{X}_1 - \overline{X}_2)$$
$$= \sigma_1^2/n_1 + \sigma_2^2/n_2$$

since $c_1^2 = c_2^2 = 1$. The variance of the estimator of the difference between the means of the second and third populations is

$$V(\overline{X}_2 - \overline{X}_3) = \frac{\sigma_2^2}{n_2} + \frac{\sigma_3^2}{n_3}$$

since here we have $c_1 = 0$, $c_2 = 1$, and $c_3 = -1$. If the three population variances are not assumed equal, then each σ_i^2 is estimated by the corresponding sample variance S_i^2, where

$$S_i^2 = \sum_j (X_{ij} - \overline{X}_i)^2/(n_i - 1)$$

If the population variances are assumed equal, $\sigma_1^2 = \sigma_2^2 = \sigma_3^2 = \sigma^2$, then the estimator for σ^2 is the *pooled* variance.

$$S^2 = \frac{(n_1 - 1)S_1^2 + (n_2 - 1)S_2^2 + (n_3 - 1)S_3^2}{n_1 + n_2 + n_3 - 3}$$
$$= \sum_i \sum_j (X_{ij} - \overline{X}_i)^2/(n_1 + n_2 + n_3 - 3)$$

This quantity is analogous to the pooled estimator of variance defined earlier in Section 9.1, using independent samples from three populations rather than two.

Example 9.3.1

Three different fertilizers are to be compared for their effects on the same hybrid corn. A total of 18 experimental plots are available for the comparison. It is assumed that the 18 plots are essentially identical in terms of expected yield, so long as they receive the same treatment. It is decided to split these resources equally among the three fertilizers to be compared. Thus, all 18 plots are planted with identical seed; of the 18 $n_1 = 6$ are treated with fertilizer 1, $n_2 = 6$ are treated with fertilizer 2, and $n_3 = 6$ are treated with fertilizer 3. The 18 observed yields are assumed then to be random samples from the corresponding conceptual populations. The three conceptual population variances, σ_1^2, σ_2^2, and σ_3^2, are assumed to be equal. The yields from the 18 plots are presented in the following table.

Fertilizer Yields

Fertilizer 1	Fertilizer 2	Fertilizer 3
16	14	10
18	18	14
18	19	13
21	15	15
15	18	12
19	16	14

We find the following values computed from these sample observations.

| | Fertilizer | | |
	1	2	3
Sample total	107	100	78
Sample sum of squares	1931	1686	1030

Then the estimates of the three population mean yields (for plots this size) are

$$\bar{x}_1 = \frac{107}{6} = 17.83 \qquad \bar{x}_2 = \frac{100}{6} = 16.67 \qquad \bar{x}_3 = \frac{78}{6} = 13$$

The three sample variances are

$$s_1^2 = \frac{1931 - (107)^2/6}{5} = 4.57$$

$$s_2^2 = \frac{1686 - (100)^2/6}{5} = 3.87$$

$$s_3^2 = \frac{1030 - (78)^2/6}{5} = 3.20$$

Since the three population variances are assumed to be equal, the pooled estimate of the common variance σ^2 is used, rather than any of the three separate ones listed above. The pooled sample variance is

$$s^2 = \frac{5s_1{}^2 + 5s_2{}^2 + 5s_3{}^2}{15}$$

$$= \frac{s_1{}^2 + s_2{}^2 + s_3{}^2}{3} = 3.88$$

because of the equal sample sizes. We estimate the difference in mean yield between fertilizers 1 and 2 to be

$$\bar{x}_1 - \bar{x}_2 = 1.16$$

the difference between mean yields of fertilizers 1 and 3 to be

$$\bar{x}_1 - \bar{x}_3 = 4.83$$

and the difference in mean yield between fertilizers 2 and 3 to be

$$\bar{x}_2 - \bar{x}_3 = 3.67$$

Each of these estimated differences has variance $s^2/6 + s^2/6 = s^2/3$, which we estimate by $s^2/3 = 1.29$. Thus the estimated standard error or each difference is $\sqrt{1.29} = 1.14$. We can also estimate other functions of the population means and their standard errors, using the results quoted earlier. For example, since there appears to be little difference in mean yields for fertilizers 1 and 2, we may want to contrast them jointly with fertilizer 3. This can be done by averaging the yields of fertilizers 1 and 2 together (since it appears quite plausible that the corresponding populations are the same), and then estimating the difference between this average and the mean yield from fertilizer 3. We would estimate this contrast by

$$\frac{\bar{x}_1 + \bar{x}_2}{2} - \bar{x}_3 = 4.25$$

The variance of this estimate is

$$\frac{\sigma^2}{6}\left(\tfrac{1}{4} + \tfrac{1}{4} + 1\right) = \frac{\sigma^2}{4}$$

(here $c_1 = 1/2$, $c_2 = 1/2$, $c_3 = -1$). The estimated standard error of this contrast, then, is $\sqrt{s^2/4} = \sqrt{(3.88)/4} = .98$.

If it is assumed that the populations sampled are normal, then confidence intervals for differences in means or of functions of the means, can be derived in much the way they were done in Section 9.1. Limitations of time and space preclude further discussion of this important topic.

With samples available from several populations it is frequently desired to test the hypothesis that all of the population means are equal versus the alternative that they are not (i.e., that some two or more population means are unequal). For example, we may want to test that the average mileage till wearout, for tires made from three different synthetic rubbers, are all equal. Or we may want to test that the average yield of corn is the same, regardless of which of the three different fertilizers is used during the growing season. Suppose five different diets could be prescribed by a doctor for overweight patients. Is the expected weight loss the same for all five?

The *analysis of variance* is used to test hypotheses that population means are equal, based on samples from normal populations. The name "analysis of variance" derives from the fact that a variance is analyzed, or split into parts; the decision to accept or reject the hypothesis of equal means hinges on the relative sizes of the parts into which the variance has been split. We shall briefly discuss the methodology of this technique and then examine an example of its use.

First, let us discuss the *one-way* analysis of variance and the assumptions behind it. (This is also referred to as the analysis of variance for a completely randomized design.) For a general notation, assume we have k populations; random samples of sizes $n_1, n_2, \ldots, n_k$ (not necessarily all equal) are selected from these populations, so that we have available $\sum_{i=1}^{k} n_i$ numbers in total. It is assumed that each population is normal, the mean of population i is μ_i, and its variance is σ^2 (thus we are assuming each population has the same unknown variance σ^2). We want to test $H_0 : \mu_1 = \mu_2 = \ldots = \mu_k$ versus the alternative that H_0 is false (some two or more means are unequal). The test proceeds as follows, letting X_{ij}, $i = 1, 2, \ldots, k$, $j = 1, 2, \ldots, n_i$ represent the sample random variables. We define

$$\overline{X}.. = \frac{\sum_i \sum_j X_{ij}}{\sum_i n_i}$$

$$\overline{X}_i. = \frac{\sum_j X_{ij}}{n_i} \qquad i = 1, 2, \ldots, k$$

to be the overall mean, and individual sample means, respectively. These individual sample means were discussed earlier in this section as the estimators for their respective population means. Intuitively, it would seem reasonable to accept H_0 (equal population means) if the sample means all have roughly the same value and to reject H_0 if they do not. This is exactly the criterion that is used, where the decision of how close together the sample means should be is determined by the estimator for σ^2. Specifically, to see why the term analysis of variances is used, let us assume H_0 is true and that in fact $\mu_1 = \mu_2 = \ldots = \mu_k$. Then all the sample values could be assumed to have come from the *same* population. Under this assumption the estimator of σ^2 is based on

$$\sum_i \sum_j (X_{ij} - \overline{X}..)^2$$

the sum of squares of all sample values about the overall mean $\overline{X}...$ It can be shown, algebraically, that

$$\sum_i \sum_j (X_{ij} - \overline{X}..)^2 = \sum_i \sum_j (X_{ij} - \overline{X}_i.)^2 + \sum_i n_i (\overline{X}_i. - \overline{X}..)^2$$

That is, we can split this "overall" variability into two parts, the first of which is used in defining the pooled estimator of σ^2 and the second of which depends on how variable the sample means are about $\overline{X}...$ The relative sizes of these two quantities determine whether H_0 should be accepted or should be rejected.

For independent samples from normal populations, all with variance σ^2, the ratio

$$\frac{\displaystyle\sum_i n_i (\overline{X}_i. - \overline{X}..)^2}{k - 1} \Bigg/ \frac{\displaystyle\sum_i \sum_j (X_{ij} - \overline{X}_i.)^2}{\sum n_i - k}$$

has the F distribution with $k - 1$ and $\Sigma n_i - k$ degrees of freedom, if $H_0 : \mu_1 = \mu_2 \ldots = \mu_k$ is true. This distribution has two parameters called degrees of freedom, the first $(k - 1)$ associated with the numerator of the ratio and the second $(\Sigma n_i - k)$ associated with the denominator. The ordering of the two degrees of freedom is of importance because, for example, an F distribution with 3 and 6 degrees of freedom is different than an F distribution with 6 and 3 degrees of freedom. If the observed value of this ratio exceeds $F_{1-\alpha}$, with the appropriate degrees of freedom, H_0 is rejected.

Table A.6 in the Appendix presents a table of the 95th and 99th percentiles of the F distributions with various pairs of degrees of freedom. The columns correspond to the numerator degrees of freedom and the rows to the denominator degrees of freedom. The 95th percentiles are given in regular type and the 99th in boldface type. Notice, then, with 2 and 6 degrees of freedom, $F_{.95} = 5.14$, $F_{.99} = 10.92$, while with 6 and 2 degrees of freedom, $F_{.95} = 19.33$, $F_{.99} = 99.33$; as already mentioned, the ordering of the degrees of freedom affects the F distribution.

Frequently a tabular array, called an analysis of variance, is used to present in an organized way the quantities needed to test the hypothesis that all the population means are equal. Table 9.3 presents such a table.

Table 9.3   Analysis of Variance for Testing Equality of Means

(Sample Size n_i from Population i, $i = 1, 2, \ldots, k$)

Source	Degrees of Freedom	Sums of Squares	Mean Squares
Between samples	$k - 1$	$B_{ss} = \Sigma n_i\,(\overline{X}_i. - \overline{X}..)^2$	$B_{ss}/(k - 1)$
Within samples	$\Sigma n_i - k$	$W_{ss} = \Sigma\Sigma\,(X_{ij} - \overline{X}_i.)^2$	$W_{ss}/(\Sigma n_i - k)$
Total	$\Sigma n_i - 1$	$T_{ss} = \Sigma\Sigma\,(X_{ij} - \overline{X}..)^2$	

Notice that there are three rows and four columns in the table. As was discussed above, in using an analysis of variance we have split a total variance into parts. The row labeled total presents the total degrees of freedom and the total sum of squares to be analyzed (in the columns labeled degrees of freedom and sums of squares, respectively). This total sum of squares is split into a part that depends on the variability between the sample means (between samples) and into a part that is used to define the pooled estimator of σ^2 (within samples). The total degrees of freedom is also split into two parts, for between samples and within samples. The mean squares for between samples and within samples are equal to their sums of squares divided by their degrees of freedom. The test statistic is equal to the ratio of the mean squares for between and within samples. The between plus within degrees of freedom equals the total degrees of freedom and the between plus within sums of squares equals the total sum of squares.

Example 9.3.2

Assume a medical doctor has devised three different diet-exercise programs to help overweight people lose weight, each of which is medically sound for the health. He then would like to compare the three regimens over, say, a one-month period to try to discover which of them might lead to a greater average weight loss. To do this he decides to find 12 volunteers among his patients who are interested in losing weight. He will allocate four patients to each of the three diets. In doing this allocation he balances the three groups so that the group receiving diet 1 is as similar as possible in initial weights to the groups receiving diets 2 and 3; in this way he can hopefully infer that any observed differences in group weight loss at the end of the month are because of the diets and not because of extraneous factors. The participants then religiously follow their respective diets and, at the end of the month, the differences between their initial weights and final weights are calculated. They are displayed in the table below, as are the average weight losses (sample means) for the three different groups.

Diet Number	Weight Losses				Mean Loss
1	23	19	17	13	18
2	19	16	12	9	14
3	20	15	19	8	13

The doctor is willing to assume that the amount of weight lost, with each diet, is a normal random variable and that the variance of the weight lost is the same for all three diets; thus, he assumes three separate conceptual populations, one for each diet regimen. The four weight losses he observes for diet one are then assumed to be a random sample from conceptual population 1; the same is true for the weight losses with the other two regimens. He wants to test the hypothesis that the population mean weight losses are the same, for these three diets with $\alpha = .05$. For the given sample results, we find the following analysis of variance. (We have $k = 3$, $n_1 = n_2 = n_3 = 4$, $\Sigma n_i = 12$.)

Source	Degrees of Freedom	Sums of Squares	Mean Squares
Between diets	2	56	28
Within diets	9	204	22.67
Total	11	260	

The value of the test statistic, then, to test that the expected weight loss is the same for all three diets, is $28/22.67 \doteq 1.24$. Since the 95th percentile of the F distribution with 2 and 9 degrees of freedom is 4.26, we accept the hypothesis that the population means are equal. Even though the sample means do obviously differ, they do not differ by a sufficiently large enough amount for us to reject the hypothesis of equal population means. Put another way, samples of size 4 from normal populations with equal population means will frequently have sample means that differ as much as these do.

Example 9.3.3

Let us use the fertilizer data, presented in Example 9.3.1, to test the hypothesis that the average crop yields are the same for the three fertilizers.

Again, then, $k = 3$ conceptual populations are involved, this time with $n_1 = n_2 = n_3 = 6$. The analysis of variance for testing that the population means are equal is given below; we shall test that the expected yields are equal with $\alpha = .01$. Thus, from Table A.6, we find we should reject H_0 if the F statistic exceeds 6.36, the 99th percentile of the F distribution with 2 and 15 degrees of freedom. The ratio of the between fertilizers and

Analysis of Variance for Fertilizer Data

Source	Degrees of Freedom	Sums of Squares	Mean Squares
Between fertilizers	2	76.30	38.15
Within fertilizers	15	58.20	3.88
Total	17	134.50	

within fertilizers mean squares is $38.15/3.88 = 9.83$; thus we reject H_0. The probability of observing sample means as variable as these, if the population means are equal, is (less than) .01.

The analysis of variance is a widely applied tool for testing hypotheses about population means. We have had space here to discuss only the simplest possible example of its use, in the completely randomized design. It is also used in a great variety of more complex situations in which samples are available from several related populations. This type of material is discussed in textbooks on the design of experiments; the material available on the analysis of variance and its use in practical problems almost has no end. For "exact" tests of hypotheses about equality of population means, it is necessary that all populations be normal and that the variances be equal. For large sample sizes, though, the central limit theorem again justifies the F distribution for a good approximate test of equality of means. Much research has been done on this topic, and many journal articles have been written about the "robustness" of the F test, how good the rule for rejecting H_0 is even if the assumptions of normality and equal variances are not satisfied.

Exercise 9.3

1. To compare the gasoline mileage for four national brands of gasoline, the same car was used over the same 100 mile course 12 times. The same driver was used each time, the car was driven at the same speed, and the car's engine was tuned (if necessary) before each run. Each

brand of gasoline was used on three separate runs; the miles per gallon for the 12 runs are summarized below.

Gasoline	Miles per Gallon		
A	15.5	16.3	15.9
B	16.1	16.5	15.9
C	14.5	15.2	15.0
D	15.7	15.6	15.7

With $\alpha = .05$ would you accept the hypothesis that the four gasolines are equally good (in terms of expected miles per gallon)?

2. Samples of five cigarettes from each of three major cigarette brands were selected; the tar content of each cigarette was measured. The following table presents the measurements made.

Brand	Cigarette Number				
	1	2	3	4	5
A	16.8	20.2	15.3	17.1	17.0
B	12.4	9.8	10.0	11.1	10.5
C	14.0	14.0	13.9	14.3	14.7

(a) Test the hypothesis that the mean tar contents are the same for the three brands, with $\alpha = .01$.

(b) Which cigarette brand would you estimate has the lowest average tar content per cigarette?

3. Twelve students volunteered to take part in a study to compare the effects of two drugs on the time necessary to perform a simple chore. Each student performed the chore before taking a drug, then consumed the drug and a short time later again performed the same chore. The differences in times required were computed for each student and are presented below. Each student used only one of the two drugs.

Drug

1	20	10	14	5	9	11
2	12	18	16	17	11	11

With $\alpha = .05$, test the hypothesis that the expected reduction in reaction time for this chore is the same for both drugs.

4. An agricultural experiment station performed an experiment to compare the effects of three different fertilizers on the yield of a crop. The following analysis of variance resulted.

Source	Degrees of Freedom	Sums of Squares	Mean Squares
Between fertilizers	2	143	
Within fertilizers			
Total	20	153	

Fill in the missing entries in this table and test the hypothesis that the fertilizers are equal in their effects, with $\alpha = .05$.

5. Show algebraically that

$$B_{ss} + W_{ss} = T_{ss}$$

That is,

$$\sum_i n_i (\overline{X}_i. - \overline{X}..)^2 + \sum_i \sum_j (X_{ij} - \overline{X}_i.)^2 = \sum_i \sum_j (X_{ij} - \overline{X}..)^2$$

[*Hint*: $X_{ij} - \overline{X}.. = (X_{ij} - \overline{X}_i.) + (\overline{X}_i. - \overline{X}..)$ and $\sum_j (X_{ij} - \overline{X}_i.) = 0$.]

6. For the data given in problem 3, use the t test to test the hypothesis that the drugs are equal in their effects, again with $\alpha = .05$. Square the t value just computed and note that this is equal to the F statistic computed for the analysis of variance in problem 3. Since this will always be true, the two tests are in fact equivalent.

7. Given a sample of size n from each of two normal populations with equal variances, show that the square of the t statistic used to test that the means are equal is identical with the F statistic used to test the same hypothesis in the analysis of variance.

8. To compare five different brands of golf balls, a professional golfer drove one ball of each type from the same tee four times on the same day. The lengths of the drives were as follows:

Brand

A	225	265	250	290
B	275	260	280	305
C	240	250	275	305
D	250	270	290	320
E	290	305	310	340

With $\alpha = .01$, test the hypothesis that the golf balls are equally good.

9. Three cakes were made with cake mix A, four with cake mix B, and six with cake mix C. The moisture contents of the 13 cakes were determined; it was desired to test the hypothesis that the average moisture content was the same for all three brands of cake mix. The

following table presents a partial summarization of the computations made.

Source	Sums of Squares
Between brands	
Within brands	50
Total	100

Given this information would you reject the hypothesis that the average moisture contents of the three brands are equal, with $\alpha = .05$?

10. Assume we have independent random samples of sizes n_1 and n_2, respectively, from two normal populations. Let σ_1^2 and σ_2^2, respectively, be the two population variances. As has been mentioned several times in this chapter, many of the standard statistical methods are based on the assumption that the population variances are equal. Thus, in many cases we would like to test $H_0: \sigma_1^2 = \sigma_2^2$ versus $H_1: \sigma_1^2 \neq \sigma_2^2$, with a probability of type I error equal to α. The following procedure provides such a test. Let S_1^2 be the sample variance for the values from population 1 and let S_2^2 be the same quantity for population 2. Then the ratio, S_1^2/S_2^2, has the F distribution with $n_1 - 1$ and $n_2 - 1$ degrees of freedom. Since we want to reject $H_0: \sigma_1^2 = \sigma_2^2$ if it appears that either $\sigma_1^2 < \sigma_2^2$ or if $\sigma_1^2 > \sigma_2^2$, we should reject H_0 if $S_1^2/S_2^2 < F_{\alpha/2}$ or if $S_1^2/S_2^2 > F_{1-\alpha/2}$. Thus, for example, to have $\alpha = .1$ we would need $F_{.05}$ and $F_{.95}$. Recall that Table A.6 presents only values for $F_{.95}$ and $F_{.99}$; thus it would not appear possible to make this two-sided test using these tables, since the tables do not give the values for $F_{.05}$. The labels (populations 1 and 2), as they are attached to the samples, are arbitrary; let S_1^2 be the larger of the two sample variances and let S_2^2 be the smaller of the two. Then S_1^2/S_2^2 is always at least 1. If we reject $H_0: \sigma_1^2 = \sigma_2^2$ only if $S_1^2/S_2^2 > F_{.95}$, this rule is equivalent to the above two-sided rule with $\alpha = .1$. Similarly, if we reject $H_0: \sigma_1^2 = \sigma_2^2$ only if $S_1^2/S_2^2 > F_{.99}$, this rule is equivalent to the two-sided rule with $\alpha = .02$. Thus, Table A.6 allows us to test

$$H_0: \sigma_1^2 = \sigma_2^2 \qquad \text{versus} \qquad H_1: \sigma_1^2 \neq \sigma_2^2$$

with $\alpha = .1$ or .02. Larger tables are necessary if we want other values for α. Assume we have selected independent samples, each of size 10, from two normal populations. We find the two sample variances to be 3.46 and 6.43. Would you accept the hypothesis the two population variances are equal, versus the alternative that they are not, with $\alpha = .1$?

11. Assume the data in problem 3 arose as two independent samples, of size 6, from two normal populations. Using the procedure described above in problem 10, would you accept the hypothesis that the population variances are equal, versus the alternative they are not, with $\alpha = .02$?

9.4   SUMMARY

Comparative experiments are used in comparing or contrasting two or more populations. When given independent samples from each of two (possibly) different normal populations, the t distribution can be used to compute confidence intervals about the difference in population means or to test hypotheses about the two population means, if the population variances are equal. Given independent large samples from each of two populations, the central limit theorem, and its extensions, can be used to compute confidence intervals and test hypotheses about the population means, even if the populations are not normal and the variances are not equal. The paired t techniques may be used for comparing related samples from two normal distributions.

In comparing three or more populations, techniques of estimation are readily generalized to estimate the difference between any pair of population means and the standard error of the estimate; any linear combination of the population means can also be estimated in a straightforward way. Granted that all the populations are normal and .their variances are equal, the analysis of variance and the F distribution may be used to test the hypothesis that all the population means are equal. This F test is quite "robust"; the rule for rejecting equality of means is quite sensible even if the populations are not normal or the variances are not equal.

Exercise 9.4

1.   Over a period of 10 years the same three hybrid grains were compared at the same experiment station. (That is, 10 separate experiments, one per year, were run.) An analysis of variance of the yields observed was computed each year, and with $\alpha = .10$ (they had a bigger F table than we do) the hypothesis of equal mean yields was rejected one time and accepted nine times. Would you conclude there was a difference in mean yields? Why or why not?

2.   A hospital selected at random the records of 10 babies born in June and those of 12 babies born in December. The mean birth weight of those selected in June was 6.9 pounds and the estimated standard deviation, computed from the 10 selected, was .7 pounds. The comparable figures for the babies born in December were 7.3 and .6 pounds respectively.

   (a)   Estimate the difference in mean weights of babies born in June versus December, for this hospital.

   (b)   What is the standard error of your estimate?

3. Using the data quoted in question 2, above, would you accept the hypothesis that the average weights of babies born in June and December are the same, with $\alpha = .05$?

4. Two different deposits of iron ore have been sampled for contents, the first in five locations, the second in seven. The following readings resulted.

   Deposit 1: 15.4, 16.7, 12.3, 13.9, 14.4

   Deposit 2: 21.6, 20.7, 18.4, 22.3, 21.0, 24.2, 17.9

   (a) Estimate the difference in the average value of the measurements for the two deposits.
   (b) Compute the standard error of the estimate in (a).

5. Thirty kindergarten children were chosen to take part in a physical dexterity test. Three different socioeconomic backgrounds had been defined and 10 of the children were from each type of background. The analysis of variance of the times needed to complete the test is presented below:

Source	Degrees of Freedom	Sums of Squares
Background	2	23
Within backgrounds	27	27
Total	29	

   Would you conclude that the mean time to complete the test was the same for children from each of the three different backgrounds, with $\alpha = .01$?

6. Fifteen plots of land in the Napa Valley were sampled; the average grape tonnage per acre harvested from the 15 was 27.4 and estimated standard deviation was 2.31. Twelve plots of land in the San Joaquin Valley were also sampled; the sample average tonnage per acre (with the same grape variety) was 30.2 and the estimated standard deviation was 3.12.
   (a) Estimate the expected difference in tons per acre for this type of grape, grown in the Napa Valley versus the San Joaquin Valley.
   (b) Compute a 95% confidence interval for the differences in yield between these two locations.

7. Use the data given in problem 4 to test the hypothesis that the average iron ore contents are the same in the two deposits with $\alpha = .01$.

8. Use the data quoted in problem 6 to test that the expected tonnage per acre for this variety of grape is the same in the two valleys, with $\alpha = .01$.

9.   Five 100-watt bulbs from manufacturer A were left on until they burned out. The average time to failure was 762 hours and the estimated standard deviation was 47 hours. Eight 100-watt bulbs from manufacturer B and 10 from manufacturer C were tested in the same way. The sample quantities were, for B, a mean lifetime of 781 hours and a standard deviation of 47 hours; the bulbs from manufacturer C had an average lifetime of 775 hours and a standard deviation of 47 hours. Would you accept the hypothesis that the bulbs made by these three manufacturers have the same average lifetime, with $\alpha = .05$?

10.   A consumer testing group was interested in comparing the average number of miles per gallon attainable with cars made by foreign manufacturer A versus the average mileage attainable with cars made by foreign manufacturer B. To make the comparison they selected five cars from each manufacturer and paired them together arbitrarily (one A with one B) giving a total of five pairs. The two cars in each pair were treated identically, driven the same distance at the same speed over the same course; the amounts of gasoline consumed were then measured. The car from manufacturer B used less gasoline in each pair. Would you conclude that B's cars give better mileage? (*Hint.* If the two types have equal expected consumption rates, we would expect the probability that B does better than A, in any pair, to be 1/2.)

CHAPTER 10
LEAST SQUARES AND REGRESSION

The great German mathematician, Gauss, is generally credited with being the first person to state a certain type of problem and to propose a criterion that gives the type of solution still in use today. He formulated the concept of applying the technique called least squares in trying to estimate population parameters from a set of data. Let us first consider examples of problems in which this criteria may be used.

In many problems two variables may be related so that the behavior of one (or the value of one) gives information about the other. For example, consider the distribution of weights of babies, between age one month and six months, say, in the United States. Undoubtedly the average weight of one-month-old babies is not the same as the average weight of the six-month-old babies. If we let x represent the age of a baby and let Y be the population random variable for the distribution of weights of all babies of age x, it is clear that the distribution for Y depends on x. (Y is a random variable, x is not; thus only Y is capitalized.) The distribution of weights of $x =$ one-month-old babies is not identical with the distribution of weights of $x =$ six-month old babies. What sort of dependence might there be?

It seems reasonable that the population mean of Y should certainly change with x; as babies grow older they get heavier, at least on the average. Furthermore, the mean of Y may very well increase linearly with x, at least over the range of one month to six months. Symbolically, then, we might assume

$$E[Y] = \beta_0 + \beta_1 x \qquad 1 \le x \le 6$$

where β_0 and β_1 are unknown parameters that might be estimated from a sample of pairs of values of Y's and x's.

As another example of the same sort, consider the distribution of grade point averages of college students at the end of their freshman year. Letting x represent the high school grade point average (or some measure of academic success in high school), we can conceive of the population of freshman-year grade point averages for all students with the same value of x (same high school performance). As x increases it would seem reasonable that very probably the population mean of freshman grade point averages also increases. That is, letting Y represent the population random variable of freshman grade point averages for all students having a fixed value of x for high school achievement, we might assume

$$E[Y] = \beta_0 + \beta_1 x$$

where again β_0 and β_1 are unknown parameters that could be estimated from a sample.

Both of these problems are examples of *simple linear regression* problems. We have, or can conceive of, a population of values of Y for each value of x; the population mean is assumed to change linearly with x. This is the type of problem considered by Gauss over 175 years ago. The procedure he suggested to estimate β_0 and β_1 is called least squares. It is of such universal appeal and produces such good estimators that it is still in general use today. We shall study this least squares criterion, and the estimators it produces, in Section 10.1. To compute confidence intervals or test hypotheses we need to assume a specific type of distribution for the Y population for a given x. These topics will be considered in Section 10.2.

10.1   LEAST SQUARES ESTIMATION

Let us keep in mind the example concerning baby weight and age in discussing the least squares problem and procedure. We assume we have a random sample of n pairs (Y_i, x_i) (weight and age, respectively); the Y values are assumed to have been selected at random from the population of all Y values for the same x. The mean values of the Y populations are assumed to depend linearly on the associated value of x:

$$E[Y] = \beta_0 + \beta_1 x$$

(This equation is called the *regression* of Y on x.) It is assumed that the variances of the Y populations are the same for all x; the variability in weights of one-month-old babies is the same as the variability of six-month-old babies. This equal variance assumption is the standard one that is made; if it is assumed that the variances also change with the value of x (and, thus, the variance of Y with $x = 1$ is not assumed to be the same as the variance of Y with $x = 6$), then the estimators for β_0 and β_1 are different.

This assumption about the mean of Y shifting linearly with x is portrayed graphically in Figure 10.1.1. For each different value of x there is a population of Y values. The mean of each population lies on the straight line $\beta_0 + \beta_1 x$. The Y measurements that occur in the sample are assumed to have been selected at random from the population of Y values for the given x.

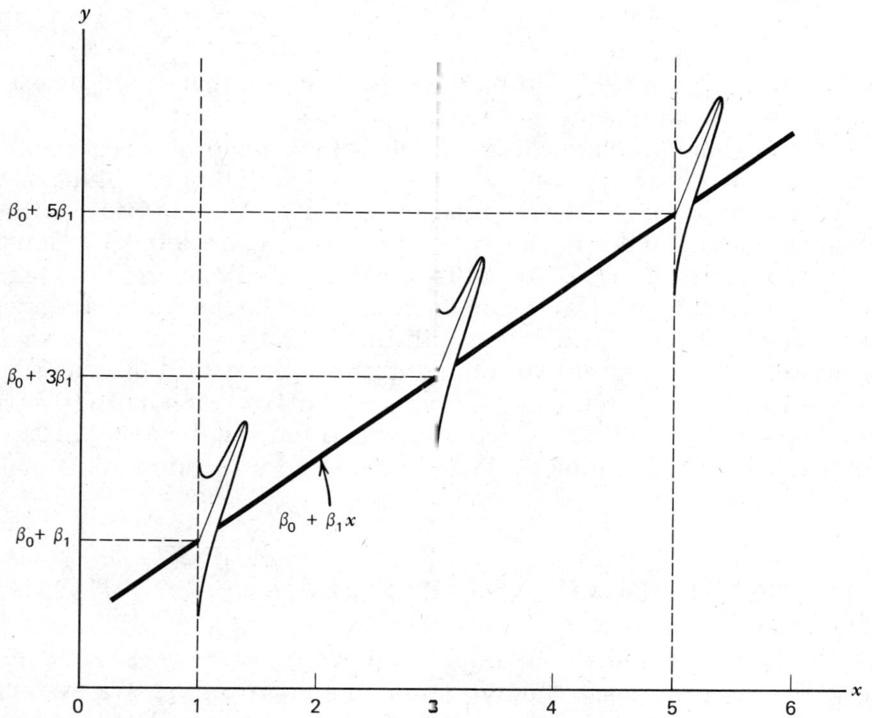

Figure 10.1.1 Populations of Y values for given values of x.

We realize, of course, that the sample pairs will not lie on a straight line, when plotted in the (x, y) plane, because of the variability of y's for each x. The particular y values we get in our sample will surely not all be exactly equal to the means of their corresponding distributions; thus, even though the population mean values may exactly lie on a straight line, the sample values will not. Figure 10.1.2 illustrates this with a sample of 10 babies' ages and weights. The weights and ages are also given in Table 10.1 in Example 10.1.1.

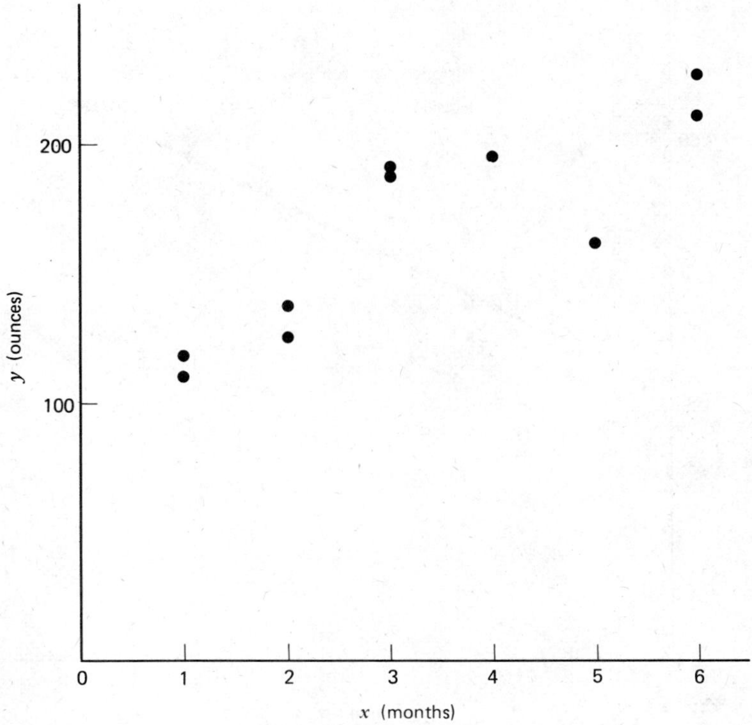

Figure 10.1.2 Sample of 10 weights and ages of babies.

Since the sample values will not lie on the true line connecting the population means, we cannot expect to discover the true values of the parameters of the line (β_0 and β_1) from a sample. We can, though, use the sample values to estimate the true values of β_0 and β_1, just as we discussed, in Chapter 7, estimating population parameters with sample values. The method used to estimate β_0 and β_1 from the sample is called *least squares*. Given a plot of the sample values such as Figure 10.1.2, there are any number of possible straight lines that more or less pass through the collection of plotted points. The least squares line in particular minimizes the sum of squares of deviations (or distances), in the y direction, between the observed sample points and the points that lie on the line. Figure 10.1.3 reproduces the same 10 ages and weights of babies, together with the resulting least squares line. The calculations needed to define this line are carried out in Example 10.1.1 below. This least squares line has the property that it gives the smallest possible value for the squares of the vertical deviations between the actual observed points and the points that lie on the line, as pictured in the figure. Thus, the y intercept and the slope of this line are our

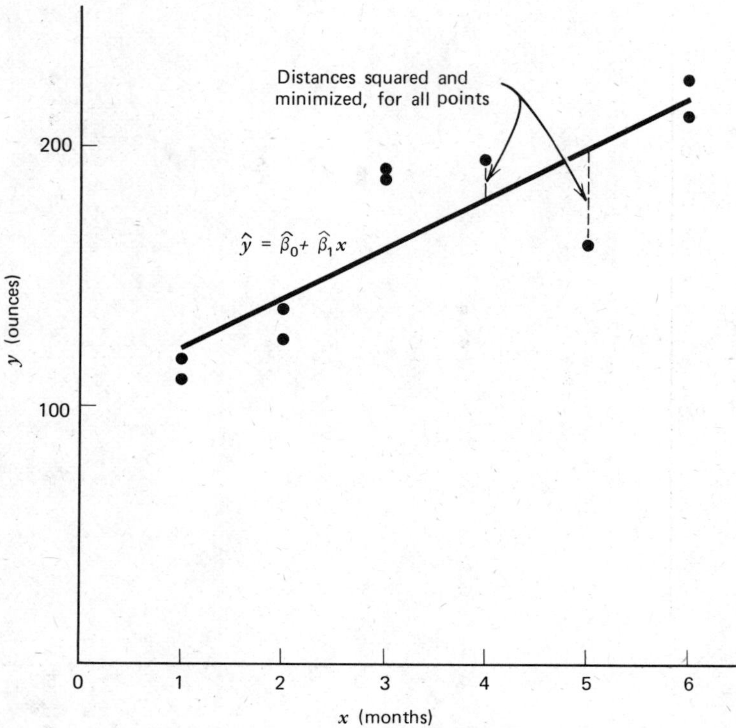

Figure 10.1.3   Sample of 10 weights and ages and the least squares line $\hat{y} = \hat{\beta}_0 + \hat{\beta}_1 x$.

estimates of β_0 and β_1, respectively; these estimates will be denoted by $\hat{\beta}_0$ and $\hat{\beta}_1$, as indicated in the figure.

Mathematically, the quantity to be minimized is

$$\sum_{i=1}^{n} (y_i - \hat{\beta}_0 - \hat{\beta}_1 x_i)^2$$

since $\hat{\beta}_0 + \hat{\beta}_1 x_i$ is the value on the line corresponding to y_i; the difference between them is $y_i - \hat{\beta}_0 - \hat{\beta}_1 x_i$. Each of these quantities is then squared and their total is minimized to determine $\hat{\beta}_0$ and $\hat{\beta}_1$. It can be shown that this procedure leads to the following two equations in $\hat{\beta}_0$ and $\hat{\beta}_1$, which then must be solved simultaneously to determine the estimates:

$$n\hat{\beta}_0 + (\Sigma x_i)\hat{\beta}_1 = \Sigma y_i$$

$$(\Sigma x_i)\hat{\beta}_0 + (\Sigma x_i^2)\hat{\beta}_1 = \Sigma x_i y_i$$

It can be verified that the solutions are

$$\hat{\beta}_0 = \bar{y} - \hat{\beta}_1 \bar{x}$$

$$\hat{\beta}_1 = \frac{\Sigma(x_i - \bar{x})(y_i - \bar{y})}{\Sigma(x_i - \bar{x})^2}$$

Just as when calculating the variance of a set of numbers, the following equivalent forms make the computation of $\hat{\beta}_1$ simpler than the formula just given:

$$\Sigma(x_i - \bar{x})(y_i - \bar{y}) = \Sigma x_i y_i - \frac{(\Sigma x_i)(\Sigma y_i)}{n}$$

$$= \Sigma x_i y_i - n\bar{x}\bar{y}$$

To compute an estimate of the variance σ^2, based on a random sample from the population, we have previously used

$$s^2 = \frac{1}{n-1} \Sigma(y_i - \bar{y})^2$$

The difference, $y_i - \bar{y}$, is actually the estimate of the difference between the individual sample value y_i and its mean. The same reasoning generalizes in estimating the population variance in a regression (or least squares) problem; the estimate of the mean of y_i is $\hat{\beta}_0 + \hat{\beta}_1 x_i$, rather than $\bar{y}$, when we make our regression assumption. Thus $y_i - (\hat{\beta}_0 + \hat{\beta}_1 x_i)$ would estimate the difference between a sample value and its mean, and our estimate of σ^2 is

$$s^2 = \frac{1}{n-2} \Sigma(y_i - \hat{\beta}_0 - \hat{\beta}_1 x_i)^2$$

Note that the denominator now is $n - 2$, since we have to estimate two quantities, β_0 and β_1, in estimating the mean of a single y_i. It can be shown that this numerator is

$$\Sigma(y_i - \hat{\beta}_0 - \hat{\beta}_1 x_i)^2 = \Sigma(y_i - \bar{y})^2 - \hat{\beta}_1^2 \Sigma(x_i - \bar{x})^2$$

which is a better computational formula than the original definition. The following example presents the baby weight-age data used in Figures 10.1.2 and 10.1.3 and applies these formulas to compute the least squares line.

Example 10.1.1

Assume that a pediatrician's records yielded the ages and weights for 10 babies given in Table 10.1. We assume that the expected weight of a baby aged x months is $E(Y) = \beta_0 + \beta_1 x$, for $1 \le x \le 6$, and that the variances of the weights are the same for each age. We shall use least squares to estimate β_0 and β_1. For the data given in Table 10.1 we find $\Sigma x_i = 33$, $\Sigma y_i = 1663$, $\Sigma x_i^2 = 141$, $\Sigma y_i^2 = 292{,}187$, and $\Sigma x_i y_i = 6104$; from these quantities we can compute $\bar{x} = 33/10 = 3.3$ and $\bar{y} = 1663/10 = 166.3$,

$$\Sigma(x_i - \bar{x})^2 = 141 - \frac{(33)^2}{10} = 32.1$$

$$\Sigma(y_i - \bar{y})^2 = 292{,}187 - \frac{(1663)^2}{10} = 15{,}630.1$$

$$\Sigma(x_i - \bar{x})(y_i - \bar{y}) = 6104 - \frac{(33)(1663)}{10} = 616.1$$

Then we find $\hat{\beta}_1 = 616.1/32.1 = 19.19$, $\hat{\beta}_0 = 166.3 - (19.19)(3.3) = 102.97$; thus the least squares line is

$$\hat{y} = 102.97 + 19.19x$$

This has already been plotted in Figure 10.1.3. Note that with this estimated line we can now estimate the mean weight of a baby for any age between one and six months, not just for those values of x that occurred in our sample. Thus, we would estimate the mean weight of a $3\frac{1}{2}$-month-old baby to be $102.97 + (19.19)(3.5) = 170.1$ ounces; our estimate of the mean weight of a $1\frac{1}{4}$-month-old baby is $102.97 + (19.19)(1.25) = 127.0$ ounces. We should restrict our predictions only to babies with ages between one and six months, since that is the range covered by our sample data. While a

Table 10.1   Ages and Weights for 10 babies

x (Age in Months)	y (Weight in Ounces)
1	110
1	118
2	125
2	137
3	187
3	191
4	195
5	162
6	211
6	227

straight line may not be a perfect assumption for ages one to six months, our sample data has not deviated to a great extent from it; if we were to go outside this range we simply have no idea whether the straight-line assumption remains at all appropriate. Our estimate of σ^2, the population variance of weights of babies all the same age, is

$$s^2 = \tfrac{1}{8}[15{,}630.1 - (19.19)^2(32.1)] = 476.13$$

If in this example we had ignored the ages of the babies and looked at our sample of weights as being selected from the larger population of weights of all babies from ages one to six months, notice that our estimate of the variance of that population would be $\tfrac{1}{9}(15{,}630.1) = 1736.68$, over 3.5 times as large as our estimate of the variance of the weights of babies all the same age. Thus, by taking the age of the baby into account, we can expect to be able to make a more accurate guess as to what its weight should be, than we are if we ignore this information.

The least squares estimated line that we have just discussed has many optimal properties. If we assume

(a) The y value associated with each x is randomly selected from the population of y's for that x.
(b) $E[Y] = \beta_0 + \beta_1 x$, for the range of x's considered.
(c) The population variances of y's are all equal to σ^2, for each x.

We can then show that the least squares estimators of β_0 and β_1 are the *best linear unbiased* estimators of these quantities. This means, first, that both $\hat{\beta}_0$ and $\hat{\beta}_1$ are unbiased estimators and, second, among all possible estimators of β_0 and β_1 that are unbiased and linear functions of the y's, $\hat{\beta}_0$ and $\hat{\beta}_1$ have the smallest possible variance over repeated samples. This is essentially a statement of the *Gauss-Markov* theorem, which is generally stated and proved in courses on mathematical statistics. Note that we have made no specific assumption about the particular form of the distribution of y's for a particular x, yet we are able to say the estimators are unbiased and have the smallest possible variance in a certain class of estimators.

The variances of the estimators $\hat{\beta}_0$ and $\hat{\beta}_1$ can be shown to be

$$\sigma_{\hat{\beta}_0}^{\,2} = \frac{\sigma^2 \Sigma x_i^2}{n\,\Sigma(x_i - \overline{x})^2}$$

$$\sigma_{\hat{\beta}_1}^{\,2} = \frac{\sigma^2}{\Sigma(x_i - \overline{x})^2}$$

where σ^2 is the common variance of the Y populations for any given value of x. We can estimate these variances by replacing σ^2 by S^2, its estimate from the sample; the estimated standard deviations of the estimators then are called the standard errors of estimate:

$$s_{\hat{\beta}_0} = s\sqrt{\Sigma x_i^2 / n\Sigma(x_i - \overline{x})^2}$$

$$s_{\hat{\beta}_1} = s/\sqrt{\Sigma(x_i - \overline{x})^2}/$$

For the baby weight–age data given in Example 10.1.1 these standard errors are

$$s_{\hat{\beta}_0} = \sqrt{476.13(141)/10(32.1)} = 14.46$$

$$s_{\hat{\beta}_1} = \sqrt{476.13}/\sqrt{32.1} = 3.85$$

These quantities estimate the standard deviations of the respective estimators over repeated samples of the same size; they indicate how close we might expect the values of $\hat{\beta}_0$ and $\hat{\beta}_1$ to be to the ones already computed, if we were to take a new random sample of $n = 10$ babies.

Example 10.1.2

Table 10.2 presents the high school grade point average (x values) and the

Table 10.2 Grade Point Averages for Twenty Students

High School gpa x	Freshman Year gpa y
2.05	2.19
2.02	2.24
2.17	2.30
2.21	2.25
2.39	2.30
3.76	3.45
3.25	3.20
3.04	2.96
2.97	3.05
2.44	2.63
2.29	2.20
2.88	3.00
3.55	3.21
3.16	3.20
3.84	3.80
3.33	3.20
3.70	3.55
2.08	2.30
2.71	2.65
2.79	2.90

college freshman year grade point averages (y values) for 20 students. In both cases, an A was scored as 4 points and a B as 3 points, and so on, and the grade points are weighted by the number of credit hours. We shall assume that the average freshman grade point average is linearly related to the same student's high school grade point average; that is, that

$$E[Y] = \beta_0 + \beta_1 x$$

Again, as in the previous example, we are assuming we have a random sample of Y's selected from the full population of Y values for the corresponding x. We shall use the method of least squares to estimate β_0 and β_1, and shall also estimate the variance σ^2 as well as the standard errors of $\hat{\beta}_0$ and $\hat{\beta}_1$. From the data given in the table we find

$$\Sigma x_i = 56.63 \qquad \Sigma y_i = 56.58$$

$$\Sigma x_i^2 = 167.4099 \qquad \Sigma y_i^2 = 164.9688 \qquad \Sigma x_i y_i = 165.9459$$

Using these quantities we compute

$$\overline{x} = 2.8315 \qquad \overline{y} = 2.8290$$

$$\Sigma(x_i - \overline{x})^2 = 7.0621 \qquad \Sigma(y_i - \overline{y})^2 = 4.9040$$

$$\Sigma(x_i - \overline{x})(y_i - \overline{y}) = 5.7397$$

From these we can compute the estimates

$$\hat{\beta}_1 = \frac{5.7397}{7.0621} = .8127$$

$$\hat{\beta}_0 = 2.8290 - (.8127)(2.8315) = .5279$$

$$s^2 = \frac{4.9040 - (.8127)^2\,(7.0621)}{18} = .0133$$

$$s_{\hat{\beta}_0} = \sqrt{\frac{(.0133)(167.4099)}{(20)(7.0621)}} = .1252$$

$$s_{\hat{\beta}_1} = \sqrt{\frac{.0133}{7.0621}} = .0433$$

The least squares line then is

$$\hat{y} = .5279 + .8127x$$

This line and the original data are both given in Figure 10.1.4. Granted this estimated line, it then is possible to use a student's high school grade point average (x) to predict what his expected freshman year grade point average will be, even before he enters college.

Any time a sample of pairs of values is available, and it is desired to assume that the mean of Y changes linearly with x, it is wise to plot the points, as done in Figures 10.1.3 and 10.1.4. As long as the

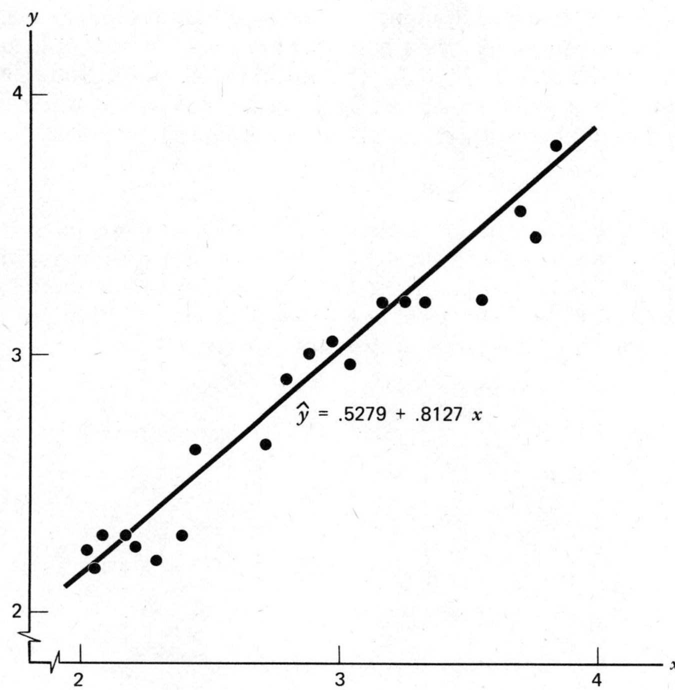

Figure 10.1.4 High school (x) and freshman (y) grade point averages for 20 students.

points do roughly lie on a straight line, then the regression model we have discussed in this section may be appropriate and may result in a useful fitted line ($\hat{y} = \hat{\beta}_0 + \hat{\beta}_1 x$) from which an expected value of Y may be computed for any given x (in the appropriate range). If, however, the plot of points again roughly lie on a line, but not a straight line, then a different sort of functional dependence between $E[Y]$ and x may be indicated. If the plotted points show no tendency to lie on any sort of line, and present a roughly circular scatter, then the mean of Y may not depend on x and least squares techniques may not be called for.

Exercise 10.1

1. Eight one-foot-tall Monterey pine trees were planted in similar controlled environments but were subjected to differing amounts of irrigation to simulate the effects of differing amounts of rainfall. At the end of one year their achieved heights were measured. The following table presents their achieved heights (y) at the end of one year, as well as the simulated amount of rainfall (x). Assume

$$E[Y] = \beta_0 + \beta_1 x$$

and compute the estimates of β_0 and β_1. Also estimate the variance of the population of growths, for a given amount of rainfall. What would you estimate as the expected height of a tree receiving 20 inches of rainfall?

Height y	Rainfall x
19	10
22	14
25	18
31	22
33	26
39	30
44	34
45	38

2. A professor gives two exams each time he teaches a course, a midterm and a final. Over the past few years, he has used his accumulated data to estimate an equation relating the score on the final exam (y) linearly to the score on the first exam (x). He has found, with $n = 100$, $\bar{x} = 67.6$, $\bar{y} = 74.3$, $\Sigma(x_i - \bar{x})^2 = 9703$, $\Sigma(y_i - \bar{y})^2 = 9947$, and

$$\Sigma(x_i - \bar{x})(y_i - \bar{y}) = 4813.$$

What is his estimated equation? Given a student that scored 78 points on the midterm, what is his expected score on the final?

3. Presented below are the weights (y) and heights (x) of the 10 Miss America contest winners between 1962 and 1971 inclusive,

Weight y (pounds)	Height x (inches)
118	65.5
115	65
124	66.5
124	66
115	67
116	66
135	69
125	67
110	65.5
121	68

(a) Assume that there is a linear relationship between the mean of y and x and compute the least squares estimates for β_0 and β_1.

(b) Using the estimates from (a), what would you estimate as the weight of a Miss America winner who was 5 feet $7\frac{1}{2}$ inches tall?

(c) Compute the standard errors for $\hat{\beta}_0$ and $\hat{\beta}_1$.

4. Presented below are the crude petroleum production totals for Kansas (y) and Louisiana (x), over the period 1965 to 1970 inclusive. Both are in millions of barrels. (A barrel is 42 gallons.)

Kansas Production y	Louisiana Production x
104.7	594.9
103.7	674.3
99.2	774.5
94.5	817.4
88.7	884.6
84.9	906.9

   (a) Assume there is a linear relationship between these two quantities and use least squares to estimate β_0 and β_1.
   (b) What are the standard errors for $\hat{\beta}_0$ and $\hat{\beta}_1$?

5. Let the Kansas production be denoted x, and the Louisiana production y, for the data given in question 4. Again assume $E[Y] = \beta_0 + \beta_1 x$ and estimate $\hat{\beta}_0$ and $\hat{\beta}_1$. Compute the standard errors of $\hat{\beta}_0$ and $\hat{\beta}_1$.

6. The following table gives the total United States gasoline production (y) and demand (x) for the years 1965 to 1970 inclusive. Both are in billions of 42-gallon barrels.

Gasoline Production y	Gasoline Demand x
1.73	1.76
1.82	1.83
1.87	1.87
1.97	1.98
2.06	2.07
2.14	2.16

   (a) Assume there is a linear relationship between these two quantities and use least squares to estimate β_0 and β_1.
   (b) Compute the standard errors of $\hat{\beta}_0$ and $\hat{\beta}_1$.

7. The total amount of ordinary life insurance policies in force for the years 1965 to 1969, inclusive, was 497.6, 539.0, 582.6, 630.4, and 678.9, respectively, all in billions of dollars.
   (a) Assume these amounts are, essentially, increasing linearly with time (the year). Use least squares to estimate β_0 and β_1. (Label 1965 year 0, 1966 year 1, etc., for ease of computation.)
   (b) Using the estimated relationship from (a), what would you guess as the amount of insurance in force in 1970, the next year? (The actual amount in force was 731.1.)

8. A random sample of 10 wage earners was selected to investigate the relationship between years of formal education (x) and salary level (y), in thousands of dollars. It was found that $\bar{x} = 12$, $\bar{y} = 10$, $\Sigma(x_i - \bar{x})^2 = 81$, $\Sigma(y_i - \bar{y})^2 = 130$, $\Sigma(x_i - \bar{x})(y_i - \bar{y}) = 97.2$.
    (a) Assume there is a linear relation between y and x and use this data to estimate β_0 and β_1.
    (b) Assume that the salary levels of all individuals with the same education are normally distributed. Estimate σ^2, the variance of each population.

9. Verify the fact that

    $$n\hat{\beta}_0 + (\Sigma x_i)\hat{\beta}_1 = \Sigma y_i$$

    $$(\Sigma x_i)\hat{\beta}_0 + (\Sigma x_i^2)\hat{\beta}_1 = \Sigma x_i y_i$$

    where

    $$\hat{\beta}_1 = \frac{\Sigma(x_i - \bar{x})(y_i - \bar{y})}{\Sigma(x_i - \bar{x})^2} \qquad \hat{\beta}_0 = \bar{y} - \hat{\beta}_1 \bar{x}$$

10. Show that the least squares line will always go through the point with coordinates $(\bar{x}, \bar{y})$.

11. Suppose we have a random sample of pairs (y_1, x_1), (y_2, x_2), ..., (y_n, x_n) and that all the x_i's are equal. Can you see any difficulties in estimating β_0 and β_1 in this case? (*Hint.* What are the values of $\bar{x}$ and $\Sigma(x_i - \bar{x})^2$?)

10.2 CONFIDENCE INTERVALS AND TESTS OF HYPOTHESES

The method of least squares may be employed to estimate β_0 and β_1, no matter what the form of the probability measure for Y, for a given x, so long as

$$E[Y] = \beta_0 + \beta_1 x$$

$$\sigma_Y^2 \text{ is constant for all } x$$

The resulting estimators have good properties, as mentioned in Section 10.1. To compute confidence intervals, or test hypotheses, about β_0 and β_1, though, it is necessary to assume that the probability measure for Y is of some definite form for each x. The most commonly assumed probability measure for Y is the normal. Under the assumption of normality it is quite easy to compute interval estimates for β_0 and β_1 and test hypotheses about their values. Throughout this section we shall assume the following:

> We have a sample of pairs (Y_i, x_i), $i = 1, 2, \ldots, n$. For each x_i, the corresponding Y_i has a *normal* probability measure with mean $\beta_0 + \beta_1 x_i$ and variance σ^2. The x_i's are given constants and the Y_i's are independent random variables. At least two of the x_i's are different.

With the above assumptions it can be shown that the probability measures of $\hat{\beta}_0$ and $\hat{\beta}_1$ are both normal as well. That is, over repeated samples of the same size, $\hat{\beta}_1$ will vary from one sample to another in accord with a normal probability measure, as will $\hat{\beta}_0$. As a consequence of this normality of $\hat{\beta}_0$ and $\hat{\beta}_1$, the results given in the following theorem can be shown to be true.

Theorem 10.1

With the assumptions quoted above

1. $\dfrac{\hat{\beta}_1 - \beta_1}{S_{\hat{\beta}_1}}$ and $\dfrac{\hat{\beta}_0 - \beta_0}{S_{\hat{\beta}_0}}$ each have the t distribution with $n - 2$ degrees of freedom; $\hat{\beta}_0$, $\hat{\beta}_1$, $S_{\hat{\beta}_1}$ and $S_{\hat{\beta}_0}$ are as defined in Section 10.1.

2. $\dfrac{\hat{Y} - (\beta_0 + \beta_1 x)}{S_{\hat{Y}}}$ has the t distribution with $n - 2$ degrees of freedom where

$$\hat{Y} = \hat{\beta}_0 + \hat{\beta}_1 x, \qquad S_{\hat{Y}} = S\sqrt{\frac{1}{n} + \frac{(x - \bar{x})^2}{\Sigma(x_i - x)^2}} .$$

The following examples illustrate the use of these results in computing confidence intervals and in testing hypotheses.

Example 10.2.1

Let us use the baby weight–age data presented in Example 10.1.1 to compute 95% confidence intervals for (a) β_1, (b) β_0, (c) the average weight of four-month-old babies. To do so we are making the assumption that the probability measure for the weights of babies, all the same age, is normal.

(a) To compute a 95% confidence interval for β_1, we can use the fact given in Theorem 10.1 that $(\hat{\beta}_1 - \beta_1)/S_{\hat{\beta}_1}$ has the t distribution with $n - 2 = 8$ degrees of freedom. From Table A.4 we find $t_{.975} = 2.306$; thus the probability is .95 that

$$-2.306 \le \frac{\hat{\beta}_1 - \beta_1}{S_{\hat{\beta}_1}} \le 2.306$$

This double inequality is equivalent to

$$\hat{\beta}_1 - 2.306\, S_{\hat{\beta}_1} \le \beta_1 \le \hat{\beta}_1 + 2.306\, S_{\hat{\beta}_1}$$

Thus, $L = \hat{\beta}_1 - 2.306\, S_{\hat{\beta}_1}$, $U = \hat{\beta}_1 + 2.306\, S_{\hat{\beta}_1}$ provide a 95% confidence interval for β_1. The data given in Example 10.1.1 gave $\hat{\beta}_1 = 19.19$ and $s_{\hat{\beta}_1} = 3.85$. The observed 95% confidence interval for β_1 then goes from

$$19.19 - 2.306(3.85) = 10.31$$

to

$$19.19 + 2.306(3.85) = 28.07$$

(b) To compute a 95% confidence interval for β_0, we again use the result given in Theorem 10.1. Since $(\hat{\beta}_0 - \beta_0)/S_{\hat{\beta}_0}$ has the t distribution with eight degrees of freedom, the interval from $L = \hat{\beta}_0 - 2.306\, S_{\hat{\beta}_0}$ to $U = \hat{\beta}_0 + 2.306\, S_{\hat{\beta}_0}$ will cover the true value of β_0 with probability .95. From the data given in Example 10.1.1, it was found that $\hat{\beta}_0 = 102.97$ and $s_{\hat{\beta}_0} = 14.46$. Thus, based on this sample we can be 95% sure that β_0 lies between

$$102.97 - 2.306(14.46) = \quad 69.63$$

and

$$102.97 + 2.306(14.46) = 136.31$$

(c) The true mean weight of four-month-old babies is, with our assumptions, $\beta_0 + 4\beta_1$; we would estimate this quantity by $\hat{\beta}_0 + 4\hat{\beta}_1 = 179.73$, the value of $\hat{y}$ on the regression line with $x = 4$. From Theorem 10.1.1, again,

$$\frac{\hat{Y} - (\beta_0 + 4\beta_1)}{S_{\hat{Y}}}$$

has the t distribution with eight degrees of freedom, where $\hat{Y}$ is the value on the regression line with $x = 4$. Manipulating this t variable just as we did above for β_1, we find the probability is .95 that $\beta_0 + 4\beta_1$ will lie between $L = \hat{Y} - 2.306 S_{\hat{y}}$ and $U = \hat{Y} + 2.306 S_{\hat{y}}$. From the data given earlier,

$$s_{\hat{Y}} = \sigma\sqrt{\frac{1}{n} + \frac{(x - \bar{x})^2}{\Sigma(x_i - \bar{x})^2}} = \sqrt{476.13}\ \sqrt{\frac{1}{10} + \frac{(4 - 3.3)^2}{32.1}}$$

$$= 7.41$$

Thus, given these sample results we are 95% sure that the true mean weight of four-month-old babies lies between $179.73 - 2.306(7.41) = 162.6$ and $179.73 + 2.306(7.41) = 196.8$.

The cases treated in the preceding example all concerned confidence intervals. As with the intervals discussed earlier for other parameters, to change the confidence coefficients $(1 - \alpha)$ we need to use different percentiles from the t distribution with the same degrees of freedom. The result given in Theorem 10.1 that was used to compute a confidence interval for the mean weight of four-month-old babies can be employed for any given value of x. The confidence interval for $E[Y]$ as x changes actually generates a "confidence band," centered about the estimated regression line that has a known probability of including the true regression line.

The following example illustrates the use of Theorem 10.1 in testing hypotheses.

Example 10.2.2

A precision scale is claimed to be accurate to the nearest 1000th of a pound, for objects whose true weight lies between 0 and 1/2 pound. Five objects, with "true" weights of .1, .2, .3, .4, and .5 pounds are available. Each is weighed twice on the scale, giving rise to 10 observed weights. These observed weights are given in Table 10.3.

Table 10.3 True Weights (x) and Scale Readings (y)

x	y
.1	.098
.1	.099
.2	.208
.2	.200
.3	.302
.3	.298
.4	.405
.4	.401
.5	.502
.5	.495

Let us assume that the scale reading (Y) is a normal random variable with mean

$$E[Y] = \beta_0 + \beta_1 x$$

where x is the true weight of the object, and with variance σ^2 for all x. For an ideal scale, of course, we should have $Y = x$, that is $\beta_0 = 0$ and $\beta_1 = 1$; in addition the variance σ^2 should be zero (i.e., every time the object is weighed, the same reading is obtained from the scale). We can see immediately that this scale is not perfect since it does not give an identical reading for repeated weighings of the same object. We shall use the observed data to test the hypotheses that $\beta_1 = 1$ and that $\beta_0 = 0$, with $\alpha = .05$. With the methodology available to us we are only able to test these two hypotheses separately. There is a procedure available to test both hypotheses at once, but limitations of space preclude our discussing it here.

To make these tests we again must make certain computations with the data. We find

$$\Sigma x_i = 3.0 \qquad \Sigma x_i^2 = 1.1 \qquad \Sigma(x_1 - \bar{x})^2 = 0.2$$

$$\Sigma y_i = 3.008 \qquad \Sigma y_i^2 = 1.1045 \qquad \Sigma(y_i - \bar{y})^2 = .19973$$

$$\Sigma x_i y_i = 1.11022 \qquad \Sigma(x_i - \bar{x})(y_i - \bar{y}) = .1998$$

Using these values, we can then compute

$$\hat{\beta}_1 = \frac{.1998}{.2} = .9990$$

$$\hat{\beta}_0 = .3008 - (.9990)(.3) = .0011$$

$$s^2 = \frac{.19973 - (.9990)^2 (.2)}{8} = .00001622$$

$$s_{\hat{\beta}_1}^2 = \frac{.00001622}{0.2} = .00008110$$

$$s_{\hat{\beta}_0}^2 = \frac{(.00001622)(1.1)}{10(.2)} = .00000892$$

Thus $s_{\hat{\beta}_1} = .00901$ and $s_{\hat{\beta}_0} = .00299$. Then, to test $H_0: \beta_1 = 1$ with $\alpha = .05$, we find $t_{.975} = 2.306$ with eight degrees of freedom. We should reject H_0 if $|\hat{\beta}_1 - 1| > ts_{\hat{\beta}_1} = (2.306)(.00901) = .0208$; since $|\hat{\beta}_1 - 1| = .001$ we would accept $H_0: \beta_1 = 1$.

To test $H_0: \beta_0 = 0$ with $\alpha = .05$, we should reject H_0 if $|\hat{\beta}_0| > ts_{\hat{\beta}_0} = (2.306) \times (.00299) = .00689$. Since $|\hat{\beta}_0| = .0011$, we also accept $H_0: \beta_0 = 0$. Thus, based on this data, even though the scale is not perfect (it does not have 0 variance) it does appear that the means of the distributions of repeated weighings shift with x as they should, that is, the sample results are consistent with $E[Y] = x$, as we would expect if the scale is to be of any use.

Exercise 10.2

1. Use the data given in Example 10.1.1 to test $H_0: \beta_1 = 16$ versus $H_1: \beta_1 > 16$, with $\alpha = .05$. If H_0 is true, it would imply that the average weight of babies increases 16 ounces (1 pound) per month between the ages of 1 and 6 months.

2. Use the data given in Example 10.1.2 to compute a 90% two-sided confidence interval for the average freshman grade point average of high school students having a 3.0 high school average.

3. Use the data given in exercise 10.1.1 to compute a 95% lower confidence interval for β_1.

4. Frequently regression techniques are used to discover whether there is a relationship between two variables. If a sample of n pairs are avail-

able, the least squares line can be computed, as well as $s_{\hat{\beta}_0}$ and $s_{\hat{\beta}_1}$. Then $H_0 : \beta_1 = 0$ versus $H_1 : \beta_1 \neq 0$ is tested. If H_0 is accepted, then based on the sample results obtained it appears that the mean of Y is $E[Y] = \beta_0 + 0 \cdot x = \beta_0$, a constant for all x, which implies that x is an extraneous variable that has no effect on the population of Y values. A sample of 11 beginning probability and statistics books yielded the following data. The number of pages in the book is Y and the number of letters in the title is x. Assume

$$E[Y] = \beta_0 + \beta_1 x$$

and test $H_0 : \beta_1 = 0$ for this data. (Also make the necessary normality assumptions and use $\alpha = .1$)

y	x
403	22
288	38
375	39
143	15
356	34
309	20
322	34
384	21
353	30
527	38
638	33

5. Before being accepted for pilot training, applicants are required to take an aptitude test that is divided into five parts each of which is scored separately. It is suspected that one part of the test is really immaterial in judging the abilities of potential pilots. To see if this was indeed the case, 20 pilots were selected and given a proficiency test; the scores of these 20 persons on the suspect portion of the aptitude test, taken before their training, were also available. Using y to denote the proficiency scores and x to denote the aptitude scores, we then have an (x, y) pair for each person. Assume, then, that $n = 20$. It was found that $\Sigma(x_i - \bar{x})^2 = 100$, $\Sigma(y_i - \bar{y})^2 = 73$, and $\Sigma(x_i - \bar{x})(y_i - \bar{y}) = 10$. Test the hypothesis that $\beta_1 = 0$, with $\alpha = .05$. Would you recommend deleting the suspect portion of the aptitude test, based on these sample results?

6. In cases similar to Example 10.2.2 (concerning data from scale readings), in which we accept $\beta_0 = 0$, it would then be logical to delete β_0 from the model and assume that

$$E[Y] = \beta x$$

a regression line that necessarily passes through the origin. It can be shown that the estimator of β becomes $\hat{\beta} = \Sigma x_i y_i / \Sigma x_i^2$; σ^2 is esti-

mated by $S^2 = \Sigma(y_i - \hat{\beta}x_i)^2/(n - 1)$ and the standard error of this new $\hat{\beta}_1$ is $S_{\hat{\beta}} = S/\sqrt{\Sigma x_i^2}$.
(a)   Use the scale data from Example 10.2.2 to estimate β.
(b)   Test $H_0: \beta = 1$ with $\alpha = .05$.

7.   You were asked in Exercise 10.1.10 to show that the usual least squares line (including β_0) goes through the point with coordinates $(\bar{x}, \bar{y})$. Show that this is not true for the regression line forced through the origin (see problem 6 above).

10.3 CORRELATION TECHNIQUES

In the previous two sections we have assumed that we had available a sample of pairs (x_i, y_i), $i = 1, 2, \ldots, n$ where the y values were selected at random from a population of Y values associated with the same x; the x's were assumed to be constants. Then, if the population mean values of the Y's changed linearly with the x values, we saw how we might estimate the linear relation. If we further assumed that each Y population is normal in form, it is straightforward to test hypotheses or compute confidence intervals.

There is a large body of problems in which it is reasonable to assume that *both* the y value and the x value in each pair are observed values of random variables. For example, suppose x is the observed number of people that smoked one or more packs of cigarettes, per day, during a given calendar year, and y is the observed number of people that died of a specific ailment, say lung cancer, during the same calendar year. Are x and y related over a series of years? Or suppose x is the measured IQ value for an individual and y is the measured mathematical aptitude score for the same individual. Are x and y related, over some population of people for which both x and y have been measured? If x is the interest rate charged prime investors by large banks, for a given week, and y is the average value of the Dow Jones Industrial Average closing values over the same week, do x and y tend to move together over a series of weeks?

In each of the cases mentioned, the question asked concerns the existence of a relationship between the two variables. The standard procedure applied to answer such questions (at least as a first cut) is to compute the *correlation coefficient* r for the pairs of measurements. This quantity is defined below.

Definition 10.3.1.   Assume (x_i, y_i), $i = 1, 2, \ldots, n$ is a set of measured pairs. The correlation between the two variables, x and y, is

$$r = \sum_{i=1}^{n} \frac{(x_i - \bar{x})(y_i - \bar{y})}{(n-1)s_x s_Y} = \frac{\Sigma x_i y_i - n\bar{x}\bar{y}}{(n-1)s_X s_Y}$$

where s_x is the standard deviation of the x measurements and s_Y is the standard deviation of the y measurements.

The following example illustrates the computation of r.

Example 10.3.1

Reproduced below are the number of letters in the title (x) and the number of pages (y) for each of 11 different beginning textbooks in probability and statistics. (This same data appeared in Exercise 10.2.4.)

x	y
22	403
38	288
39	375
15	143
34	356
20	309
34	322
21	384
30	353
38	527
33	638

We find, for this data

$$\Sigma x_i = 324 \qquad \bar{x} = 29.45 \qquad \Sigma x_i^2 = 10,260$$

from which we find $s_X = 8.47$. Similarly

$$\Sigma y_i = 4098 \qquad \bar{y} = 372.55 \qquad \Sigma y_i^2 = 1,689,166$$

so $\qquad s_Y = 127.47$. Then

$$\Sigma x_i y_i = 125,546 \text{ and}$$

$$r = \frac{\Sigma x_i y_i - n\bar{x}\bar{y}}{(n-1)s_X s_Y} = \frac{125,546 - 11(29.45)(372.55)}{10(8.47)(127.47)} = .449$$

The correlation coefficient r is a measure of linear association only; to help see why it is called a measure of linear association, notice that

$$r = \frac{\Sigma(x_i - \overline{x})(y_i - \overline{y})}{(n-1)s_X s_Y}$$

$$= \frac{\Sigma(x_i - \overline{x})(y_i - \overline{y})}{(n-1)s_X s_Y} \cdot \frac{s_X}{s_X}$$

$$= \frac{\Sigma(x_i - \overline{x})(y_i - \overline{y})}{\Sigma(x_i - \overline{x})^2} \cdot \frac{s_X}{s_Y}$$

when we recall that $s_X{}^2 = \Sigma(x_i - \overline{x})^2/(n-1)$. Thus, we have

$$r = \hat{\beta}_1 \frac{s_X}{s_Y}$$

where $\hat{\beta}_1$ is the least squares estimate of the slope of the relationship between $E[Y]$ and x, discussed in the preceding two sections. That is, if we were to assume that $E[Y]$ is *linearly* related to x, the correlation coefficient is equal to the estimated slope of the straight line, times the ratio of the two estimated standard deviations.

In particular, then, we can see that the signs of r and $\hat{\beta}_1$ must be identical, since s_X/s_Y must be positive; if $\hat{\beta}_1$ is negative so is r, if $\hat{\beta}_1$ is positive so is r. If $\hat{\beta}_1$ is positive, there appears to be a direct relation between $E[Y]$ and x; as x increases so does $E[Y]$. If $\hat{\beta}_1$ is negative, there appears to be an inverse relation between $E[Y]$ and x; as x increases, $E[Y]$ apparently decreases. Thus, a positive value for r indicates a direct linear relationship, and a negative value for r indicates an inverse linear relationship. A zero value for r indicates no linear relationship between the two variables, since $r = 0$ if and only if $\hat{\beta}_1 = 0$.

Example 10.3.2

The table below gives two recorded grades, x and y, for each of 10 students in a probability course. The x score is the midterm grade (50 is the maximum possible) and the y score is the final exam grade made by the same student (again 50 is the maximum possible).

Student Number	x	y
1	30.2	26.8
2	47.0	42.5
3	16.7	11.8
4	49.5	49.5
5	40	23.8
6	49	42.5
7	39	32
8	38.5	38
9	43.5	32.5
10	38	32

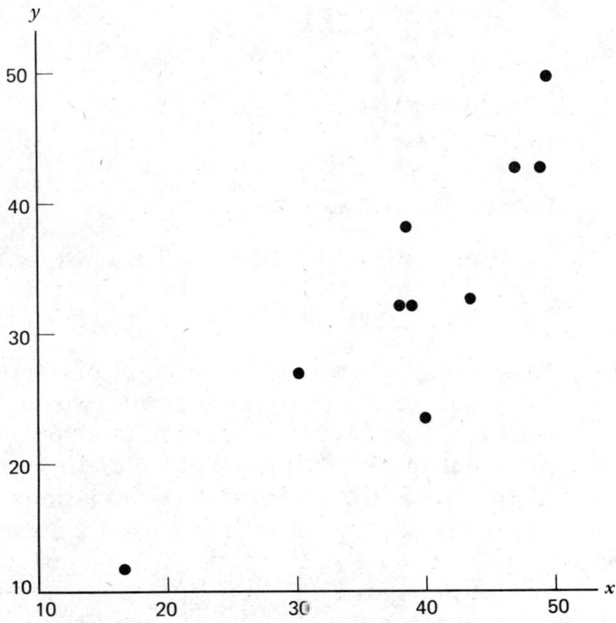

Figure 10.3.1 Exam scores for 10 students.

These values are also plotted in Figure 10.3.1. Notice from the figure that there is a tendency for y to be larger as x increases as we would expect; thus we might anticipate that r will be positive, indicating a direct relationship. The magnitude of r provides a measure of the strength of the relationship. We find, for this data,

$$\Sigma x_i = 391.4 \qquad \bar{x} = 39.14 \qquad \Sigma x_i^2 = 16{,}190.68$$

so
$$s_X^2 = \frac{16{,}190.68 - 10(39.14)^2}{9} = 96.81 \qquad s_X = 9.84$$

Similarly, $\Sigma y_i = 331.4$, $\bar{y} = 33.14$, $\Sigma y_i^2 = 12{,}034.92$, and

$$s_Y^2 = \frac{12{,}034.92 - 10(33.14)^2}{9} = 116.92 \qquad s_Y = 10.81$$

and $\Sigma x_i y_i = 13{,}829.42$. Thus,

$$r = \frac{13{,}829.42 - 10(39.14)(33.14)}{9(9.84)(10.81)} = .897$$

As we expected, r is positive because of the tendency for y to increase as x increases.

In both Examples 10.3.1 and 10.3.2 the value of r was a fraction. We might inquire then about the possible range of r. Can it ever exceed 1? The answer to this question is no; it is possible to show that in all cases $r^2 \leq 1$, which implies that r itself must also lie between -1 and 1 inclusive. It can also be shown that the only time $r = 1$ is when all the observed values (x, y) lie exactly on a straight line with positive slope; the only time that $r = -1$ is when all the observed values (x, y) lie exactly on a straight line with negative slope. As mentioned earlier, $r = 0$ indicates no linear relationship between x and y. Figure 10.3.2 presents graphs illustrating these facts. The closer that r gets to ± 1, the "closer" the points are to lying on a straight line.

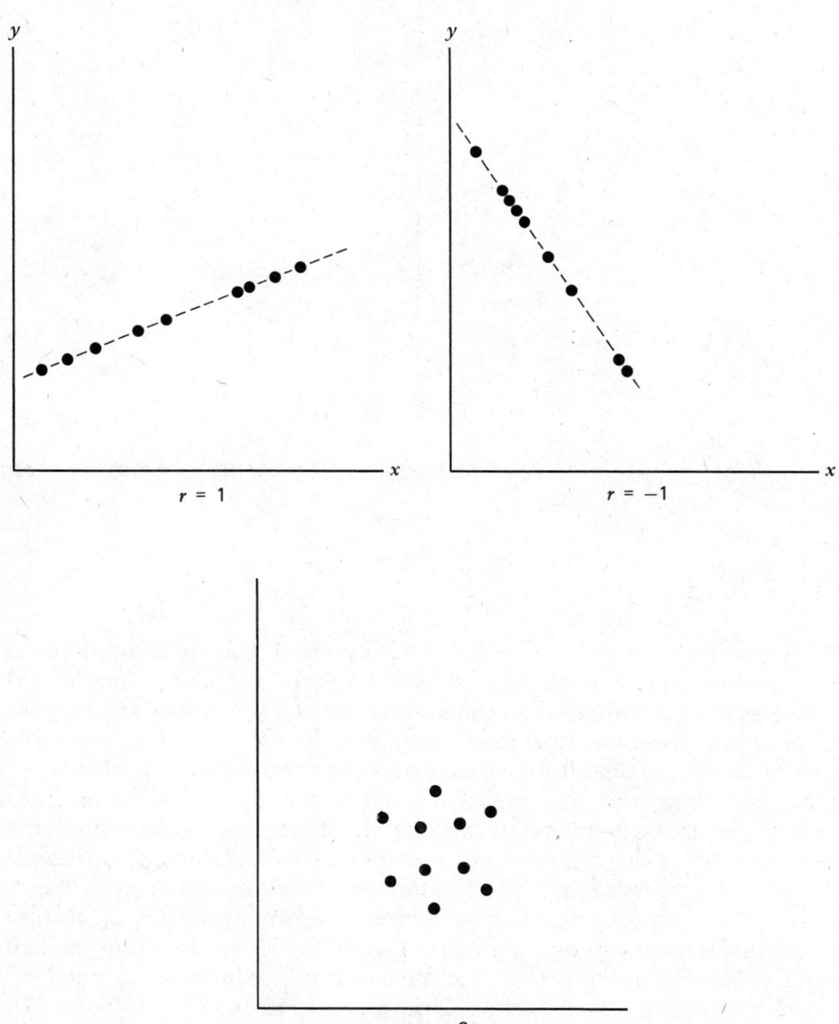

Figure 10.3.2   The extreme values for r.

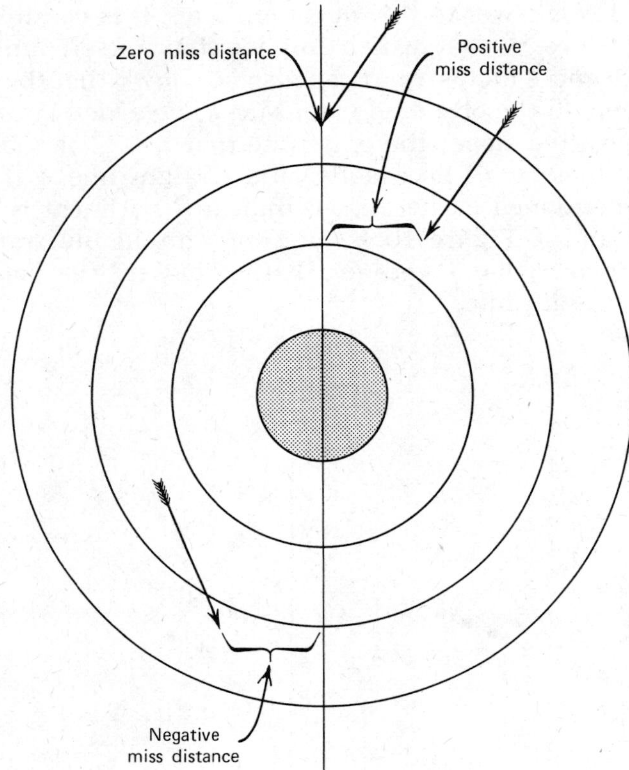

Figure 10.3.3 Miss distances are measured horizontally from the center line. (See Example 10.3.3.)

Example 10.3.3

This example shows that r is a measure only of linear association between two variables; it is possible for x and y to be perfectly related (nonlinearly), yet r may equal 0. Suppose we shoot an arrow at a bull's-eye; the target has a vertical line from top to bottom through its center. If the arrow lands to the right of the vertical line, we measure the miss distance positively (perpendicularly from the line) and if it lands to the left of the center line we measure the miss distance negatively. If the arrow lands right on the center line the miss distance is zero (Figure 10.3.3). Imagine we shoot three arrows at this target and find their miss distances to be $x_1 = 1$, $x_2 = 0$, and $x_3 = -1$, respectively. Let us define a second variable, y_i, to be the square of the corresponding x_i, for $i = 1, 2, 3$. Then certainly the y measurement is perfectly related to the x measurement, since we derived the y measurement from the x value. This gives the set of three pairs $(1, 1)$, $(0, 0)$, and $(-1, 1)$. It is easily verified that $r = 0$ when computed from these values, even though the y values were completely determined by the x

values. The relationship is, however, nonlinear, which is why we find $r = 0$. It is important to realize that r measures only the degree of *linear* association between two variables.

As discussed up to now, the correlation coefficient is a measure of the amount of linear association between two variables. In many social science applications, it is assumed that the (x, y) pairs that occur are a random sample selected from a population (the exam scores in Example 10.3.2 could be considered a sample from the population of all possible students for that course).

When we think of a population of pairs, it is also possible to define a population correlation coefficient ρ (computed in the same way as r, over the whole population rather than over just the sample). It would be natural then to use r, computed from the sample, to make inferences about the value of ρ in the population. Textbooks specializing in statistics for the social sciences describe many types of inference problems concerning ρ and their solutions (generally depending on r in some way).

We have the space here to discuss only the most basic of possible problems of inference about ρ; namely, how to test the hypothesis that ρ equals zero. If the underlying population of pairs has the bivariate normal distribution and $\rho = 0$ (this says that both the X population and the Y population are normal, with X and Y independent), then

$$\frac{r}{\sqrt{(1 - r^2)/(n - 2)}}$$

has the t distribution with $n - 2$ degrees of freedom. (n is the number of pairs in the sample.) Thus, if we reject $H_0: \rho = 0$ and accept $H_1: \rho \neq 0$ when

$$\frac{|r|}{\sqrt{(1 - r^2)/(n - 2)}} \geq t_{1-\alpha/2}$$

our probability of type I error is α.

Example 10.3.4

Let us assume the pairs of exam scores given in Example 10.3.2 are a random sample from the population of all students to take this course. Further, we assume the population of midterm scores is normal, as is the population of final exam scores. We want to test

$$H_0: \rho = 0 \qquad \text{versus} \qquad H_1: \rho \neq 0$$

with $\alpha = .05$. Since we have a sample of $n = 10$ pairs, we find $t_{.975} = 2.306$, with eight degrees of freedom. We previously computed $r = .897$; thus

$$\frac{r}{\sqrt{(1 - r^2)/(n - 2)}} = \frac{.897}{\sqrt{[1 - (.897)^2]/8}} = 5.739 > 2.306$$

and we reject H_0. Our probability of type I error using this rule is .05.

Exercise 10.3

1.   The examination scores discussed in Example 10.3.2 were selected from a class of 23 students. The 10 pairs given in the example were for the first 10 students in alphabetical order. The remaining 13 students in the class had the following scores:

x	y
38	44
41.5	24.5
41.5	38.5
29.8	16.5
39	38.8
44.5	38
37.5	33
41.5	24
30	34.8
24	20.8
39	47
21.5	15.8
29.7	19.8

(a)   Plot these pairs on a piece of graph paper and see how similar this plot is to Figure 10.3.1.

(b)   Compute r, the correlation coefficient for these 13 pairs.

2.   Combine the data from Example 10.3.2 with the data given above in problem 1 and compute the correlation coefficient r for the whole class.

3.   Use the data given in Example 10.3.1 to test $H_0: \rho = 0$ versus $H_1: \rho \neq 0$, with $\alpha = .05$, where ρ is the population correlation coefficient between the number of letters in the title and the number of pages in the book for the population of beginning textbooks in probability and statistics.

4.   Use the data in Exercise 10.1.4 to compute the correlation between the crude petroleum production of Kansas and Louisiana. Does r have the sign you would expect?

5.   In Section 10.2, Theorem 10.1, we saw that $(\hat{\beta}_1 - \beta_1)/S_{\hat{\beta}_1}$ has the t distribution with $n - 2$ degrees of freedom. Thus, to test $H_0: \beta_1 = 0$ versus

H_1: $\beta_1 \neq 0$, our test statistic would be $\hat{\beta}_1/S_{\hat{\beta}_1}$. Show, algebraically, that this test statistic is identical with

$$\frac{r}{\sqrt{(1 - r^2)/(n - 2)}}$$

the statistic used to test H_0: $\rho \neq 0$. Thus these two tests are identical; any sample for which we would accept $\beta_1 = 0$ we would also accept $\rho = 0$; conversely, if one is rejected, so is the other.

6. Use the data given in problem 3 of Exercise 10.1 to estimate the correlation between weight and height of Miss America contest winners.

7. Make the appropriate normality assumptions and test the hypothesis that heights and weights of Miss America contest winners are uncorrelated, versus the alternative that they are correlated, with $\alpha = .01$; use the data given in problem 3, Exercise 10.1.

8. Given a sequence of values, $x_1, x_2, \ldots, x_n$, define a new sequence $y_1, y_2, \ldots, y_n$ by

$$y_i = a + bx_i \qquad i = 1, 2, \ldots, n$$

We now have n pairs, $(x_1, y_1), (x_2, y_2), \ldots, (x_n, y_n)$.

Show that

$$r = 1 \quad \text{if} \quad b > 0$$

$$= -1 \quad \text{if} \quad b < 0$$

10.4 SUMMARY

One of the oldest, and still most frequently used, statistical techniques is the method of least squares. The procedure is appropriate when

(a) We have a population of Y values for each of several different values of an independent variable x.
(b) $E[Y] = \beta_0 + \beta_1 x$ (for the formulas presented in the text.)
(c) σ_Y^2 is the same for all x.
(d) The sample Y values are random samples from the populations.

The criterion then says that the values used to estimate β_0 and β_1 define the line having the smallest value for the sum of squares of the deviations between the observed y values and the values on the line. The resulting estimators, $\hat{\beta}_0$ and $\hat{\beta}_1$, are the best linear unbiased estimators for β_0 and β_1, no matter what the probability measure(s) may be for the Y populations.

If, in addition to the assumptions stated above, we assume that the probability measure for each Y population is normal, we may use the t distribution to construct confidence intervals for β_0 and β_1 and to test hypotheses about their values. We may also use the t distribution to construct confidence intervals for the mean value of Y for any given value of x, as well as test hypotheses about this mean value.

The correlation coefficient, r, is used as a measure of linear association between two variables. If the two variables are linearly related, perfectly, then $r = 1$ or $r = -1$, depending on whether y increases or decreases as x increases. If $r = 0$, the two variables are uncorrelated. Assuming the (x, y) values are actually an observed random sample from a bivariate normal population, a t statistic is available for testing that the population correlation equals 0. Other procedures of inference about ρ are also available.

Exercise 10.4

1. A random sample of $n = 9$ employed males was selected. Each was asked the number of years of formal education he had (x) and the amount of his annual salary (y). The responses are summarized below.

y (Annual Salary/$1000)	x (Years of Education)
12	7
11	10
14	14
15	16
27	18
32	17
19	8
8	9
20	15

   Assuming there is a population of salaries (Y's) for each x and that $E[Y] = \beta_0 + \beta_1 x$, σ_Y^2 is the same for all x, use this data to estimate β_0 and β_1.

2. Estimate σ_Y^2 for the case discussed in problem 1 above; also compute the standard errors for $\hat{\beta}_0$ and $\hat{\beta}_1$.

3. Making the appropriate normality assumption, compute a 90% confidence interval for β_1, for the data given in question 1.

4. The same automobile was run over the same measured 10-mile course a total of $n = 8$ times. Each time the auto was maintained at a constant speed over the 10-mile course; three times the speed was held at 40 miles per hour, two times at 50 miles per hour, and the remaining three times at 60 miles per hour. The amount of gasoline consumed was measured for each run (to the closest 1/100th of a gallon). The results are summarized below.

y (Gasoline Consumed)	x (Speed)
.61	40
.63	40
.63	40
.72	50
.71	50
.78	60
.78	60
.79	60

Assume the amount of gasoline consumed on one of these runs is a normal random variable with $E[Y] = \beta_0 + \beta_1 x$ and variance σ_Y^2, for all x.
(a)   Estimate β_0 and β_1.
(b)   Estimate σ_Y^2.
(c)   Estimate $\sigma_{\hat{\beta}_0}$ and $\sigma_{\hat{\beta}_1}$.

5. Use the data given in problem 4 to test $H_0: \beta_1 = 0$ versus $H_1: \beta_1 \neq 0$ with $\alpha = .01$.

6. Use the data given in problem 4 to compute a 95% confidence interval for the expected amount of gasoline this car will consume, over this course, at a speed of 55 miles per hour.

7. A random sample of $n = 7$ males were selected. The height (x) and weight (y) were recorded for each; these measurements are given below.

y (Weight in Pounds)	x (Height in Inches)
145	68
160	70
160	69
165	68
190	74
152	66
170	71

    (a) Estimate the regression of Y on x (i.e., estimate β_0 and β_1 assuming $E[Y] = \beta_0 + \beta_1 x$).

    (b) Compute the standard errors of $\hat{\beta}_0$ and $\hat{\beta}_1$.

8. Interchange the roles of x and Y in problem 7 and answer the questions posed there.

9. If we solve the equation

$$y = \beta_0 + \beta_1 x$$

for x in terms of y, we find

$$x = \frac{-\beta_0}{\beta_1} + \frac{y}{\beta_1} = \beta_0{}' + \beta_1{}'y$$

Why is not the value for $\hat{\beta}_0{}'$, computed in problem 8, equal to $-\hat{\beta}_0/\hat{\beta}_1$ where $\hat{\beta}_0$ and $\hat{\beta}_1$ are the values from problem 7?

10. Use the data given in problem 6, Exercise 10.1 to compute the correlation between United States gasoline production and demand, over the years given there.

11. Use the data given in problem 1 above to compute the correlation between annual salary and years of formal education, for the nine employed males discussed there.

12. Use the data given in problem 4 above to compute the correlation between the gasoline consumption and the driving speed for the eight trials discussed there.

CHAPTER 11
SOME NONPARAMETRIC PROCEDURES

All of the statistical procedures we have studied thus far in this book either made some assumptions about the population distribution that we sampled from, or depended on the central limit theorem for a normal approximation (with the exception of least squares estimation). If the actual form of the distribution was not the same as the one assumed, or if the sample size was not large enough, these techniques may be less accurate than desired. For example, the 95% confidence interval for a population mean may have a probability smaller than .95 of including the population mean if the underlying assumptions are not met; instead of a probability .01 of rejecting a hypothesis when it is true, using one of these procedures with $\alpha = .01$, we may in fact have a larger probability of making a type I error and, at the same time, the probability of a type II error may be larger than warranted. The techniques that we shall study in Section 11.1 generally make fewer assumptions about the form of the population distribution and, thus, can be expected to be more widely applicable. This greater applicability, though, is obtained only at a price; if, for example, the population distribution is really normal in form, then the techniques we have already discussed, where a normal population was assumed, are more efficient. They provide the greatest amount of information and the greatest sensitivity for the given sample size.

The procedures we have discussed in the previous chapters are called *parametric procedures*; they are generally concerned with estimating or testing hypotheses about underlying population parameters, assuming in general that the population has certain properties, such as a normal distribution. The techniques that we shall examine in Section 11.1 are called *nonparametric* or *distribution-free*. They are not necessarily concerned with estimating or testing hypotheses about means or variances and, within rather broadly defined limits, are appropriate no matter what the underlying distribution may be. The past 20 years has seen a great deal of effort expended in the study of these techniques. A very large number of them are available, having been designed for very specific purposes. We shall look briefly at a few of them that involve frequently occurring problems.

11.1 ESTIMATION OF PERCENTILES

Research workers in the social sciences (and other areas) are often concerned with estimating the *percentiles* of distributions. (The reader might like to review the discussion of percentiles in Sections 2.1 and 4.3.) If the population can be assumed to be *normal* in form, a sample can be used to estimate μ and σ^2; these estimates could then be used to estimate any percentile of interest since, for a normal distribution, the $100k$th percentile is given by $\mu + \sigma z_k$, where z_k is the $100k$th percentile of the standard normal distribution and may be found directly from Table A.2. If one is not willing to assume that the population distribution is normal, then this procedure, based on z_k from the standard normal table, should not be used. What, then, would be a reasonable procedure to use in estimating population percentiles?

Suppose we are given a random sample of n observations from a continuous population. It will be recalled that the $100k$th percentile of the population is the number t_k, such that $100k\%$ of the population values are smaller than it. Thus, using this notation, $t_{.5}$ is the population median, and $t_{.25}$, $t_{.5}$, $t_{.75}$ are the population quartiles, for example. The original sample values would, as usual, be denoted by $X_1, X_2, \ldots, X_n$, where the subscript indicates the order in which the values were drawn from the population. Let $X_{(1)}$ denote the smallest of the n numbers, $X_{(2)}$ the second smallest, etc., up to $X_{(n)}$, which denotes the largest of them. Note that here the subscript is used to denote the numerical magnitude, from smallest to largest, of the sample values, rather than the order in which the values were drawn. The quantities, $X_{(1)}, X_{(2)}, \ldots, X_{(n)}$, are called a *ranked array*, and are also referred to as the *order statistics* of the sample. It can be shown that on the average, over repeated samples, the proportion of population values less than or equal to $X_{(1)}$ is $1/(n + 1)$, *no matter what the form of the continuous distribution* in the population; similarly the proportion of population values no larger than $X_{(2)}$ is $2/(n + 1)$, the proportion no larger than $X_{(3)}$ is $3/(n + 1)$, up to the proportion no larger than $X_{(n)}$ is $n/(n+1)$. In short, the expected proportion of population values no larger than $X_{(j)}$, is $j/(n+1)$, for $j = 1, 2, \ldots, n$. (This does *not* say $E[X_{(j)}] = t_{j/(n+1)}$. Can you see the difference? Study this difference using a *triangular* density, such as that studied in Section 4.3). Thus, when sampling from a continuous population, a natural estimator to use for the $[j/(n+1)]$th population percentile is $X_{(j)}$. To estimate intermediate percentiles, we can use linear interpolation between the order statistics of the sample. The following example illustrates this.

Example 11.1.1
Assume that it is desired to estimate the median family income of families living in a large city. A random sample of $n = 9$ family incomes from this city are found to be 5562, 4164, 2031, 10,153, 164,761, 17,621, 14,300, 13,225, 6994 in the order in which they were collected. (Thus these are the values for $x_1, x_2, \ldots, x_9$, respectively.) Then the ranked array for the sample is 2031, 4164, 5562, 6994, 10,153, 13,225, 14,300, 17,621, and 164,761. (These are the values of the order statistics $x_{(1)}, x_{(2)}, \ldots, x_{(9)}$, respectively.) On the average for samples of size 9, the proportion of population values to the left of $x_{(5)}$ is $5/(9+1) = .5$; thus the middle sample value, $x_{(5)} = 10,153$, is our point estimate of the median family income for this city. Let us also use the sample values to estimate $t_{.25}$ and $t_{.75}$, the lowest and highest population quartiles. The estimate of $t_{.2}$ is $x_{(2)} = 4164$ and the estimate of $t_{.3}$ is $x_{(3)} = 5562$; using linear interpolation between these two quartiles our estimate of $t_{.25}$ is $\frac{1}{2}(4164 + 5562) = 4863$. Similarly, our estimate of $t_{.75}$ is

$$\frac{x_{(7)} + x_{(8)}}{2} = \frac{14,300 + 17,621}{2} = 15,960.50$$

To construct interval estimates of the percentiles of a continuous population we can reason as follows. Suppose we want a confidence interval for t_p, the $100p$th population percentile. (We are using p instead of k for a reason that will soon be apparent.) Each value in our random sample either does or does not exceed t_p. (Granted t_p is unknown, we do not know which specific ones do and which ones do not exceed t_p, but we can be assured that each value must be either on one side of t_p or the other.) The probability is p that a value selected at random from the population is smaller than t_p, the subscript on t. Thus, from the definition of percentiles, each element in the random sample is a Bernoulli trial with probability p of success (calling success the outcome that the element is smaller than t_p). The kth order statistic, $X_{(k)}$, is less than or equal to t_p if and only if at least k sample values are less than or equal to t_p; if fewer than k sample values are smaller than t_p, then $X_{(k)} > t_p$. Thus, the probability that X_k is smaller than t_p is given by the probability that a binomial random variable, with parameters n and p, is at least as big as k. That is,

$$P(X_{(k)} \le t_p) = \sum_{j=k}^{n} \binom{n}{j} p^j q^{n-j}$$

Thus $X_{(k)}$ provides a lower confidence limit for t_p with the confidence coefficient given by the value of the above sum. For specific values of p and k this can be evaluated from our binomial tables.

Example 11.1.2

Let us consider the income data given above in Example 11.1.1 and use it to evaluate some lower confidence limits for $t_{.5}$ ($p = .5$), the population median. From the reported data, $x_{(2)} = 4164$; from Table A.1 with $n = 9$, $p = .5$, we find the probability that the binomial variable is 2 or greater is $.0703 + .1641 + \ldots + .0176 + .0020 = .9804$. Thus, based on this sample we can be 98% sure that the median income in this city is at least \$4,164. We also have $x_{(3)} = 5562$ and, from Table A.2, the probability that a binomial random variable with $n = 9$, $p = .5$ is 3 or greater is

$$.1641 + .2461 + \ldots + .0176 + .0020 = .9101$$

We are 91% sure the median income is at least as large as $x_{(3)} = \$5562$. By using different order statistics we get different values for the confidence coefficient.

The order statistics can also be used to construct two-sided confidence intervals for percentiles. Again using t_p to represent the $100p$th percentile, we can show that

$$P(X_{(k)} \le t_p < X_{(m)}) = \sum_{i=k}^{m-1} \binom{n}{i} p^i q^{n-i}$$

where $X_{(k)}$ is the kth order statistic and $X_{(m)}$ is the mth order statistic from the same sample and $m > k$.

Example 11.1.3

Let us use the sample data from Example 11.1.1 again, this time to examine two-sided confidence intervals for $t_{.4}$, the 40th percentile of the population of incomes. The sample size again is $n = 9$ and to construct intervals for $t_{.4}$, we use $p = .4$; from Table A.1 we find, for example,

$$\sum_{i=1}^{8} \binom{9}{i} (.4)^i (.6)^{9-i} = .9896$$

$$\sum_{i=2}^{7} \binom{9}{i} (.4)^i (.6)^{9-i} = .9257$$

$$\sum_{i=2}^{6} \binom{9}{i} (.4)^i (.6)^{9-i} = .9045$$

Then we can be 99% sure the interval from $x_{(1)} = 2031$ to $x_{(9)} = 164{,}761$ includes $t_{.4}$; we are also 93% sure that the interval from $x_{(2)} = 4164$ to $x_{(8)} = 17{,}621$ covers $t_{.4}$, and we are 90% sure the interval from $x_{(2)} = 4164$ to $x_{(7)} = 14{,}300$ includes $t_{.4}$. By taking different pairs of order statistics, we get different values for the confidence coefficient.

Exercise 11.1

1. The population *terciles* cut the population distribution into thirds; denote them by $t_{.33}$ and $t_{.67}$. For the family income data in Example 11.1.1, estimate these two terciles. What are the estimates of the population *deciles* $t_{.1}, t_{.2}, \ldots, t_{.9}$, for this data?

2. For the income data given in Example 11.1.1 compute $\bar{x}$ and s. Estimate $t_{.25}$, $t_{.5}$ and $t_{.75}$ by $\bar{x} - .67s$, $\bar{x}$ and $\bar{x} + .67s$, the quantities that would be appropriate if we assumed the incomes in the population were normally distributed. Both $\bar{x}$ and s are greatly affected by the one extremely large income in the sample.

3. The number of tornadoes reported in the United States from 1966 to to 1970, respectively, were 570, 912, 660, 604, and 647. Assuming these values to be a random sample from a conceptual (idealized) continuous population, what would you estimate as the median number of tornadoes in the United States, per year? Construct a 93.76% confidence interval for the population median number of tornadoes per year.

4. In a random sample of 11 college students each was given a physical dexterity test. The lengths of time required for them to complete the test were 22.1, 15.8, 26.4, 20.2, 17.3, 18.1, 22.3, 24.7, 24.0, 19.3, and 16.5 minutes, respectively.
    (a)   Use this data to estimate the median time required to complete this test, in the population of college students.
    (b)   Estimate $t_{.1}$ and $t_{.9}$.
    (c)   What is an upper 84% (approximately) confidence limit for $t_{.2}$?

5. In five successive years, the numbers of days until the first hurricane struck the United States (counting from January 1) were 173, 192, 161, 169, and 154, respectively.

(a) What would you estimate as the median number of days until the first hurricane strikes the United States?

(b) Give a 94% confidence interval for the median number of days.

6. Assume you and 19 other students take a mathematics achievement test and you score 527, the best of your group. What percentile is your score an estimate of, for the population at large, assuming the scores made by the 20 people in your group constitute a random sample from the population?

7. Table 2.3 presents a ranked array of the weights of a sample of 60 male college students. Assume these students were randomly selected from the population of all male college students.

(a) Estimate the median weight in the population.

(b) Estimate the 10th and 90th percentiles in the population.

(c) Construct a 90% two-sided confidence interval for $t_{.5}$. (*Hint*. With $n = 60$ and $p = .5$, remember that the binomial distribution is well approximated by the normal.)

8. The sample mean of the weights of the 60 male college students presented in Table 2.1 is $\bar{x} = 192.9$ and the standard deviation is $s = 20.8$. Assume that the weights of male college students are normally distributed.

(a) What is your estimate of $t_{.5}$?

(b) Estimate $t_{.1}$ and $t_{.9}$ and compare these values with those derived in problem 7(b) above.

9. To derive an upper confidence limit for t_p, show that

$$P(t_p < X_{(m)}) = \sum_{i=0}^{m-1} \binom{n}{i} p^i q^{n-i}$$

(*Hint*. The mth order statistic exceeds t_p if and only if $m - 1$ or fewer sample observations are less than or equal to t_p.)

11.2 SOME NONPARAMETRIC TESTS

There are quite a few nonparametric tests of hypotheses. They have been designed to test a large variety of different hypotheses in a variety of contexts. We have the space to discuss only a few.

11.2.1 Sign Test

The *sign test* is one of the simplest examples of a nonparametric test and can even be applied without numerical measurements in certain circumstances, as illustrated in Example 11.2.1 below. The sign test is useful in comparing two populations, based on a sample of pairs, where one element from each pair is from population 1 and the other is from population 2. (This situation is just like the one assumed in the paired t test, discussed in problem 12, Exercise 9.2). Given two sample values, either the value from population 1 exceeds the value from population 2 or vice versa; granted the two population medians are equal, these two possibilities should occur equally frequently. Thus, if the two populations have the same median, the probability is 1/2 that the value from population 1 will be larger. We replace each pair by a plus sign (+) if the population 1 value is larger, and replace the pair by a minus sign (−) if it is not. Then Y, the number of pluses, is a binomial random variable with parameters n (the number of pairs) and $p = 1/2$ under the assumption that the two medians are equal. If Y is either very large or very small, the hypothesis of equal medians is rejected (for a two-sided test). The following example illustrates this test.

Example 11.2.1

Two exterior metal paints are to be compared for durability. To make the comparison 15 metal rods, each two feet long, are used. One foot of each rod is painted with paint A and the other half of the same rod is painted with paint B. Then all 15 rods are exposed to a variety of weather conditions and, at the conclusion of the experiment, a judgment is made about which paint has provided the best protection for each rod. If paint A is judged better, a plus is scored and if paint B is judged better, a minus is scored. Following the judgments we have 13 +'s and 2 −'s (A was judged better on 13 of the 15 rods). Should we accept $H_0: m_A = m_B$ versus $H_1: m_A \neq m_B$ with $\alpha = .05$? The random variable Y, the number of +'s in the sample, is binomial with $n = 15$, $p = 1/2$ if H_0 is true. From Table A.1 we see that

$$P(Y \leq 3) + P(Y \geq 12) = .0352$$

$$P(Y \leq 4) + P(Y \geq 11) = .1186$$

Thus we cannot use $\alpha = .05$ with a two-tailed test because of the discreteness of the binomial distribution. Thus, we shall use the value for α that is closest to, but no larger than .05, namely .0352. Since we observed $y = 13$, we would reject H_0; the probability we are wrong in this decision, using this rule, is .0352.

11.2.2   Population Percentile Test

In the previous section we saw a method for deriving confidence intervals for population percentiles. This same procedure can be used to test a hypothesis about a population percentile, t_p. Suppose, based on a sample of size n from a continuous population, we can construct a $100(1 - \alpha)\%$ confidence interval for t_p. Then, to test the hypothesis that t_p equals some specific value, we need merely take our sample, construct the $100(1 - \alpha)\%$ interval and, if the hypothesized value for t_p falls within the interval, accept the hypothesis; otherwise we should reject the hypothesis. This test has a probability of type I error equal to α, 1 minus the confidence coefficient used in constructing the interval. Thus, the intervals described in Section 11.1 for percentiles can, in fact, be used to test hypotheses about the values for the population percentiles. Let us examine this test in greater detail in the following example.

Example 11.2.2

The nationwide median score on an abstract reasoning ability test is known to be 200. A random sample of 11 students selected from a large school district made scores of 159, 202, 137, 144, 169, 192, 213, 180, 162, 152, and 171, respectively. Should we accept the hypothesis that the median score in the school district is also 200, versus the alternative that it is less than 200? We can make this type of one-sided test by computing a one-sided confidence interval for $t_{.5}$, the school district median score and then checking to see whether the hypothesized value 200 is covered by the interval. From Table A.1, with $n = 11$ and $p = .5$, we find that $X_{(9)}$ provides an upper confidence limit for the median with confidence coefficient .9673; that is, that probability is .9673 that $t_{.5} < X_{(9)}$. For this sample we find $x_{(9)} = 192$ and, thus, since 200 does not belong to this one-sided interval, we would *reject* the hypothesis that the school district median is 200 and conclude it is some smaller number (our point estimate of it is 169). The probability we are wrong (type I error) is $1 - .9673 = .0327$.

11.2.3 Wilcoxon-Mann-Whitney Test

In Section 9.1 we studied a t test of the hypothesis that two population means are equal; this test was based on the assumption that both of the two populations were normal and that their variances were equal. The commonly used nonparametric test for this same case was first described by Wilcoxon and later extended by Mann and Whitney. Thus it is commonly called the *Wilcoxon-Mann-Whitney test*. This test proceeds as follows. Suppose we are given a random sample of size n_1 from one continuous population and a sample of size n_2 from a second continuous population; we want to test that the two population distributions are the same (which would imply that their means and variances are equal, in particular). Then, let Z_1 be the number of values from population 1 that are smaller than the *smallest value* from population 2, let Z_2 be the number of values from population 1 that are smaller than the *second smallest value* from population 2, etc.; Z_{n_2} then, is the number of values from population 1 that are smaller than the *largest sample value* from population 2. The statistic used to test equality of the two distributions is U, the sum of these Z_j's $U = \sum_{j=1}^{n_2} Z_j$. Amazingly, if n_1 and n_2 are both larger than 8, the exact distribution of U is well approximated by a normal distribution with mean

$$\mu_U = (n_1 n_2 - 1)/2$$

and variance $\sigma_U^2 = n_1 n_2 (n_1 + n_2 + 1)/12$; to have a type I error equal to α, the hypothesis that the two population distributions are the same is rejected if $|U - \mu_U|/\sigma_U \geq z_{1-\alpha/2}$. The following example illustrates this test.

Example 11.2.3

To compare the effectiveness of two different methods of teaching the same material, two classes, with 15 students each, have been selected so that the previous experience and abilities of the two classes are essentially equal. One class will be taught using method I and the other taught using method II; at the end of the period of instruction, each student from both classes will take an achievement test; (thus, we are assuming the scores made by the students in class 1 are a random sample of the achievements recorded for the conceptual population of all students taught using method I, with a similar assumption about the scores made by the students in class 2). The Wilcoxon-Mann-Whitney test can then be used to test the hypothesis that these two conceptual populations are equal. The achievement scores are recorded in the following table.

Class 1   125, 97, 143, 100, 89, 74, 94, 115, 78, 52, 111, 91, 85, 70, 104

Class 2   82, 116, 101, 93, 79, 110, 98, 138, 121, 127, 107, 88, 90, 132, 120

The U statistic, defined above, will be approximately normally distributed with mean $\mu_U = [(15)(15) - 1]/2 = 112$ and variance $\sigma_U^2 = (15)(15)(31)/12 = 581.25$; thus $\sigma_U = \sqrt{581.25} = 24.1$. To have $\alpha = .05$, we will reject the hypothesis if $|U - 112|/24.1 \geq 1.96$. Perhaps the easiest way to compute the observed value of U for these samples is to combine the two sets of observed scores in a ranked array, where those scores from class 2 have been underlined. This is presented below:

52, 70, 74, 78, 79, 82, 85, 88, 89, 90, 91, 93, 94, 97, 98, 100, 101, 104, 107, 110, 111, 115, 116, 120, 121, 125, 127, 132, 138, 143

Once the combined ranked array is constructed, it is a simple matter to count the number of scores from class 1 that are smaller than the smallest class 2 score of 79 (this number is 4), the number of class 1 scores smaller than the second smallest class 2 score of 82 (also 4), etc., up to the number of class 1 scores smaller than the largest class 2 score of 138 (14); we then simply add these numbers together to get the observed value of U. Thus, for these two samples,

$$u = 4 + 4 + 5 + 6 + 7 + 9 + 10 + 11 + 11 + 13 + 13 + 13 + 14 + 14 + 14$$

$$= 148$$

Since we find, then, $|148 - 112|/24.1 = 1.49$, we would accept the hypothesis, with $\alpha = .05$, that the two conceptual distributions are the same. Had we chosen $\alpha = .15$, we would have rejected equality since $z_{1-\alpha/2} = z_{.925} = 1.44$.

11.2.4 Test for Randomness

We have, in discussing every technique covered in this book, assumed that the sample values available were a random sample of a population random variable. There are many ways in which the "randomness" of the sample values may be tested. Let us discuss the distribution of the number of *runs* in a random sample. This distribution is frequently used to test the hypothesis that the sample values, in the order in which they are chosen, are consistent with the assumption that their given ordering occurred by chance.

First, let us define a *run*. Given any sequence, each of whose positions is filled by one of two different symbols, an unbroken succession of the same symbol is called a run. For example, in the sequence

$$s\,s\,s\,f\,f\,s\,s\,s\,s\,s\,f\,s$$

we have three runs of s's and two runs of f's, for a grand total of five runs. Given a sequence of $n_1 + n_2$ positions filled with n_1 symbols of one type and n_2 symbols of a second type, it is possible to derive the distribution of the total number of runs that will occur, if the symbols are arranged in a random order. If we find either an extremely large number of runs or an extremely small number of runs, we may want to reject the assumed randomness of the sequence.

Thus, suppose we have a sequence of $n_1 + n_2$ positions, filled with

n_1 symbols of type 1 and n_2 symbols of type 2. Let R be the total number of runs to occur in the sequence, if the symbols have been randomly allocated to the positions. As long as both n_1 and n_2 are at least 10, then the distribution of R is well approximated by a normal distribution with mean $\mu_R = 2n_1 n_2/n$ and variance $\sigma_R^2 = 4n_1^2 n_2^2/n^3$, where $n = n_1 + n_2$. Then, if we wanted to reject randomness with either too many or too few runs, we should reject the randomness of the sequence if

$$\frac{|R - \mu_R|}{\sigma_R} \geq z_{1-\alpha/2}$$

where $z_{1-\alpha/2}$ is chosen from the standard normal table; the (approximate) probability of a type I error then is α. The following example illustrates this test.

Example 11.2.4

One of the most frequent uses of this test of randomness is in the control of industrial production lines. Thus, successive items selected from the production line are either nondefective (s) or defective (f), and we would expect these two symbols to be randomly intermixed; if they are not randomly intermixed, it may indicate a problem in production that can and should be corrected. Thus, suppose we inspect 50 successive items, of which we find $n_1 = 40$ are nondefective and $n_2 = 10$ are defective; furthermore, we observe that the first 20 were nondefective, followed by 5 defectives and then again by 20 nondefectives, followed by 5 defectives, for a grand total of $2 + 2 = 4$ runs. With $\alpha = .05$, should we reject the randomness of this sequence? Here we have $n_1 = 40$, $n_2 = 10$, and $n = 40 + 10 = 50$. Thus

$$\mu_R = \frac{2(40)(10)}{50} = 16 \qquad \sigma_R^2 = \frac{4(40)^2(10)^2}{(50)^3} = 5.12$$

and $\sigma_R = \sqrt{5.12} = 2.26$. Then to have $\alpha = .05$, we should reject randomness if $|R - \mu_R|/\sigma_R \geq 1.96$. Since we find $|4 - 16|/2.26 = 5.31$, we should reject the hypothesis that the defectives and nondefectives in the sample were randomly intermixed; our probability of type I error is .05.

Exercise 11.2

1. For the data given in problem 4, Exercise 11.1, would you accept the hypothesis that the median time to complete the test is 24.5 seconds, with $\alpha = .066$, versus the alternative that $t_{.5} \neq 24.5$?

2. Eight trees of the same kind were toppled in the same forest by the same storm. By counting the annual rings on each, their ages were estimated to be 15, 7, 8, 12, 26, 24, 18, and 22 years, respectively. With $\alpha = .2890$, would you accept the hypothesis that the median age of trees of this type toppled by storms of this magnitude is 20 years?

3. It has often been said that the atomic bomb disturbs the upper atmosphere and that it has changed the weather patterns in the world, for the worse in many cases. In the 10 years from 1936 to 1945 the numbers of tornadoes reported in the United States were 151, 147, 213, 152, 124, 118, 167, 152, 169, and 121, respectively; in the 11 years from 1946 to 1956 the numbers of tornadoes were 106, 165, 183, 249, 199, 272, 236, 437, 549, 593, and 532, respectively. Assume these are two random samples, one of size $n_1 = 10$ from one population and the other of size $n_2 = 11$ from a second. Use the Wilcoxon-Mann-Whitney statistic to test the hypothesis that the two population distributions are equal, with $\alpha = .01$. Can you think of any other reasons for the differences here, besides the fact that the samples were before and after the bomb?

4. (Continuation of question 3). The numbers of deaths reported, caused by tornadoes, in the United States from 1936 to 1945 were 552, 29, 183, 87, 65, 53, 384, 58, 275, and 210, respectively. From 1946 to 1956 the numbers of deaths were 78, 313, 140, 212, 70, 34, 230, 516, 35, 125, and 83, respectively. Again use the Wilcoxon-Mann-Whitney statistic to test the hypothesis that these two samples were selected from the same population with $\alpha = .01$. Why might there be such a big difference between questions 3 and 4?

5. Twenty pairs of brothers were selected for a study of smoking habits and its effect on age at death. In each pair, one of the two brothers smoked cigarettes and the other did not. It was found that the smoker lived to an older age in four of the pairs and that the nonsmoker lived to an older age in the remaining 16. Would you accept $H_0: m_S \geq m_N$ versus $H_1: m_S < m_N$ with $\alpha = .01$, where m_S is the median age at death for smokers and m_N is the median age at death for nonsmokers? (Adapt the sign test mentioned in the text for a one-sided alternative.)

6. The distribution of the number of runs can also be used for testing the randomness of samples selected from populations of other than the defective-nondefective Bernoulli type. Given a sample of size n from a population, let m be the sample median. With the observations kept in the order in which they were received, label each value that exceeds m with an s and all the others by an f, thus generating again a sequence of s's and f's. If the sample is really random, then again the number of runs observed should be neither too small nor too large. Question 3 presents the numbers of tornadoes reported in the United States over a 21-year period, in their order of occurrence. Use this run

test to test the hypothesis that these values are a random sample from a population, with $\alpha = .01$.

7. The distribution of the number of runs can also be used to test the hypothesis that two samples were selected from the same population. To do this, the values in one sample are all represented by s's and those in the other sample by f's. The two samples are combined, ranked in order of magnitude, and then their numerical values are replaced by these two symbols. If the two samples were selected from the same population, we would expect quite a bit of intermingling of s's and f's in this combined ranking; thus we would expect a rather large number of runs to occur. On the other hand, if the two populations are not the same, we would expect the s's to congregate together and the f's to congregate together, leading to a small number of runs. Thus we would only want to reject the hypothesis that the two populations are equal if we find a small number of runs in the combined ranking (this is a one-tailed test). Use this test of the hypothesis that the numbers of tornadoes reported in the United States from 1936 to 1945, has the same distribution as the numbers reported from 1946 to 1956, with $\alpha = .01$. (The data is given in question 3 above.)

8. Use the run test of randomness, described in problem 6, to test that the weights of the winning Miss America contestants, given in question 10, Exercise 2.1, are a random sample from a population with $\alpha = .05$.

9. Using the weights of the 60 male college students, presented in Section 2.1 in the order in which they were collected, test the hypothesis that they do constitute a random sample from a population, using the run test described in problem 6 above with $\alpha = .1$.

10. Apply the run test of the hypothesis that the distributions are the same, with $\alpha = .1$, described in problem 7 above, to the data collected for the two teaching methods described in Example 11.2.3.

11.3 SUMMARY

Nonparametric procedures may be used for random samples from any continuous population. They do not depend on the underlying population probability measure being normal. If, however, the population probability measure is normal, then techniques based on that assumption are more sensitive and will do a better job.

The order statistics of a sample provide natural estimates for population percentiles. They may also be used in the construction of interval estimates of population percentiles and in testing hypotheses about population percentiles.

The binomial probability measure occurs naturally in many non-parametric procedures. It is used for constructing interval estimates of population percentiles and for testing hypotheses about these percentiles; it is also the basis for the sign test and occurs in many other procedures we do not have the space to describe. The Wilcoxon-Mann-Whitney statistic provides an excellent test that two population distributions are equal. The number of runs observed in a sample can be used to test the hypothesis that the sample elements were selected at random from the population.

Exercise 11.3

1. To compare how well its brand sells in two different geographic areas (east and west), a toothpaste manufacturer selects 10 retail stores in each area. The stores are purposely paired together based on the store location and the size of the store, thus giving 10 pairs of stores, one in the east and one in the west. Over the same 30-day period it was found that nine out of the 10 western stores sold more of this manufacturer's brand than did their eastern counterparts. Would you accept the hypothesis that the median number of its sales (for a 30-day period) in the west is no greater than its median sales for the east, with $\alpha = .02$?

2. A sample of 12 school children from families on welfare were selected, as was a sample of 15 school children from families not on welfare; all these families lived in the same school district. During the school year the number of days the 12 welfare children were absent were 18, 22, 14, 12, 13, 21, 20, 19, 27, 41, 25, and 28, respectively. The number of days the other 15 children were absent during the same year were 27, 10, 19, 6, 4, 53, 22, 31, 11, 2, 12, 15, 20, 29, and 30, respectively. Would you accept the hypothesis the two populations of days absent are the same, with $\alpha = .05$? (Use the Wilcoxon-Mann-Whitney test, after removing all ties between a person in the welfare group with a person in the other group.)

3. Using the run test, would you accept the hypothesis that the number of days absent from school for welfare children in question 2 is, in the order reported, a random sample from a population, with $\alpha = .1$?

4. Answer question 3, using the data for the nonwelfare children.

5. Estimate the median number of days absent for welfare children using the data given in question 2. Do the same for the nonwelfare children.

6. The number of sales made by a large bookstore of a novel on the best-seller's list, per day, for 10 days were 125, 210, 117, 194, 176, 125, 188,

142, 203, and 154, respectively. Assume these values are a random sample from a continuous (idealized) population and compute a two-sided 97.84% confidence interval for the median sales per day of a bestselling book, by such a large store.

7. If the conceptual population mentioned in problem 6 has a symmetric probability measure, then the population median is the same as the population mean. Assuming the values reported in problem 6 are a random sample from a normal population, compute a 97.84% confidence interval for the population mean and compare this interval to the answer given in that problem.

APPENDIX

PROBABILITY THEORY

There is much evidence to indicate that men have played and enjoyed many types of games for thousands of years. Some of the most popular of these games have involved special equipment that inject elements of randomness into the play. Dice, playing cards, and spinners, for example, have long been used in such games; they insert a certain unpredictability into the play, adding to the interest of the participants. Indeed, as long as games have been around, so has gambling or wagering on who will win the game. More than likely, the person who enjoys a record of frequently winning such wagers has more than luck on his side. He may very well have tried to analyze the chance mechanism used in the play and discovered that certain regularities are sure to occur over repeated plays, in spite of the unpredictable behavior on individual games. This regularity may then be used to his own advantage.

Over the years casinos have emerged as places where anyone is free to gamble on certain games of chance, placing bets against the owners of the casino. The men who ran (and still run) such places had noted that certain bets could be counted on to make money over long series of repetitions of the game played. As long as they were on the right side of the bet, they could essentially be sure of coming out ahead in the long run. This sort of discovery apparently predates probability theory, the mathematical study of uncertainty, by hundred of years; casinos and gambling houses have been around much longer than have any records of a theory of probability.

A series of letters between the two great French mathematicians,

Pascal and Fermat, over 300 years ago, is generally taken as the beginning of the theory of probability. They both knew a French nobleman of the time, the Chevalier de Meré. The Chevalier enjoyed games of chance at Monte Carlo and had tried unsuccessfully himself to apply mathematics in describing his chances of winning at certain games. One game that he tried to analyze involved repeated rolls of a pair of dice. The bettor was allowed to roll a pair of dice repeatedly. If the bettor got one or more double-6's in 24 rolls, he won the bet; if he did not get one or more double-6's in the 24 rolls, he lost the bet. The Chevalier's analysis led him to believe that the bettor should win this game more than half the time, but his practical experience in playing it did not bear this out. He asked Pascal and Fermat to examine his analysis of the game; this (and other questions) started a famous exchange of letters between the two mathematicians, who correctly concluded that the bettor should win this game of dice slightly less than half the time. Indeed, the chances of winning are quite close to 1/2 and, should the bettor be allowed 25 rolls instead of 24, the bettor would then win more than half the time. Gambling games that remain popular in casinos for long periods of time generally share this feature. The bettor's chances of winning are close enough to 1/2 so that he will win quite frequently, but over long repetitions he will not come out ahead.

The mathematical theory employed by Pascal and Fermat has evolved into what is today called "probability theory." Probability is essentially synonomous with likelihood or chance. It provides a measure of the chance that a phenomenon will be observed when an experiment is performed. We shall examine this theory and see some of its applications in the following sections. The language of set theory is an ideal medium of communication for the study of probability. Thus, we shall first briefly study some concepts of set theory and then see their use and application in the study of probability theory.

A.1 SETS AND OPERATIONS

A set is any collection of objects. The objects that belong to the set, called the elements of the set, can be anything. We can talk of sets of people, sets of buildings, sets of boats, sets of numbers, etc. A set is completely determined by the elements that belong to it. The important thing is that we must be able to determine whether or not any object we may think of is an element of the set. Thus, if A is the set of people in the same room with you while you eat dinner this evening, the people in the room, including you, are the elements of the set.

We shall use capital letters, generally from the start of the alphabet (A, B, C, etc.) to denote sets and shall use lower case letters from the end of the alphabet (x, y, z, etc.) to denote elements of sets. Printer's braces, { } (also called set builders), are used in defining sets. The things that belong to the set are specified between the two braces. Two different methods are commonly useful in defining sets, the roster method (the full list of elements is given between the set builders) or the rule method (a rule is given between the set builders, which identifies the elements of the set). Thus, if we want to define B as the set of whole numbers, or integers, between 1 and 5, inclusive we may write

$$B = \{1, 2, 3, 4, 5\} \qquad \text{(roster method)}$$

or, equivalently,

$$B = \{x : x = 1, 2, \ldots, 5\} \qquad \text{(rule method)}$$

The order used in listing the elements in the roster is unimportant. Thus, we could also write $B = \{2, 5, 1, 3, 4\}$ or $B = \{5, 3, 1, 4, 2\}$.

To indicate that a specific element, represented say by x, belongs to a set A, we shall write $x \epsilon A$; the symbol ϵ is read "belongs to." Thus for the set B defined above $1 \epsilon B$ and $4 \epsilon B$. To indicate "does not belong to" we shall use $\notin$. Thus $8 \notin B$, $a \notin B$, $102 \notin B$, etc.

Given two numbers, a and b, there are several operations we could perform, each of which yields another number. For example, we could add a and b together; this operation yields the number $a + b$. Or, we could multiply a by b; this operation yields the number ab. Addition and multiplication are called operations, appropriate for things called numbers. The result of these operations is again a number ($a + b$ or ab). Given two sets, A and B, we can also define operations that will then result in sets. Two set operations that prove quite useful are called the *union* of A and B and the *intersection* of A and B. First, the union of A and B is defined as follows:

Definition A.1.1. The *union* of A and B is

$$A \cup B = \{x : x \epsilon A \text{ or } x \epsilon B \text{ or both}\}$$

The union of two sets thus is a merging together of the rosters of the two sets. $A \cup B$ consists of all elements that belong to A or to B or to both.

Consider, for example, the set A of female students at your school and the set B of male students at your school; $A \cup B$ then is the set of all students at your school. If

$$C = \{1, 2, 3, 4, 5\}$$
$$D = \{3, 6, 9, 12, 15\}$$

then

$$C \cup D = \{1, 2, 3, 4, 5, 6, 9, 12, 15\}$$

Notice that, in general, every element belonging to C will then belong to the union of C with *any* other set D. We say that C is a *subset* of $C \cup D$.

The intersection of two sets is given in the following definition.

Definition A.1.2. The intersection of A and B is

$$A \cap B = \{x : x \epsilon A \text{ and } x \epsilon B\}$$

The intersection, then, has as elements only those elements that belong to *both* A and B; any element that belongs only to A but not to B will not belong to $A \cap B$. For the particular sets A and B mentioned above (female and male students at your school), there is no element belonging to both A and B. Thus, for this case, $A \cap B$ is *empty* or *null*, the set with no elements. The Greek letter ϕ (phi) will be used to denote this null set. For the two sets of numbers, C and D, defined above, notice that 3 belongs to C and to D; no other element belongs to both so

$$C \cap D = \{3\}$$

Venn diagrams provide a very useful representation of sets and operations with sets. Figure A.1.1 gives Venn diagrams of two sets,

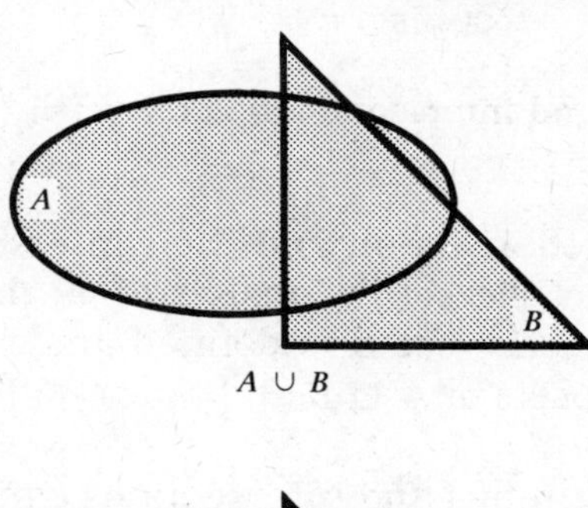

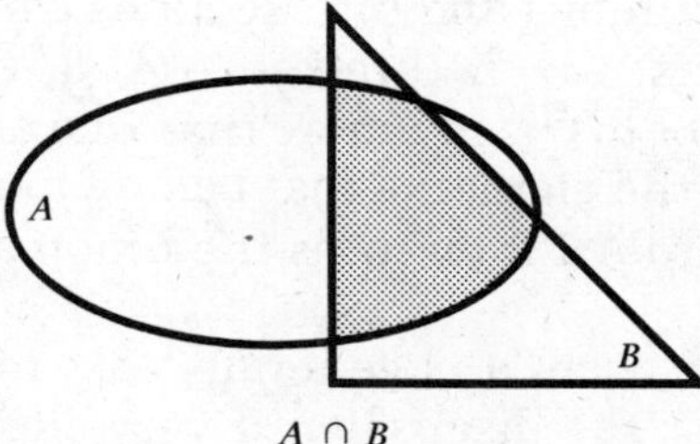

Figure A.1.1 Union and intersection of two sets.

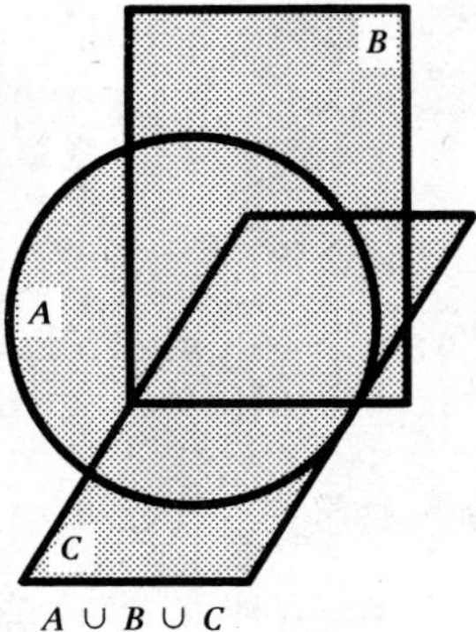

$A \cup B \cup C$

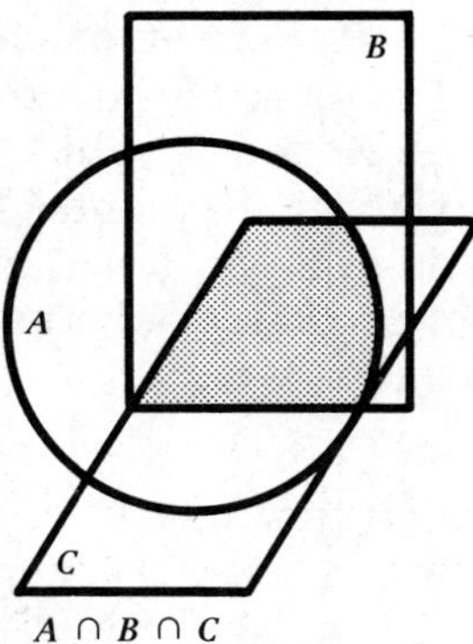

$A \cap B \cap C$

Figure A.1.2 Union and intersection of three sets.

A and *B*, as well as their union and their intersection. In such pictorial representations the set *A* consists of all the points interior to the figure labeled *A*; the set *B* is defined analogously. Notice that both *A* and *B* are subsets of $A \cup B$ and that $A \cap B$ is a subset of both *A* and *B*.

The operations of unions and intersections can easily be extended to three or more sets. Given three sets, *A*, *B*, *C*, any element that belongs to *A* or to *B* or to *C* (or to more than one of them) also belongs to $A \cup B \cup C$. Only the elements that belong to all three sets belong to $A \cap B \cap C$. Figure A.1.2 pictures the union and the intersection of three sets.

Whenever sets are used, it is generally easy to define an appropriate *universal set* for the discussion. Every element of every set in

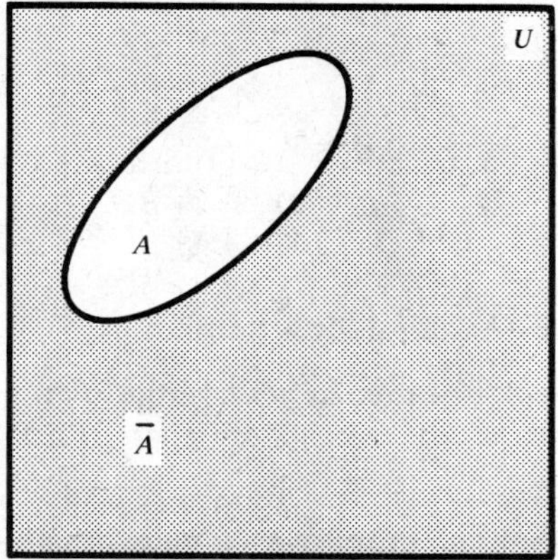

Figure A.1.3 A and its complement, $\overline{A}$.

the discussion belongs to the universal set. For example, if we wanted to discuss various sets of college students we could use the set of all college students as our universal set. If we wanted to discuss sets of dishes, our universal set could be the set of all dishes. If we wanted to discuss sets of numbers, our universal set could be the set of all numbers. As we shall see in our discussions of probability, the sample space for an experiment (set of all possible outcomes which could be observed) is a natural universal set.

Granted we have a well-defined universal set in mind, every time we define a set (call it A) we also have defined a second set called the complement of A (denoted by $\overline{A}$). The complement of A is defined below:

> **Definition A.1.3.**   The complement of a set A is defined to be the set of elements *not* belonging to A (but still in the universal set).

It is plain why the universal set must be well defined; if it were not, the elements not belonging to A would not be well specified. Figure A.1.3 pictures a universal set U, a set A and its complement $\overline{A}$.

A number of interesting relationships involving complements can be shown always to hold true. For example, suppose B is a subset of A; as mentioned earlier, this says then that every element that belongs to B must also belong to A. Then, necessarily $\overline{A}$ must be a subset of $\overline{B}$ since every element not belonging to A (then it does belong to $\overline{A}$) must also certainly not belong to B (it does belong to $\overline{B}$). Draw a Venn diagram showing this fact.

Exercise A.1

1. Define $A = \{1, 2, 3, 4\}$ and $B = \{3, 4, 5, 6, 7\}$ and find $A \cup B$ and $A \cap B$.

2. Let $U = \{1, 2, 3, 4, 5, 6, 7, 8, 9, 10\}$ and find
   (a) $\overline{A}$ (d) $\overline{A} \cap \overline{B}$
   (b) $\overline{B}$ (e) $\overline{A} \cup B$
   (c) $\overline{A \cup B}$ (f) $A \cap \overline{B}$,
   where A and B are as defined in problem 1.

3. Let A be the set of firemen in your hometown and let B be the set of policemen in your hometown. Define, in words, the two sets $A \cup B$ and $A \cap B$.

4. Letting ϕ represent the empty set, what is the set $A \cap \phi$?

5. Again, letting ϕ represent the empty set, what is the set $A \cup \phi$?

6. Is the set of all United States citizens identical with the set of people living in the United States?

7. Define $A = \{-1, 1\}$, $B = \{0, 1\}$, $C = \{-1, 0\}$ and find
   (a) $A \cap (B \cup C)$.
   (b) $(A \cap B) \cup (A \cap C)$.

8. Notice that the two sets asked for in problem 7 are equal. The fact these two sets are always equal is called a *distributive law*. Draw a Venn diagram showing the equality of these two sets. (This distributive law is analogous to the fact that $a(b + c) = ab + ac$, where a, b, and c are numbers.)

9. For A, B, and C as defined in problem 7, find
   (a) $A \cup (B \cap C)$
   (b) $(A \cup B) \cap (A \cup C)$

10. The two sets asked for in problem 9 are also always equal; this is a second distributive law for set operations. Draw a Venn diagram illustrating this law.

11. Can you define a set A such that $\overline{A} = A$?

12. Use a Venn diagram to show that
$$\overline{A \cup B} = \overline{A} \cap \overline{B}$$
for any two sets A and B. (This is one of De Morgan's laws.)

A.2   EXPERIMENTS AND SAMPLE SPACES

The word probability is in fairly common usage today. As has been discussed, the theory of probability has its origins in the description of games of chance; this same theory has progressed a great deal, in terms of breadth, depth, and scientific respectability. Today probability theory is used in virtually every field of science and in many other endeavors from practical business decision making to descriptions of likely locations for successful oil wells to measuring the chances you may require an umbrella tomorrow.

In all the applications we shall discuss, probability is a measure of relative frequency. This means, first of all, that we are referring to an experiment or operation that could be repeated any desired number of times (at least conceptually). On each repetition the event of interest either does occur or it does not; the probability of the event is the relative proportion of the time that the event should be observed over repeated experiments. For example, suppose our operation is the flip of a penny. Most people would agree that the probability of observing a head when the penny is flipped is 1/2. By this they mean that if the coin were flipped a large number of times, half the flips should result in the coin resting heads up; 1/2 is the measure of the relative frequency of occurrence of heads over the repeated flips. It is quite useful to keep in mind that the probabilities we shall discuss are in fact relative frequencies.

Let us begin with some concepts that are quite useful in discovering what the relative frequencies, or probabilities, of certain events should be.

> **Definition A.2.1.**   An *experiment* is any operation whose outcome cannot be predicted in advance, with certainty.

Examples of experiments are a flip of a coin, a roll of a pair of dice, taking an examination (the number of questions answered correctly), length of time required for a person to run a mile, the effect of a drug on an illness, etc. In each of these cases it is not possible to say in advance exactly what the outcome will be; thus each of these is an experiment.

Even though it is not possible to know in advance the particular outcome that will be observed if the experiment is performed, it is in general fairly straightforward to list the full set of possible outcomes, any one of which might be observed. This set of possible outcomes is called the sample space for the experiment, as is given in the following definition.

> **Definition A.2.2.**   The *sample space* for an experiment is the set of all possible outcomes that could be observed when the experiment is performed.

Let us examine possible sample spaces for the experiments mentioned earlier. As will become apparent, the sample space for an experiment is not, in general, unique; that is, we shall frequently be able to specify more than one possible sample space for the same experiment. In such a case we shall want to use the sample space that allows the easiest evaluation of probabilities of interest.

Example A.2.1

Suppose our experiment consists of a flip of a coin. Then, assuming the coin will not be lost in some manner (grabbed in flight by a passing bird, fall into a crack and disappear, etc.) and that it will not end up standing on edge, it seems reasonable to assume that we will observe either a head (H) or a tail (T) as the outcome of the experiment. The sample space for the experiment then could be taken to be the set

$$S = \{H, T\}$$

Example A.2.2

Suppose our experiment is one roll of two dice. A die (singular of dice) is a six-sided cube whose six faces are marked, respectively, with one spot, two spots, etc., up to six spots. There are several different sample spaces that could be employed for the roll of two dice, depending on the level of detail desired (and the probabilities one wants to compute). For example, if we roll the two dice, either the two dice will show the same face (s) or they will show different numbers of spots (d). Thus, we might use as our sample space

$$S_1 = \{s, d\}$$

Alternatively, we might reason that the smallest pair of numbers that could occur is given by ones on both dice and the largest pair by sixes on both dice; thus the sum of the two numbers could range anywhere from 2 to 12. We might then use as a sample space

$$S_2 = \{2, 3, 4, 5, \ldots, 12\}$$

since one of the integers between 2 and 12 must occur as the sum when the two dice are rolled. Yet another alternative might also occur to us, most easily visualized if one die is red and the other is green. Surely the red die will end up with one spot uppermost, or two spots uppermost, or three spots uppermost, etc., and so will the green die. Thus, we might use a pair of numbers to represent a possible outcome, the first number from the red die and the second number from the green die. For example, then, $(1, 4)$ would stand for the outcome "one on red, four on green"; $(3, 2)$ would stand for "three on red, two on green." Using this sort of system we could take as our sample space

$$S_3 = \{(1, 1), (1, 2), \ldots, (1, 6), (2, 1), (2, 2), \ldots, (2, 6), \ldots, (6, 6)\}$$

or, equivalently,

$$S_3 = \{(x, y): x = 1, 2, 3, 4, 5, 6; \ y = 1, 2, 3, 4, 5, 6\}$$

Notice that, for this particular experiment, we can find three different ways of listing the set of possible outcomes and thus can find three different possible sample spaces S_1, S_2, and S_3. For most experiments we shall be able to think of several ways of specifying the list of possible outcomes (several sample spaces). We shall want to use the one that makes our computations the easiest to carry out. For the roll of two dice we shall find that S_3, somewhat surprisingly, is the easiest to use, even though it is the most complex of the three. This is because S_3 has retained all the information from the experiment itself. We shall be able to compute, using S_3, the probability that the two faces are alike, or that the sum of the two numbers is 5, or that the red die shows a one. It would not be possible to compute the probability of this latter event (red die shows a one) if we were using either S_1 or S_2.

Example A.2.3

Assume Rachel takes an examination with 15 true-false questions. The answer she gives will be either right or wrong for each question. The number of questions she will answer correctly will be 0 or 1 or 2, etc., out of 15. We might use as our sample space, then,

$$S = \{0, 1, 2, \ldots, 15\}$$

the number indicating the number of questions she answers correctly. (As with Example A.2.2 there are other possible sample spaces for this experiment, one of which will be examined later.)

Example A.2.4

Assume Douglas is on a college track team and participates in the mile run. Our experiment is the length of time it will take him to run the mile on a specific day, say at a track meet with another school. The sample space must list all possible outcomes, that is, the collection of times that it may take him to complete the mile distance. Using a second as a unit we might then use as our sample space

$$S = \{x: 210 \leq x \leq 600\}$$

if we are sure it will take him at least $3\frac{1}{2}$ minutes (210 seconds) to complete the mile and that it will take him no more than 10 minutes (600 seconds) to complete it. This is an example of a *continuous* sample space, one that has as elements all the numbers in a continuous interval. This type of sample space is natural to use when the experiment consists of measuring the length of time to accomplish something, since time is thought of as varying continuously. The general study and use of continuous sample spaces involves calculus techniques; since it is not assumed that the reader is familiar with calculus, we shall concentrate our efforts on experiments having discrete sample spaces, such as those given in Examples A.2.1 through A.2.3 above.

Example A.2.5

Assume our experiment consists of treating a person afflicted with a certain illness (e.g., a skin infection) with a given dosage of a drug (say 250 mg doses of tetracycline). This type of situation is, of course, quite complex; there are many aspects of such a real-life experiment that could be of interest and that should be reflected in the list of possible outcomes. We might crudely assume the infection will be cured (c_1), controlled (c_2), not affected (n), or made worse (w) by the treatment. With these assumptions, then, our sample space would be

$$S = \{c_1, c_2, n, w\}$$

Exercise A.2

1. A single die is rolled one time. Give a reasonable sample space for this experiment.

2. Two coins, a penny and a nickel, are each flipped one time. Give a reasonable sample space for this experiment.

3. Suppose the same penny is flipped two times. Is the sample space you gave in problem 2 above also appropriate for this experiment?

4. A roulette wheel has 38 spots on it, numbered 00, 0, 1, 2, 3, . . . , 36. Both 00 and 0 are green; half the remaining spots are red; the other half are black. Assume the roulette wheel is spun once. What is a good sample space for the outcome of this single spin? (Give at least two sample spaces.)

5. A standard 52-card deck of playing cards contains 13 spades, 13 clubs, 13 hearts, and 13 diamonds. The spades and clubs are black; hearts and diamonds are red. Assume such a deck is well shuffled and then the top card is turned over. Give two good sample spaces for this experiment.

6. Suppose, in problem 5, the top two cards are turned over, instead of just the top one. What would be a good sample space for this experiment?

7. Eight horses are entered in a race, and are numbered 1 to 8. The first horse across the finish line is the winner. What is a good sample space for this experiment?

8. Hugh takes a true-false examination that contains 10 questions. After the examination the number answered correctly will be determined. Give a sample space for this experiment.

A.3 PROBABILITY, THE EQUALLY LIKELY CASE

An experiment has many possible outcomes; the sample space for the experiment lists the possible outcomes. Granted that the experiment will be performed, people are interested in computing the probabilities of occurrence of things called "events." For example, if a penny is flipped one time, we may want to know the probability that a head will occur as the outcome. The event we are interested in, then, is called "Head occurs." Or suppose a die is rolled one time. We might be interested in the possibility that four spots are uppermost when the die stops rolling; this event is "4 occurs." Or, again, for the roll of the die we might be interested in the probability that an even number of spots is uppermost when the die stops rolling; this event would be labeled "even number occurs." In any case, when the experiment is performed, the event of interest may occur or it may not. The probability of the event occurring should be a measure of the relative frequency with which the event will occur, over repetitions of the experiment.

Notice, then, that an event is a collection of elements from the sample space for the experiment. The event may have a single element belonging to it (head for the flip of a coin, four spots uppermost when die is rolled) or it may have more than one element belonging to it (even number of spots when die is rolled). Since an event is a collection of elements from the sample space, it is in fact a subset of the sample space; an event is itself also a set. The formal definition of an event is as follows:

> **Definition A.3.1.** An *event* is a subset of a sample space. An *event occurs* if any one of its elements is the observed outcome of the experiment.

Since events are sets, they will be denoted by capital letters, the convention introduced earlier.

Example A.3.1

Assume a die is rolled one time; the sample space for the experiment is

$$S = \{1, 2, 3, 4, 5, 6\}$$

Let us define several events:

    A: 2 occurs
    B: 6 occurs
    C: odd number occurs
    D: number less than 4 occurs

Then, as sets, these events are

$$A = \{2\} \qquad B = \{6\} \qquad C = \{1,3,5\} \qquad D = \{1,2,3\}$$

Notice, as mentioned above, that an event may have only one element of S belonging to it, or it may have more than one. Furthermore, for any given outcome of the experiment, several events may have occurred. For example, if the die is rolled and the number of spots uppermost is 2, then events A and D defined above both are said to have occurred. If 3 spots are uppermost, then events C and D are said to have occurred. You are cautioned not to confuse outcomes (elements of the sample space) with events (subsets of the sample space). When the experiment is performed, exactly one outcome will be observed, but many events will have occurred.

Example A.3.2

It was mentioned earlier that the sample space for an experiment is not unique, that the set of possible outcomes could be specified in different ways (this was specifically illustrated in Example A.2.2.) Since an event is a subset of the sample space, different choices of the sample space will have different subsets and different events. Suppose two dice are rolled and we use the sample space

$$S_2 = \{2,3,4,5,\ldots,12\}$$

(Recall from Example A.2.2 that the possible sums of the two numbers that may occur are the elements of S_2.) Using this sample space, let us define the events.

    A: the sum is 7.
    B: the sum is at least 10.

Then, as subsets, these events are

$$A = \{7\} \qquad B = \{10,11,12\}$$

Notice, though, that with this sample space there are things we may observe that cannot be called events, since they are not subsets of S_2. For example, we might be interested in the probability that the two dice have the same number of spots uppermost. If we use this sample space, this is not an event, since the statement "two dice show the same face" is not a subset of S_2. If both dice have 3 spots uppermost, then the sum is 6; but the sum is also 6 if 1 and 5 occur or if 2 and 4 occur. Thus, the occurrence of a 6 does not imply the two faces were the same. Suppose, on the other hand, we use as our sample space

$$S_3 = \{(x,y): x = 1,2,3,4,5,6; \; y = 1,2,3,4,5,6\}$$

(The elements of S are the different possible pairs of numbers that may occur.) Then the events A (sum is 7) and B (sum is at least 10) are still events; they have different representations since S_3 has different elements. As subsets of S_3 these events are

$$A = \{(1,6),\, (2,5),\, (3,4),\, (4,3),\, (5,2),\, (6,1)\}$$

$$B = \{(4,6),\, (5,5),\, (6,4),\, (6,5),\, (5,6),\, (6,6)\}$$

Now we can define

C: the two faces are the same

C is also an event since

$$C = \{(1,1),\, (2,2),\, (3,3),\, (4,4),\, (5,5),\, (6,6)\}$$

S_3 is a more "basic" sample space and has a richer collection of events than does S_2 (or S_1 defined in Example A.2.2). The interested reader can verify that any event using S_2 can also be defined using S_3, but the converse is not true. Thus, the choice of the sample space determines the possible events. If we want to compute the probabilities of occurrence of specific things, we must be sure to adopt a sample space that includes these things as subsets. In essentially all cases we shall want to adopt the most basic sample space possible, the one that keeps track of *all* the information to be observed. S_2 above records only the sum of the two numbers that will occur; the sum being constant does not uniquely identify the two individual numbers rolled (except for a sum of 2 and 12). S_2 explicitly records both numbers rolled. As we shall see, the more basic sample space allows the easier computations, in general.

We have seen the concept of a sample space, and that we shall be able to compute probabilities for events, subsets of the sample space. How can we in fact compute the probability of an event and how will probabilities be denoted? First, the probability of an event A will be denoted by $P(A)$, read probability of A. Since A is an event it is a set; $P(A)$ is the long-term relative frequency of A and, thus, is a number. We are in fact then asking how we can assign numbers to subsets when we want to compute probabilities. A probability measure is a special kind of real-valued function defined over a class of subsets.

The actual computation of the probability of an event must necessarily rest on some basic assumptions concerning the experiment being modeled. We shall in this section restrict our discussion to a special, and very important, type of sample space, ones in which we are willing to assume *equally likely outcomes*. This assumption of equally likely outcomes means that we are willing to grant that any element of the sample space is as likely to occur as any other;

if the experiment were repeated indefinitely, we expect to see the different elements of S occur with equal frequencies. If this equally likely assumption is made, the probability of any event A is defined to be

$$P(A) = \frac{\text{number of elements in } A}{\text{number of elements in } S} = \frac{\#(A)}{\#(S)}$$

Let us immediately examine some examples using this definition.

Example A.3.3

Suppose a dime is to be flipped two times. Let us adopt the sample space

$$S = \{(H,H), (H,T), (T,H), (T,T)\}$$

where the first position in each pair describes the results of the first flip (head-H or tail-T) and the second position describes the results of the second flip. Since a dime seems to be fairly symmetric, we might reason that these different pairs should be observed with equal frequencies; thus we would make the equally likely assumption. Once we are willing to make this assumption, we can then compute the probabilities of occurrence of any events of interest. The computed probabilities then represent the correct relative frequencies, so long as the equally likely assumption is valid. For example, let us define

A :   we get a head on the second flip.
B :   the two faces are the same.
C :   heads occur on both flips.

Then

$$A = \{(H,H), (T,H)\}, \qquad B = \{(H,H), (T,T)\},$$

$$C = \{(H,H)\}, \#(S) = 4, \#(A) = 2, \#(B) = 2, \#(C) = 1$$

so

$$P(A) = \frac{2}{4} = \frac{1}{2}, \; P(B) = \frac{2}{4} = \frac{1}{2}, \; P(C) = \frac{1}{4}$$

These are the expected relative frequencies of occurrence of these events over repetitions of the experiment, given the assumption of equally likely outcomes.

Example A.3.4

Let us again refer to the experiment of rolling two dice. We shall use as our sample space the basic one discussed above (now denoted simply S):

$$S = \{(x,y): x = 1,2,3,4,5,6; \quad y = 1,2,3,4,5,6\}$$

If it appears that each die is symmetric and well balanced, again it would seem reasonable to make the equally likely assumption for this sample space S. We then can compute probabilities as ratios of the number of elements belonging to the event, divided by 36, the number of elements in S. Let us examine the probabilities of the following events with this equally likely assumption:

A: The first die has a one on it.
B: The two dice show the same face.

Then

$$A = \{(1,1), (1,2), (1,3), (1,4), (1,5), (1,6)\}$$

$$B = \{(1,1), (2,2), (3,3), (4,4), (5,5), (6,6)\}$$

and $\#(A) = 6$, $\#(B) = 6$. Therefore,

$$P(A) = \frac{6}{36} = \frac{1}{6}, \quad P(B) = \frac{6}{36} = \frac{1}{6}$$

Let us also compute the probabilities of occurrence of the different possible sums that could occur. We define

C_i: sum of the two numbers is i for $i = 2,3,4,\dots,12$

The probabilities of occurrence of these different sums then are gotten by counting the number of elements belonging to C_i and dividing by 36. The following table presents these values.

i	2	3	4	5	6	7	8	9	10	11	12
$\#(C_i)$	1	2	3	4	5	6	5	4	3	2	1
$P(C_i)$	1/36	2/36	3/36	4/36	5/36	6/36	5/36	4/36	3/36	2/36	1/36

Notice then, that if we were to use

$$\{2,3,4,\dots,12\}$$

as the sample space for this experiment, it would not be reasonable to assume that these elements are equally likely to occur.

Computing probabilities for equally likely sample spaces involves counting the numbers of elements in various sets. Thus, when using such sample spaces, it is beneficial to have some knowledge of counting techniques, procedures that are useful in counting the number of elements that belong to a set without going through a complete enumeration of the roster of the set. We shall examine some counting techniques now.

The factorials of whole numbers occur frequently in counting problems. By definition $n!$ (read n factorial) is the product of all the whole numbers from 1 up to n, inclusive; thus

$$n! = n(n-1)(n-2)\ldots 2 \cdot 1$$

For example, then

$$3! = 3 \cdot 2 \cdot 1 = 6$$

$$4! = 4 \cdot 3 \cdot 2 \cdot 1 = 4 \cdot 3! = 24$$

$$5! = 5 \cdot 4 \cdot 3 \cdot 2 \cdot 1 = 5 \cdot 4! = 120$$

$$n! = n(n-1)!$$

$n!$ gets large very fast as n increases. As a convention that proves quite useful, we shall define $0! = 1$ so that both $0!$ and $1!$ are equal to 1.

The multiplication principle is frequently useful in counting the number of elements in a set. This principle is stated below.

Multiplication Principle:
Each element of a set contains k symbols. If the first symbol can be any one of n_1 letters and then the second symbol can be any one of n_2 letters, etc., out to the kth symbol can then be any one of n_k letters, the total number of elements in the roster of the set is $n_1 n_2 \cdots n_k$, the product of the number of letters available for the various positions.

Thus, in Example A.3.4 the sample space S consisted of the possible pairs of numbers that could be observed in rolling two dice. In that case, each element contained $k = 2$ symbols; the first symbol could be any of $n_1 = 6$ different letters (actually the integers $1,2,3,4,5,6$), as could the second. Thus, the number of elements in that sample space is $n_1 n_2 = 6 \cdot 6 = 36$ as mentioned in the example. If three dice were rolled, the sample space consisting of the various possible trios of numbers (or letters) that could occur would have $6 \cdot 6 \cdot 6 = 216$ elements.

Suppose 6 cards are in a bowl, numbered $1,2,3,4,5,6$ and that two cards are to be drawn without replacement. (This means the first card drawn is not replaced before the second is chosen). The list of possible pairs of numbers that could occur would have $6 \cdot 5 = 30$ elements, since the first card could bear any of the 6 integers and, regardless of which number occurs in the first position, five different numbers are still available to fill the second position.

A *permutation* of a group of objects is an arrangement of those objects in a row, in a definite order. Different orderings are different permutations. Suppose we have n objects and want to count the total number of permutations possible. This is identical with counting the number of elements in the set consisting of all possible permutations of the n objects (each element of the set is a permutation of the objects). The leftmost object could be any of the n. Regardless of which one is chosen for that position, the second leftmost object could be any one of the remaining $n - 1$, the third leftmost object could then be any of the remaining $n - 2$, etc. Thus, the total number of permutations possible is $n(n - 1)(n - 2) \ldots 2 \cdot 1 = n!$

Three books may be placed side by side on a desk in $3! = 6$ ways. If five horses are in a race there are $5! = 120$ different possible orderings for them to be in when they cross the finish line.

Suppose n objects are available. From these n, k objects are to be selected and placed in a definite order. (k, the number selected, is of course no larger than n.) This resulting ordered collection of k objects is called a permutation of n things k at a time. Using the multiplication principle the number of permutations of n things k at a time is $n(n - 1)(n - 2) \ldots (n - [k - 1]) = n(n - 1)(n - 2) \ldots (n - k + 1)$ since the number of objects available after $k - 1$ positions have been filled is $n - [k - 1]$. If, for example, 8 people enter a race and 3 prizes are to be awarded for the 1st, 2nd, and 3rd place fiinishers, the number of different possible allotments of prizes to people is given by the number of permutations of 8 things 3 at a time: $8 \cdot 7 \cdot 6 = 336$. This number includes all the different possible selections of 3 people that could be made from the 8 in the race, as well as the permutations of prizes among the three individuals chosen.

As a second example, suppose a firm has three reserved parking spaces available and five people eligible for reserved spaces. The number of different ways in which the spaces could be allocated to three of the eligible individuals is given by the number of permutations of 5 things 3 at a time: $5 \cdot 4 \cdot 3 = 60$. Again, as is always true with permutations, the allocation is counted as being different if the same 3 people receive spaces, but the spaces are permuted among the people.

We have seen that the number of permutations of n things k at a time is $n(n - 1)(n - 2) \ldots (n - k + 1)$. This quantity is unchanged if we multiply it by $(n - k)!/(n - k)!$ giving

$$\frac{n(n - 1) \ldots (n - k + 1)(n - k)!}{(n - k)!} = \frac{n!}{(n - k)!}$$

This formula is easier for many people to remember than is the original product. Notice, then, that the number of permutations of n things n at a time (which is just the number of ways n objects could be arranged in a row) is $n!/(n - n)! = n!/0! = n!$, with our convention that $0! = 1$, the same result mentioned earlier.

In many counting problems we shall want to know how many

different selections of k objects can be made from n objects, regardless of the order of selection; we want to count only the different groups of size k that could be selected. The answer to this type of problem, where order does not count, is given by the number of *combinations* of n things k at a time. By definition the number of different subsets, each of size k, which can be selected from a set with n elements, is called the *number of combinations of n things k at a time*. We shall denote this quantity by $\binom{n}{k}$. What, then, is the value of $\binom{n}{k}$? We saw above that the number of permutations of n things k at a time is $n!/(n-k)!$. This quantity counts two things: the number of distinct selections of size k that could be made from n objects and the number of ways in which the k objects selected can be rearranged or permuted among themselves. Any collection of k objects can be permuted among themselves in $k!$ different ways. Thus, if we divide $n!/(n-k)!$ by $k!$, we shall be left with only the number of distinct collections of size k that could be selected from the n objects, that is, we have the value for $\binom{n}{k}$. Thus

$$\binom{n}{k} = \frac{n!}{k!(n-k)!}$$

is the number of subsets, each of size k, which can be selected from a set with n elements.

Suppose we are given a set with $n = 4$ elements. This set has $\binom{4}{1} = \frac{4!}{1!\,3!} = 4$ subsets, each with 1 element, as we know it should. It has $\binom{4}{2} = \frac{4!}{2!\,2!} = 6$ subsets each with 2 elements. (It might be instructive to list the 6 subsets of size 2 that can be constructed from $\{1,2,3,4\}$.) We are now able to evaluate the number of different subsets, each of the same size, that any given set has; this number depends both on the size of the original set (n) and on the number of elements the subset is to have (k).

Let us examine some examples. First, the game of bridge is played with a standard 52-card deck of cards; thus this deck can be thought of as a set with $n = 52$ elements (the individual cards). An individual bridge hand contains 13 cards, selected from the 52 (an individual hand is a subset of size 13). We might inquire, then, how many different bridge hands could be dealt from the deck? The answer to this question is given by the number of combinations of 52 things at a time:

$$\binom{52}{13} = \frac{52!}{13!\,39!} = 635{,}013{,}559{,}600$$

To get an idea of the magnitude of this number, we could give a different 13 card hand to each person in the world, every day for more than 6 months, and never repeat a hand. The game of poker is played with the same 52-card deck. A poker hand contains 5 cards instead of 13. How many different poker hands can be dealt? Again, the order in which the cards are dealt does not matter and should not be counted so the answer is

$$\binom{52}{5} = \frac{52!}{5!\,47!} = 2{,}598{,}960$$

a considerably smaller number. In fact, there are almost 1/4 million bridge hands for every poker hand. A set with 52 elements has a lot more subsets of size 13 than it does of size 5.

A primitive tribe has developed a printing machine. They have only two different symbols that can be used, which we will denote by + and −. How many different 5 letter words can they write each with 2+'s and 3−'s? (How many different sequences of 5 symbols are possible, when two symbols must be + and the remaining three must be −?) The +'s could occupy any two of the positions, and then the remaining positions are filled with −'s. We can choose two positions from five (order does not count) in $\binom{5}{2} = 10$ ways; the two chosen positions are filled with +'s, and the other three with −'s. Thus the answer is 10 different words may be found, each using 2+'s and 3−'s.

Let us now examine two examples of computing probabilities from equally likely sample spaces.

Example A.3.5

Suppose you are at a party with 29 other people. Two door prizes are to be given by lot (at random). Let us make two different assumptions about the values of the prizes and compute the probability that you receive a prize.

(*a*) Assuming the prizes are of equal value, what is the probability you receive one? First, we must decide on the sample space for the experiment. Since the prizes are of equal value, let us use as S, the sample space, the collection of all distinct pairs of people that could be selected from the 30 at the party; the order of listing within the pair plays no special role. Under the assumption that the two prize recipients are chosen at random, all the elements of S are equally likely to occur and probabilities are computed as ratios of the number of elements in the event divided by the number of elements of S. Since S has as elements the distinct unordered pairs that could be selected from the 30, $\#(S) = \binom{30}{2} = 435$. Define

$A:$ You win a prize.

Then A consists of all pairs of names from S, such that your name is one of the two. If A contains only the pairs that include you, each element of A has one more position to be filled (the other person to receive a prize) and $\#(A) = \binom{29}{1} = 29$, since any of the remaining 29 people could be the recipient of the other prize. Then

$$P(A) = \frac{\#(A)}{\#(S)} = \frac{29}{435} = \frac{1}{15}$$

(b) Assume the two prizes are not of equal value, that one cost \$10 and the other \$5. Under this assumption it would seem reasonable to let S, the sample space, consist of the collection of ordered pairs of names that could be selected, where the first name in each pair receives the more expensive prize. Again, if the prizes are awarded by lot, it is reasonable to assume all the elements of S are equally likely to occur. We find the number of elements in S, then, to be the number of *permutations* of 30 things 2 at a time; that is,

$$\#(S) = \frac{30!}{28!} = 870$$

Notice that the number of elements in this sample space is exactly twice the number we had in part (a); this is because each possible distinct pair is listed twice, depending on which of the two persons gets the more expensive present. Define the two events.

$A:$ You get a prize.
$B:$ You get the more expensive prize.

Then to compute the probabilities of occurrence for these two events, we need to count the number of elements each one has. First of all, $\#(A) = 2(29) = 58$, since there are 29 pairs in S with your name listed first and also 29 pairs with your name listed second. Thus $P(A) = 58/870 = 1/15$, just as in part (a); this result does not seem surprising since the probability that you get a prize should be independent of whether the two are of equal value. The event B consists of all pairs of names in S each of which have your name listed first. Clearly the second listed name could be any of the other 29, so $\#(B) = 29$ and $P(B) = 29/870 = 1/30$.

Example A.3.6

Ms. Jones claims to have a talent for mental telepathy. In fact, if you look at a card, whose face is either red or black, she claims she can identify which color it is simply by reading your mind. To test her powers, a deck of 8 cards is composed, of which 4 are red and 4 are black; Ms. Jones is told that I will examine 8 cards successively, of which 4 are red and 4 are black. She is to tell me their colors, without being able to see them. I arbitrarily look at the 4 red cards first and then at the 4 black cards. If Ms. Jones is guessing at their identities, what is the probability she correctly identifies the color of all eight cards? To discuss the possible outcomes of this experiment, let us assume the cards are numbered from 1 to 8; cards numbered 1 to 4, then, are the red ones and 5 to 8 are the black ones. Let us use as a sample space the various collections of 4 numbers that she might respond were the red cards. Thus $(1,3,5,6)$ means she said cards $1,3,5$, and 6 were red (and $2,4,7$, and 8 were black). If she is guessing, then the various possible collections of four red card numbers are equally likely to occur. The number of elements in the sample space is

$$\#(S) = \binom{8}{4} = 70$$

exactly one of these (she says cards, $1,2,3$, and 4 are red) corresponds to her correctly identifying the colors of all 8 cards. Thus P(she is right on all $8) = 1/70$. If she does in fact identify them all correctly, we might feel she does have some extraordinary talent for mind reading, since it is quite unlikely she could do it just by guessing.

Exercise A.3

1. Two cards are drawn from a well-shuffled, standard 52 card deck. Compute the probability that
   (a) Both are red.
   (b) Two colors occur.
   (c) Both are the same denomination.
   (d) Both are from the same suit.
   (e) Both are face cards (king, queen and jack are the only face cards).

2. Assume that eight horses (numbered 1 to 8) are entered in a race and that you have bet on number 1. Assuming the horses are all of equal ability, compute the probability that
   (a) Number 1 wins.
   (b) Number 1 comes in second.
   (c) Number 1 comes in third.
   (d) Number 1 finishes in the money (i.e., he is first or second or third).

3. California license plates contain three digits (each digit can be $0, 1, 2, \ldots, 9$) followed by three letters (each letter can equal any of the 26 in our alphabet).
   (a) With no other restrictions, how many license plates can be made?
   (b) If 196 three-letter permutations have been termed undesirable in one or more languages, and thus will not be used, how many license plates are possible?

4. Each integer between 100 and 999, inclusive, has three digits. How many three-digit integers are there? How many of these do not contain the digit 5? How many contain one or more 5's?

5. A prison has 50 inmates, but adequate facilities for only 47. The warden has decided to release three, chosen at random, from the 50.
    (a) What is the probability the three oldest persons are chosen?
    (b) What is the probability the oldest and youngest are among the chosen three?

6. Ten students, eight boys and two girls, are to be seated in a row, at random.
    (a) What is the probability the girls get the end seats?
    (b) What is the probability the two girls sit in adjacent seats?

7. Show algebraically that

$$\binom{n}{r} = \binom{n}{n-r}$$

for all n and r. (That this must be so can be seen because $\binom{n}{r}$ is the number of different subsets of size r that can be selected from n; every time we choose a subset of size r, we leave behind a subset of size $n - r$. Thus there must be equal numbers of different subsets of these two different sizes.)

8. Show that $\binom{n}{0} = \binom{n}{n} = 1$ for all n. (This says there is only one way to either select none from the n or to select all n.)

9. A boys' prep school has 100 students in the senior class. Two students are to be chosen by lot (at random) to represent the school at a prep school conference.
    (a) What is the probability that the oldest and youngest boys are chosen?
    (b) What is the probability that the two students with the best scholastic records are chosen?
    (c) What is the probability that the two students with the worst scholastic records are chosen?
    (d) What is the probability that the student with the best scholastic record is one of those chosen?
    (e) What is the probability that the student with the worst scholastic record is one of those chosen?

10. A national magazine runs a sweepstakes, where the winners are chosen at random from those that respond to a mailed form. Assume that 10,000,000 such forms are mailed out and that 80% of those receiving one will respond and send it back. You are one of those that send back a response. Assume that 2000 prizes are awarded.

    (a) What is the probability that you win first prize?

    (b) What is the probability you win one of the first 10 prizes?

    (c) What is the probability you win a prize?

11. A beauty contest has seven judges; after several preliminary com-
petitions only two contestants remain, A and B. Assume four of the
judges prefer A (and thus she will be the winner) while the remaining
three judges prefer B. If three judges are selected at random from
the seven, what is the probability that a majority of the three favor A?
that a majority favors B?

12. For the equally likely case show that

$$P(A \cup B) = P(A) + P(B) - P(A \cap B)$$

where A and B are any two events.

13. A psychology examination contains 10 true-false questions. To pass
the examination a student must get at least seven of the questions
correct. Assume that Josef is absolutely unprepared to take the examination and will quess at the answer to every question. What is the
probability that Josef passes the examination?

14. Verify the values for $P(C_i)$, $i = 1, 2, \ldots, 12$, given in Example A.3.4.

15. Five people park their cars on the same side of the same street every
night. How many different orderings are possible for the five cars?

16. A standard double-6 domino set contains 28 dominos. In playing the
game "Five-Up" each player selects five dominoes for his hand. How
many distinct five domino hands can be selected?

17. Given an experiment and a sample space S whose elements are assumed equally likely to occur, show that

$$P(A) = 1 - P(\overline{A})$$

where $\overline{A}$ is the complement of A with respect to the sample space S.
This formula is useful in cases in which it is easier to evaluate $P(\overline{A})$
than to compute $P(A)$ directly.

18. Assume 40 people are at a party and four prizes are to be given by lot.
What is the probability that the first person to arrive at the party does
not get a prize?

19. Assume the same experiment as is described in Example A.3.6. If
Ms. Jones is guessing, what is the probability she correctly identifies
at least six of the eight cards?

20. Assume 25 people are in a room. What is the probability that two or
more of them have the same birthdate (same month and day of the
month)? (*Hint.* Ignore February 29 and use as a sample space the possible lists of 25 birthdates that could occur; assume all possible col-

lections of birthdates are equally likely to occur. Consider the complement of the event of interest and question 17.)

21. Six dice are rolled one time. What is the probability that each of the six different faces occur?

22. Assume a "signal" consists of five symbols, each symbol is either + or −. How many different signals can be made? (*Hint.* Count the number that have 0 +'s, exactly 1 +, exactly 2 +'s, etc., and add them together.)

A.4 PROBABILITY AXIOMS

In the last section we examined experiments that had equally likely outcomes. If the equally likely assumption is made, then probabilities of events are computed as ratios of the number of elements in the event divided by the number of elements in the sample space. This is the type of computation made by the early writers in probability theory. They may not have always mentioned it, but they would adopt sample spaces for which the equally likely assumption was natural.

More recently, the Russian mathematician A. N. Kolmogorov published an axiomatic system for probability measures that is widely used today. Theoretical advances in probability have been greatly enhanced by this more formal system. The reader may have had Euclidean geometry in high school, which is generally presented as an axiomatic system. Contrary to the many axioms required in geometry, probability theory is generated from just three axioms.

An axiom is a statement regarded as being true. It could be considered an assumption that everyone agrees to. Given a collection of axioms, then, a theory is concerned with investigating consequences and derivatives of the axioms; theorems are proved that give additional facts, which are implied by the axioms, although not directly stated by the axioms.

What, then, are the Kolmogorov axioms for probability theory? As we have seen, probability should be a measure of the relative frequency of occurrence of an event, over repetitions of the experiment. The axioms, then, are merely a succinct statement of the properties of relative frequencies. We insist that the numbers we compute, called probabilities, satisfy the axioms so that our probabilities will behave like relative frequencies.

Relative frequencies are never negative; it makes no sense to say that an event occurs -15% of the time. Thus, the first axiom we shall insist is satisfied is that the probability of any event must be non-negative. That is, if A is any event and $P(A)$ is its probability, we shall assume $P(A) \geq 0$.

Something we are sure will occur, every time the experiment is performed, should have relative frequency 100%. It happens or occurs on every performance of the experiment. If we have really listed all the possible outcomes in our sample space S, then we are sure to observe some element belonging to S every time the experiment is performed; S is sure to occur. Thus, the probability of S must be 1. Our second axiom, then, is $P(S) = 1$.

The only remaining axiom is an additivity requirement, common to all things called measures. The measure weight, for example, possesses this additivity property (at least ideally). If we have a pound of butter and cut it into two pieces, the sum of the weights of the pieces must be one pound. This is the additivity requirement. If we cut a "whole" into nonoverlapping pieces, the sum of the measures of the pieces must be identical with the measure of the whole.

In terms of events, assume A and B are mutually exclusive, meaning they have no elements in common ($A \cap B = \phi$). Then the additivity requirement says

$$P(A \cup B) = P(A) + P(B)$$

This will be our third axiom. It says, for example, that if we repeatedly roll a die the relative frequency of the event "1 or 2 occurs" must be equal to the sum of the relative frequencies of the events "1 occurs" and "2 occurs."

To summarize, then, assume we are given an experiment and a sample space S for the experiment. Our probability model for the experiment is completed by any rule assigning numbers to events (the subsets of S) such that

1. $P(A) \geq 0$ for all $A \subset S$. (A is a subset of S.)
2. $P(S) = 1$.
3. $P(A \cup B) = P(A) + P(B)$ if $A \cap B = \phi$.

Mathematically, a probability function is a real-valued set function, defined on the class of all subsets of S, which satisfies the above three axioms. (It is a rule associating real numbers with every subset of S.)

The actual values of the probabilities for specific events could be anything so long as these three rules are satisfied.

Example A.4.1.

Assume a bowl contains three slips of paper, numbered $1, 2, 3$. We are going to draw one slip. A reasonable sample space then is

$$S = \{1, 2, 3\}$$

since the slip of paper drawn must have one of these three numbers on it. The different possible events (subsets of S) and some proposed probabilities for these events are given below.

Event A	$P(A)$
ϕ	0
$\{1\}$	1/2
$\{2\}$	1/3
$\{3\}$	1/6
$\{1, 2\}$	2/3
$\{1, 3\}$	7/12
$\{2, 3\}$	1/2
$\{1, 2, 3\} = S$	1

Does this assignment of probabilities satisfy the axioms?

Since every event has a nonnegative probability axiom 1 is certainly satisfied. From the final line we see $P(S) = 1$, so the second axiom is also satisfied. Notice, though, that $\{1, 2\} = \{1\} \cup \{2\}$ and $\{1\} \cap \{2\} = \phi$, and yet

$$P(\{1, 2\}) = 2/3$$
$$P(\{1\}) + P(\{2\}) = 1/2 + 1/3 = 5/6$$

Since $2/3 \neq 5/6$, the third axiom is not satisfied with $A = \{1\}$, $B = \{2\}$; therefore the numbers given are not legitimate probabilities. You might search for other violations of axiom three using the assigned numbers.

In Section A.3 we studied the special case of equally likely outcomes and used a special rule for computing probabilities of events. We might inquire, then, whether the rule used there will satisfy the three Kolmogorov axioms. The following theorem assures us that it does.

Theorem A.4.1

Assume we are given an experiment with a sample space S that has n elements. Then the rule

$$P(A) = \#(A)/\#(S) = \#(A)/n \quad \text{for } A \subset S$$

where #(A)=number of elements in A, satisfies the Kolmogorov axioms.

Proof.

Any subset $A \subset S$ has a nonnegative number of elements. Thus, for any $A \subset S$,

$$P(A) = \#(A)/n \geq 0$$

and axiom 1 is satisfied. Also

$$P(S) = \#(S)/n = n/n = 1$$

Thus axiom 2 is also satisfied. If A and B are any two subsets with no elements in common ($A \cap B = \phi$), then

$$\# (A \cup B) = \#(A) + \#(B)$$

from the definition of the union. Thus

$$\begin{aligned}
P(A \cup B) &= \#(A \cup B)/n \\
&= [\#(A) + \#(B)]/n \\
&= \#(A)/n + \#(B)/n \\
&= P(A) + P(B)
\end{aligned}$$

Thus axiom 3 is also satisfied.

The special computational formula appropriate for equally likely outcomes thus does possess the properties of relative frequencies and can be used in any case in which we are willing to make the necessary assumption. You were asked in Exercises A.3.17 and A.3.12 to show that two special formulas were valid for the equally likely case, namely,

$$P(\overline{A}) = 1 - P(A)$$

and

$$P(A \cup B) = P(A) + P(B) - P(A \cap B)$$

These two results are actually consequences of the three axioms and thus will always be true, no matter whether we make the equally likely assumption or not. The following two theorems establish these results for the general case.

Theorem A.4.2

If $\overline{A}$ is the complement of A (with respect to S)

$$P(\overline{A}) = 1 - P(A)$$

Proof. From the definition of $\overline{A}$

$$\overline{A} \cup A = S$$

and thus $P(\overline{A} \cup A) = P(S) = 1$ from axiom 2.

Furthermore, $A \cap \overline{A} = \phi$. Therefore

$$P(\overline{A} \cup A) = P(\overline{A}) + P(A) \qquad \text{from axiom 3.}$$

Thus, we have

$$P(\overline{A}) + P(A) = 1$$

so

$$P(\overline{A}) = 1 - P(A)$$

as was to be shown.

An interesting corollary to this result is obtained by letting $A = S$ above. Then $\overline{A} = \overline{S} = \phi$, and we have

$$P(\overline{A}) = P(\phi) = 1 - P(S) = 1 - 1 = 0$$

We know ϕ is a subset of any sample space S; thus it is an event and must have some number as its probability. The above result shows that the probability of ϕ occurring must be 0, no matter what the experiment may be.

Theorem A.4.3

If A and B are any two events,

$$P(A \cup B) = P(A) + P(B) - P(A \cap B)$$

Proof.

We can see from Figure A.4.1 that a set B can be written as the union of two sets whose intersection is empty:

$$B = [A \cap B] \cup [\overline{A} \cap B]$$

Since the two sets are equal,

$$P(B) = P([A \cap B] \cup [\overline{A} \cap B])$$

Furthermore, since $A \cap B$ and $\overline{A} \cap B$ have no elements in common, we see from axiom 3 that

$$P(B) = P([A \cap B] \cup [\overline{A} \cap B])$$
$$= P(A \cap B) + P(\overline{A} \cap B)$$

from which it follows that

$$P(\overline{A} \cap B) = P(B) - P(A \cap B)$$

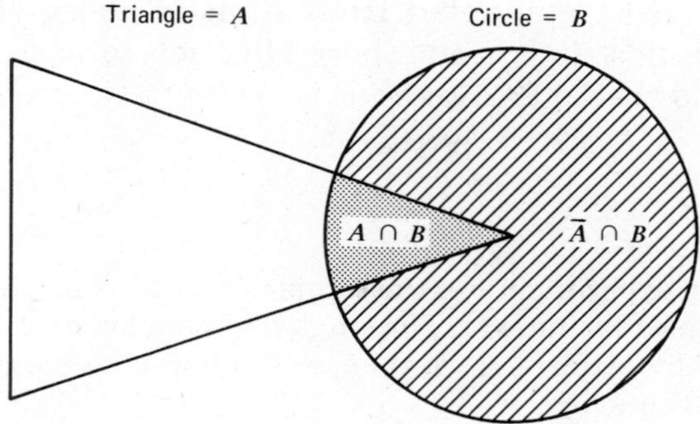

Figure A.4.1 B is the union of two disjoint sets: $[A \cap B] \cup [\overline{A} \cap B]$.

We can also see from Figure A.4.1 that

$$A \cup B = A \cup [\overline{A} \cap B]$$

and that $A \cap [\overline{A} \cap B] = \phi$. Thus

$$P(A \cup B) = P(A \cup [\overline{A} \cap B])$$
$$= P(A) + P(\overline{A} \cap B)$$

from axiom 3. Thus we have

$$P(A \cup B) = P(A) + P(B) - P(A \cap B)$$

after replacing $P(\overline{A} \cap B)$ by $P(B) - P(A \cap B)$; this is the desired result.

The result given above in Theorem A.4.3 holds for the union of any two events A and B, regardless of whether they have elements in common. Notice that if $A \cap B = \phi$, then

$$P(A \cup B) = P(A) + P(B) - P(A \cap B)$$
$$= P(A) + P(B) - P(\phi)$$
$$= P(A) + P(B)$$

since we saw earlier that $P(\phi) = 0$. Thus, this formula reproduces axiom 3 if $A \cap B = \phi$.

The equally likely formula is just one special way of computing probabilities that will satisfy the axioms. It might seem that to have any more general assignment, then, that we would have to specify individually the probabilities for all subsets, similar to the listing given in Example A.4.1. It can be shown that if S has n elements, then it has 2^n subsets. As n gets large, then, 2^n gets big very fast, and the assignment of 2^n probabilities for the individual events is an overwhelming task. Luckily, the following theorem says that probabilities only need to be assigned to the n single-element events (those subsets having only one element) and then the probability of any other event can be computed from sums of these single-element event probabilities. It is a much smaller job to assign n numbers than to have to assign 2^n numbers.

Theorem A.4.4

Assume an experiment has the sample space $S = \{1, 2, 3, \ldots, n\}$, where n is some positive integer. The single-element events then are $\{1\}$, $\{2\}$, $\ldots$, $\{n\}$. Any event $A \subset S$, then, is a union of some of these single-element events and

$$P(A) = \sum_{i \in A} P(\{i\})$$

that is, $P(A)$ is given by adding together the probabilities of the single-element events, over all single-element events whose union is A. This rule satisfies the Kolmogorov axioms.

The proof of this theorem will be left to the reader. It says that we only have to know, or decide on, the values of n events, and from these we can derive the probabilities of all remaining events. [Actually, because of the requirement $P(S) = 1$, the sum of the n single-element event probabilities must be 1 and we really only have to choose n-1 of them.]

Example A.4.2

Suppose that a bowl contains 20 slips of paper, all of identical size, color, and texture. Ten of the slips have the number 1 printed on them, six have the number 2 printed on them, and the remainder have the number 3 printed on them. We are going to select one slip from this bowl without looking. The slip we select, then, will have 1, 2, or 3 printed on it, and a reasonable sample space would be

$$S = \{1, 2, 3\}$$

What would be reasonable values for the probabilities of the single-element events $\{1\}$, $\{2\}$, $\{3\}$ for this case? Because of the differing numbers of slips in the bowl having the given numbers, the following assignments would seem reasonable:

$$P(\{1\}) = \frac{10}{20} = \frac{1}{2} \quad P(\{2\}) = \frac{6}{20} = \frac{3}{10} \quad P(\{3\}) = \frac{4}{20} = \frac{1}{5}$$

Notice that

$$P(S) = P(\{1\}) + P(\{2\}) + P(\{3\})$$

$$= \frac{1}{2} + \frac{3}{10} + \frac{1}{5} = 1$$

as, of course, it must. This really means that once we had decided on the values for $P(\{1\})$ and $P(\{2\})$, the value of $P(\{3\})$ was also necessarily specified to be 1/5, the value that makes the three single-element probabilities sum to 1. By using the result given in Theorem A.4.4, we see that the probabilities of all the other possible subsets are gotten by summing the probabilities of the appropriate single-element events. Thus

$$P(\{1,2\}) = \frac{1}{2} + \frac{3}{10} = \frac{4}{5}$$

$$P(\{1,3\}) = \frac{1}{2} + \frac{1}{5} = \frac{7}{10}$$

$$P(\{2,3\}) = \frac{3}{10} + \frac{1}{5} = \frac{1}{2}$$

Example A.4.3

A die has been "loaded" in such a way that the probabilities of occurrence of the different faces are proportional to the numbers on the faces. Thus, a two is twice as likely to occur as is a one, a six is six times as likely as a one, etc. If this die is rolled one time, what is the probability that an even number occurs? What is the probability that the number that occurs is less than three?

The sample space for our experiment is $S = \{1, 2, 3, 4, 5, 6\}$. To answer the questions posed above, we must first evaluate the probabilities of the single-element events. If we set

$$P(\{1\}) = k$$

then we must have

$$P(\{2\}) = 2k \qquad P(\{3\}) = 3k \qquad \text{etc.}$$

because of the proportionality requirement. Since the sum of the single-element event probabilities must give us 1, we have

$$k + 2k + 3k + 4k + 5k + 6k = 1$$

or

$$21k = 1$$

so

$$k = \frac{1}{21}$$

Thus, the probabilities of the single-element events are

$$P(\{1\}) = \frac{1}{21} \qquad P(\{2\}) = \frac{2}{21} \qquad P(\{3\}) = \frac{3}{21}$$

$$P(\{4\}) = \frac{4}{21} \qquad P(\{5\}) = \frac{5}{21} \qquad P(\{6\}) = \frac{6}{21}$$

Now that we have the single-element event probabilities, let A be the event that an even number occurs and let B be the occurrence of a number less than 3. Then

$$A = \{2, 4, 6\} \quad B = \{1, 2\}$$

and we have

$$P(A) = P(\{2\}) + P(\{4\}) + P(\{6\})$$
$$= \frac{2}{21} + \frac{4}{21} + \frac{6}{21} = \frac{12}{21} = \frac{4}{7}$$
$$P(B) = P(\{1\}) + P(\{2\})$$
$$= \frac{1}{21} + \frac{2}{21} = \frac{3}{21} = \frac{1}{7}$$

Exercise A.4

1. Assume $S = \{1,2,3\}$, $A = \{1\}$, $B = \{3\}$, $C = \{2\}$, $P(A) = \frac{1}{3}$, and $P(B) = \frac{1}{4}$. Find each of the following.
   (a) $P(C)$ (c) $P(\overline{A})$
   (b) $P(A \cup B)$ (d) $P(\overline{A} \cap B)$

2. Let S, A, B, C be as defined in question 1 above, but now assume $P(A) = 1/2$, $P(B) = 1/5$ and answer the question given there.

3. Show that $P(A) \leq 1$.
   [*Hint.* $P(A) + P(\overline{A}) = 1$ and $P(\overline{A}) \geq 0$.]

4. Given an experiment such that $P(A) = 1/2$, $P(B) = 1/2$, $P(A \cup B) = 2/3$. Compute these probabilities.
   (a) $P(\overline{A})$ (c) $P(A \cap B)$
   (b) $P(\overline{B})$ (d) $P(\overline{A} \cup B)$

5. A tetrahedron is a solid with four sides; each side is an equilateral triangle. Assume the four sides are numbered $1,2,3,4$; the tetrahedron is "loaded" in such a way that the probability of a given side being on the bottom, when it is rolled, is proportional to the number on the side. Thus the side numbered 2 is twice as likely to be on the bottom as is the side numbered 1, the side numbered 3 is three times as likely, and so on. What are the probabilities of the different sides being on the bottom?

6. Assume we are given an experiment and $P(A) = 1/2$, $P(B) = 1/3$, $P(A \cap B) = 1/4$. Compute the following.
   (a) $P(A \cup B)$ (c) $P(A \cap \overline{B})$
   (b) $P(\overline{A} \cup B)$ (d) $P(\overline{A} \cup \overline{B})$

7. If you know that $P(A \cup B) = 2/3$ and $P(A \cap B) = 1/2$, can you determine the values of $P(A)$ and $P(B)$?

8. A bowl contains several slips of paper. Each slip has either a 1 or a 2 or a 3 printed on it. One slip is to be drawn. The probability is $1/2$ that the number drawn is at least 2 and the probability is $5/6$ that the number drawn is 2 or less. Find the probabilities that the number drawn is
   (a) 1   (b) 2   (c) 3.

A.5   CONDITIONAL PROBABILITY AND INDEPENDENCE

In some cases we may be given partial information about the outcome of an experiment. For example, two dice may be rolled and a person observing may tell us that the sum is not 7. What can we do to use this information in computing the probability that one of the dice has one spot uppermost or that the sum is equal to 12? Or, if two prizes are to be given out by lot to the people at a party, having observed that the first prize went to someone else, what is the probability you will receive the other one? Granted that you know the answer to the first question on an exam, but must guess the answers to the remaining questions, what is the probability you will pass the exam?

In each of the situations above, we are given that an event did in fact occur and want then to compute the probability of an additional event also occurring. These questions are answered by conditional probabilities, relative frequencies of occurrence of the desired event over repetitions of the occurrence of the given event. To describe this situation more exactly, let S be the sample space for the experiment as a whole, and let A and B be two events. We are given that the experiment has been performed and that event B did occur. What is the probability that A occurred, given that B occurred? To distinguish this probability from $P(A)$, the probability of A occurring given no special information, we shall denote the probability of A occurring, given B has occurred, by $P(A|B)$.

Recall that an event occurs if any one of its elements is the outcome observed when the experiment is performed. If we are given that B occurred, then, we know that one of the elements belonging to B must have been the one observed; any element in $\overline{B}$ must not have occurred. Thus, given that B occurred, the set of elements belonging to B has effectively become the sample space. The event A may or may not have also occurred. It did occur if one of the elements belonging to both A and B ($A \cap B$) was observed; otherwise it did not occur. With the equally likely assumption for the original sample space, each of the individual elements of B is equally likely to have been the one observed. Thus, to measure the probability that A occurred, given B has occurred, it would seem reasonable to define

$$P(A|B) = \frac{\#(A \cap B)}{\#(B)}$$

If we divide both numerator and denominator of this quantity by $\#(S)$, we have

$$P(A|B) = \frac{\#(A \cap B)/\#(S)}{\#(B)/\#(S)} = \frac{P(A \cap B)}{P(B)}$$

This is in fact the definition of the conditional probability of A occurring, given that B has occurred. (It is the definition for all situations, not just the equally likely case.) It may or may not be the case that $P(A|B) = P(A)$. Let us examine some specific examples.

Example A.5.1

Two dice are rolled. Given that the sum of the two numbers is not 7, compute (a) the probability that one of the dice had one spot uppermost; and (b) the probability that the sum is 12.
Our sample space is

$$S = \{(x,y): x = 1,2,3,4,5,6; \quad y = 1,2,3,4,5,6\}$$

and we assume these elements are equally likely. The event given to have occurred is

$B:$ the sum is not 7

Thus

$$B = \{(x,y): x + y \neq 7\}$$

and, since there are six pairs in S such that $x + y = 7$, there must be 30 pairs such that $x + y \neq 7$. The set B is our conditional sample space. We find then $P(B) = 30/36 = 5/6$.

Define:

$A:$ One of the dice has one spot uppermost.
$C:$ The sum is 12.

Then, to compute $P(A|B)$ and $P(C|B)$, we must get $P(A \cap B)$ and $P(C \cap B)$; this means we must evaluate $\#(A \cap B)$ and $\#(C \cap B)$. We find

$$A \cap B = \{(1,1), (1,2), (1,3), (1,4), (1,5), (2,1), (3,1), (4,1), (5,1)\}$$

Thus, $\#(A \cap B) = 9$, $P(A \cap B) = 9/36 = 1/4$ and

$$P(A|B) = \frac{1/4}{5/6} = \frac{6}{20} = \frac{3}{10}$$

It is easily verified that $P(A) = 11/36$, so for this case $P(A|B) \neq P(A)$. Similarly

$$C \cap B = \{(6,6)\}$$

Thus,

$$P(C \cap B) = \frac{1}{36} \qquad P(C|B) = \frac{1/36}{5/6} = \frac{1}{30}$$

Since $P(C) = 1/36$, notice that $P(C) \neq P(C|B)$; conditional probabilities can in general be different (sometimes considerably so) than unconditional probabilities. If we define D as the first die that has one spot on it, then $P(D \cap B) = 5/36$,

$$P(D|B) = \frac{5}{36} \Big/ \frac{5}{6} = \frac{1}{6}$$

the same value as $P(D)$, the unconditional probability; conditional and unconditional probabilities *may* be equal.

It is easy to make errors in intuition in using conditional probabilities. Some classical "paradoxes" in probability theory are based on subtle changes in the given event. The following example discusses one of these paradoxes.

Example A.5.2

Assume that a married couple will have two children. Let us view the order of birth and the sexes of the two children as an experiment. There are four possibilities, and we shall use as a sample space

$$S = \{(b,b), (b,g), (g,b), (g,g)\}$$

where b stands for boy, g for girl, and the order in the pair represents the order of birth. Thus, (b,b) represents both children are boys, (b,g) that a boy is born first and a girl second, etc. Let us assume that the elements of S are equally likely to occur; birth records indicate that the relative frequencies of occurrence of the four cases listed in S are about .2704, .2496, .2496, and .2304, respectively. With our equally likely assumption, we are assuming the four elements to each have relative frequency 1/4. This slight discrepancy has no bearing on the point to be made. Let us define

   B: At least one child is a boy.
   A: The second child is a boy.

and we shall compute $P(A|B)$. We find $P(B) = 3/4$ and $P(A \cap B) = 1/2$. Thus,

$$P(A|B) = \frac{1/2}{3/4} = \frac{2}{3}$$

Now suppose you are married, you have one child (a boy) and are expecting your second child; since your first child was a boy, then surely at least one of your two children (after the birth of the second) will be a boy. To some people, the above conditional probability seems to imply that the probability your second child will be a boy is 2/3, rather than 1/2, which equally likely assumptions would imply. Can you see the flaw in this type of reasoning?

An important use of conditional probabilities is in the computation of the probabilities of intersections of events. Since

$$P(A|B) = \frac{P(A \cap B)}{P(B)}$$

notice that $P(A \cap B) = P(B)P(A|B)$, so that $P(A \cap B)$ may be computed directly from knowledge of $P(B)$ and $P(A|B)$. Interchanging the roles of A and B above, we also find

$$P(A \cap B) = P(A)P(B|A)$$

In certain types of problems it is easy to evaluate $P(A \cap B)$ using one or the other of these two formulas. The following example is of this type.

Example A.5.3

Assume that the dairy case in a grocery store contains 10 cartons of milk, 3 of which are sour. Two people, one after the other, each purchase a carton of milk. What is the probability that both get fresh cartons? Define

A: The first person gets a fresh carton.
B: The second person gets a fresh carton.

Then the event that both get fresh cartons is $A \cap B$. It is easy to see that $P(A) = 7/10$, $P(B|A) = 6/9$ since after the first person selects a fresh carton, there would be 9 left, 6 of which are fresh. Thus

$$P(A \cap B) = P(A)P(B|A)$$

$$= \frac{7}{10} \cdot \frac{6}{9} = \frac{7}{15}$$

Let us also compute $P(B)$, the unconditional probability that the second person gets a fresh carton. Using a Venn diagram it is easy to see that

$$B = (A \cap B) \cup (\overline{A} \cap B)$$

(Figure A.5.1). Furthermore, since no element of S could belong to both $A \cap B$ and $\overline{A} \cap B$, we see then that

$$\#(B) = \#(A \cap B) + \#(\overline{A} \cap B)$$

and thus

$$P(B) = P(A \cap B) + P(\overline{A} \cap B)$$

$$= P(A)P(B|A) + P(\overline{A})P(B|\overline{A})$$

$$= \frac{7}{10} \cdot \frac{6}{9} + \frac{3}{10} \cdot \frac{7}{9}$$

$$= \frac{7}{10}$$

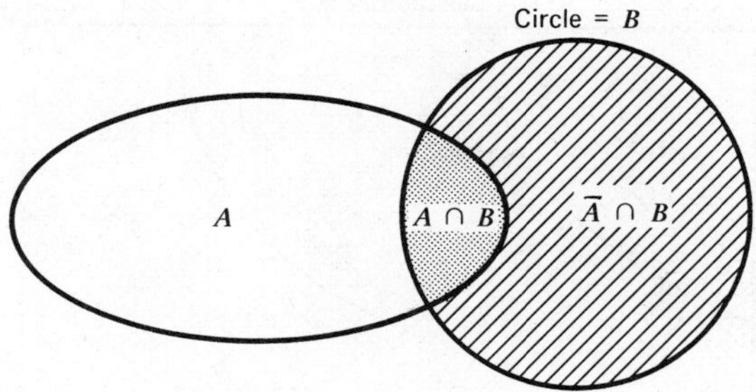

Figure A.5.1 $B = (A \cap B) \cup (\overline{A} \cap B)$.

That is, the unconditional probability the second person gets a fresh carton of milk is 7/10, the same as the probability that the first person gets a fresh carton. This result does not seem so strange if we recall the relative frequency interpretation of probability. Over repetitions of this experiment, the second person to buy will get a fresh carton 70% of the time, just as the first person will.

The Reverend Thomas Bayes was an early contributor to the theory of probability. One of his works, published after his death, contained a theorem that today bears his name. It concerns a manipulation of conditional probabilities that is fairly straightforward to establish. It was looked on with suspicion and not used until quite modern times. Today, it is the basis for a system of statistical inference generally referred to as Bayesian techniques. These procedures are quite similar to the classical procedures described in this book, but are distinguished by the fact that they require the specification of subjective or personal probabilities for their use. These subjective probabilities, describing his "degree of belief," must be supplied by the person attempting to make a statistical inference. Bayes' theorem is presented below.

Theorem A.5.1

Assume that a sample space S has been partitioned into k parts, $A_1, A_2, \ldots, A_k$. (This means the intersection of any two A_i's is empty, they have no elements in common, and $A_1 \cup A_2 \cup \ldots \cup A_k = S$. See Figure A.5.2). Then, if E is any event

$$P(A_1|E) = \frac{P(E|A_1)P(A_1)}{\sum_{j=1}^{k} P(E|A_j)P(A_j)}$$

Similar equations hold for $P(A_2|E), P(A_3|E), \ldots, P(A_k|E)$.

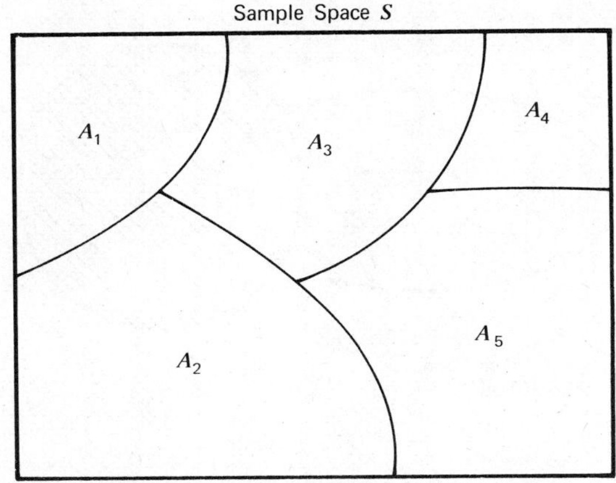

Figure A.5.2   Partition of S into five pieces.

Proof.
Granted $A_1, A_2, \ldots, A_k$ is a partition of S, these events also partition E into nonoverlapping pieces (Figure A.5.3). Thus we have

$$E = (A_1 \cap E) \cup (A_2 \cap E) \cup \ldots \cup (A_k \cap E)$$

and, since the A_j's have no elements in common

$$P(E) = P[(A_1 \cap E) \cup (A_2 \cap E) \cup \ldots \cup (A_k \cap E)]$$
$$= P(A_1 \cap E) + P(A_2 \cap E) + \ldots + P(A_k \cap E)$$

Recalling that the probability of an intersection can be written as the product of an unconditional and a conditional probability, we have then

$$P(A_1 \cap E) = P(A_1)P(E|A_1)$$

with similar expressions for $P(A_2 \cap E), \ldots, P(A_k \cap E)$.
Thus

$$P(E) = P(A_1)P(E|A_1) + P(A_2)P(E|A_2) + \ldots + P(A_k)P(E|A_k)$$
$$= \sum_{j=1}^{k} P(E|A_j)P(A_j)$$

From the definition of conditional probability

$$P(A_1|E) = \frac{P(A_1 \cap E)}{P(E)}$$

$$= \frac{P(E|A_1)P(A_1)}{\sum_{j=1}^{k} P(E|A_j)P(A_j)}$$

after substituting the relations derived above; this is the desired result and the theorem is proved.

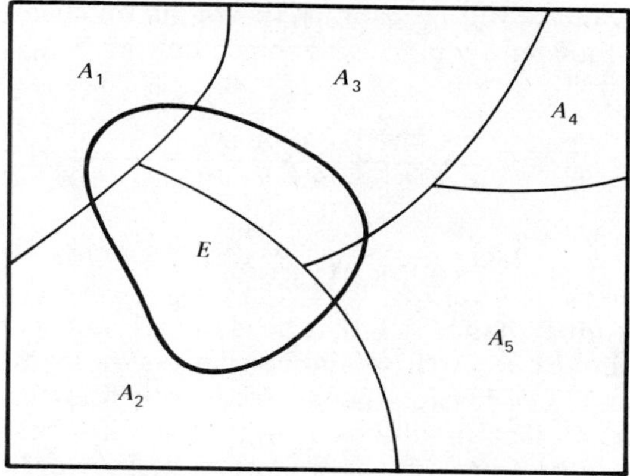

Figure A.5.3   Any event $E \subset S$ is also partitioned.

The following example applies Bayes' theorem to a particular problem.

Example A.5.4

Assume the probability is .95 that a jury selected to try a criminal case will arrive at the correct verdict, regardless of whether the defendant is guilty or not. That is, given a guilty defendant on trial, the probability is .95 that the jury finds him guilty; given an innocent defendant on trial, the probability is also .95 that the jury will find him innocent. We shall also assume that the local police force is diligent in its duties and that 99% of all people brought to court are in fact guilty. Granted these assumptions, let us compute the probability that a defendant is in fact innocent, given that the jury found him innocent.

This computation can be made using Bayes' theorem as follows. Let A_1 be the event that the defendant is in fact innocent and let A_2 be the event he is in fact guilty. Then A_1 and A_2 form a partition of the sample space, the set of possibilities regarding this defendant. Let E be the event that he is found innocent by the jury; $\overline{E}$ is the event, then, that he is found guilty. The numerical information given above, then, is

$$P(E|A_1) = .95 \qquad P(\overline{E}|A_2) = .95 \qquad P(A_1) = .01$$

A conditional probability measure behaves just like any other, and we also find

$$P(\overline{E}|A_1) = 1 - P(E|A_1) = .05$$

$$P(E|A_2) = 1 - P(\overline{E}|A_2) = .05$$

$$P(A_2) = 1 - P(A_1) = .99$$

We want to evaluate the probability that he is innocent, given that the jury found him innocent, which is denoted then by $P(A_1|E)$. From Bayes' theorem we have

$$P(A_1|E) = \frac{P(E|A_1)P(A_1)}{P(E|A_1)P(A_1) + P(E|A_2)P(A_2)}$$

$$= \frac{(.95)(.01)}{(.95)(.01) + (.05)(.99)} = .161$$

Thus, there is about 1 chance in 6 that he is in fact innocent, given that the jury finds him innocent. At first glance this result seems much smaller than it should be. This is because people frequently feel that $P(E|A_1)$ (which is .95) should be similar in value to $P(A_1|E)$ (which turns out to be .161). These two probabilities are quite different since they are conditioning on different events. Before his trial, the probability that the defendant was innocent was only .01 (since 99% of those brought to trial are guilty). After the defendant was found innocent, this probability has increased to .161; we are 16 times as sure he is innocent after the trial as we were before the trial.

As we have seen in the examples above, it may be that $P(A|B)$ and $P(A)$ are equal, or they may be considerably different. If the two are equal, then the knowledge that B had occurred did not in any way change the probability that A occurred. Such events will be called independent. The formal definition of independence is given below.

Definition A.5.1. The events A and B are independent if and only if

$$P(A \cap B) = P(A)P(B)$$

Notice immediately, then, that if A and B are independent, then

$$P(A|B) = \frac{P(A \cap B)}{P(B)} = \frac{P(A)P(B)}{P(B)} = P(A)$$

Thus the conditional probability is necessarily identical with the unconditional if the two events are independent.

Example A.5.5

Let us reexamine Example A.5.2, in which we considered the sexes and orders of birth of the two children in the family. The sample space adopted there was

$$S = \{(b,b), (b,g), (g,b), (g,g)\}$$

and we assumed the elements were equally likely to occur. Let A be the event that the second child is a boy, B is the event that at least one child is a boy, and C is the event that the first child is a boy. Thus, as subsets, these events are

$$A = \{(b,b), (g,b)\}$$
$$B = \{(b,b), (b,g), (g,b)\}$$
$$C = \{(b,b), (b,g)\}$$

With our equally likely assumption then

$$P(A) = \tfrac{2}{4} = \tfrac{1}{2} \qquad P(B) = \tfrac{3}{4} \qquad P(C) = \tfrac{2}{4} = \tfrac{1}{2}$$

We also find

$$P(A \cap B) = \tfrac{2}{4} = \tfrac{1}{2} \qquad P(A \cap C) = \tfrac{1}{4} \qquad P(B \cap C) = \tfrac{2}{4} = \tfrac{1}{2}$$

Since $P(A \cap C) = 1/4 = 1/2 \cdot 1/2 = P(A)P(C)$, the events A and C are independent; however since $P(A \cap B) = 1/2 \neq P(A)P(B)$ the two events A and B are not independent. Similarly, it is easily seen that B and C are not independent.

A frequent use of the concept of independence is to assign probabilities to intersections of two (or more) events. If A and B are independent and we know $P(A)$ and $P(B)$, then we can compute $P(A \cap B)$ as the product $P(A)P(B)$.

Example A.5.6

Suppose we flip a fair coin; the adjective fair means we expect to observe heads and tails equally frequently. Using H for heads then, we assume

$$P(H) = \tfrac{1}{2}$$

Now suppose we flip the same fair coin two times; it seems logical that the face we observe on the first flip should have no affect on what we get on the second flip. The events

$$H_1 \text{: Head on first flip}$$
$$H_2 \text{: Head on second flip}$$

should be independent. This means, then,

$$P(\text{heads on both flips}) = P(H_1 \cap H_2) = \tfrac{1}{2} \cdot \tfrac{1}{2} = \tfrac{1}{4}$$

The probability we would get three heads in a row, then, is

$$\frac{1}{2} \cdot \frac{1}{2} \cdot \frac{1}{2} = \frac{1}{8}$$

four heads in a row has probability

$$\frac{1}{2} \cdot \frac{1}{2} \cdot \frac{1}{2} \cdot \frac{1}{2} = \frac{1}{16}$$

and so on. The assumed independence from one flip to another implies that probabilities of intersections of independent events can be obtained by multiplying together the probabilities for the individual events. We shall explore this type of idea to a greater extent in the next section.

Exercise A.5

1. One red die and one white die are rolled. Given that the sum of the two numbers is 7, compute the probabilities that the red number was

    (a) 1   (b) 2   (c) 6

2. Two fair coins are tossed. What is the conditional probability there were two heads, given that at least one head occurred?

3. A nickel and a dime are each tossed (one time). What is the conditional probability that the nickel showed a head, given that at least one head occurred?

4. A man has four keys in his pocket, one of which fits his front door. On a dark night he cannot see to choose the correct one and will thus try them one at a time. Compute the probabilities he finds the right key on his first, second, third, and fourth tries.

5. Urn I contains three red and three blue balls. Urn II contains one red and five blue balls. A fair die is rolled one time; if a one occurs on the die, a ball will be selected at random from urn I; otherwise a ball will be selected at random from urn II. What is the probability of getting a red ball with this procedure?

6. In Example A.5.4 compute the conditional probabilities the defendant is truly guilty:
    (a) Given that he is found innocent.
    (b) Given that he is found guilty.

7. A chest contains two drawers. Drawer 1 has four gold and one silver coin in it; drawer 2 has three gold and three silver coins. Blindfolded, you select a drawer at random, remove a coin and after your blindfold is removed, you find it is gold. What is the probability you selected it from drawer 1?

8. According to 1970 data from the U.S. Department of Health, Education and Welfare, the probability a 20-year old United States citizen will die before age 25 is .0074. Assume you and a friend are both 20 and that you will die independently.
    (a) What is the probability neither of you survive to age 25?
    (b) What is the probability both of you survive to age 25?

9. To decide who will pay for their coffee after class each morning, two friends "match" coins (if HH or TT occur, A pays, and if HT or TH occur, B pays).

   (a) What is the probability that A must pay on any given day?

   (b) What is the probability that A must pay on each of 10 successive days?

10. In the game of "odd man out," three people, A, B, C, each flip a coin one time. If two of the faces match and the third does not, the third person is called the odd man (and he loses). Assuming the coins are fair, and independent, what is the probability of there being an odd man on a given try (one flip of all three coins)?

A.6   INDEPENDENT TRIALS AND THE BINOMIAL RANDOM VARIABLE

Many applied examples concern experiments that consist of repeated independent Bernoulli trials. A Bernoulli trial is any experiment whose possible outcomes have been grouped into only two classes, called success and failure. Examples of Bernoulli trials are a flip of a coin (head = success, tail = failure), a student's performance in a course (pass = success, fail = failure), firing a rifle at a target (hit = success, miss = failure), outcome of a basketball game (win = success, lose = failure), response of a sick person to a drug (cured = success, anything else = failure), and so on. The actual experiment may be anything; the only requirement that it be a Bernoulli trial is that any outcome that could be observed must be labeled either a success or a failure.

The sample space for a single Bernoulli trial, then, is

$$S = \{s,f\}$$

where s represents success and f represents failure. The probability measure for a Bernoulli trial is especially simple. The probability of the event success is

$$P(\{s\}) = p$$

from which it follows that the probability of the event failure must be

$$P(\{f\}) = 1 - p = q$$

The quantity p is called a parameter; since it is a probability, it must lie between 0 and 1. Depending on the nature of the Bernoulli trial being studied, the value of p could be 1/2 or 1/4 or .1 or any number in the interval from 0 to 1.

Suppose, now, we have an experiment that consists of a sequence of Bernoulli trials, one after another. If we are willing to assume that the outcome of any single trial has no effect on the outcome of any other trial, then we say the Bernoulli trials are *independent*. For example, suppose we are going to flip the same coin five times. Each individual flip is a Bernoulli trial and, as long as we assume that the occurrence of a head on any flip has no effect on whether or not a head will occur on any other, these would be called five independent Bernoulli trials. Suppose each of 11 people have the same sickness and that each of them is treated with the same dosage of the same drug; if we assume that the fact any one of them may be cured has no effect on whether or not any other one is also cured, these treatments are called independent Bernoulli trials.

Whenever we have an experiment that consists of independent Bernoulli trials, the natural sample space to use is one that records the possible occurrences of successes (s) and failures (f) and their orders of occurrence. For example, if our experiment consists of two independent trials, S has $2^2 = 4$ elements;

$$S = \{(s,s), (s,f), (f,s), (f,f)\}$$

If our experiment consists of three independent trials, the sample space has $2^3 = 8$ elements.

$$S = \{(s,s,s), (s,s,f), (s,f,s), (s,f,f),$$
$$(f,s,s), (f,s,f), (f,f,s), (f,f,f)\}$$

More generally, if the experiment consists of n independent trials, then the sample space S will have 2^n elements, recording the possible occurrences of successes and failures on the n trials:

$$S = \{(x_1, x_2, \ldots, x_n): x_i = s \text{ or } f, i = 1, 2, \ldots, n\}$$

If the experiment consists of *repeated* independent trials, we mean that the trials are independent and the probability of success is p, the same value, for every trial. In such a case the independence assumption then implies a very simple rule for computing probabilities of the single-element events: we multiply together the probabilities of observing the appropriate outcomes on the corresponding trials. Suppose a Brazilian cruzeiro (worth about 15¢) is to be flipped two times; we assume that the probability of heads (success) occurring is $p = .6$ for each flip and that the flips are independent. We are also assuming, then, that the probability of tails occurring (failure) is $q = 1 - p = .4$, for each flip. The sample space has four elements and, because we assume the flips to be independent, the probabilities of occurrence of the different elements of S are as given in the following table.

Event	$\{(s,s)\}$	$\{(s,f)\}$	$\{(f,s)\}$	$\{(f,f)\}$
Probability	$(.6)(.6) = .36$	$(.6)(.4) = .24$	$(.4)(.6) = .24$	$(.4)(.4) = .16$

As always, the probability of occurrence of any other event, then, is given by adding together the probabilities of the single-element events making it up. For example, the probability that we get the same face on both flips is

$$P(\{(s,s)\}) + P(\{(f,f)\}) = .36 + .16 = .52$$

the probability we get at least one head is

$$P(\{(s,s)\}) + P(\{(s,f)\}) + P(\{(f,s)\}) = .36 + .24 + .24 = .84$$

Suppose we perform three repeated independent Bernoulli trials, and the probability of success is p for each trial; the probability of failure is $q = 1 - p$ for each trial. Then, again, the probabilities for the different single-element events in S are given by multiplying together the probabilities for the appropriate outcomes on the individual trials. These are displayed below.

Event	Probability
$\{(s,s,s)\}$	$p \cdot p \cdot p = p^3$
$\{(s,s,f)\}$	$p \cdot p \cdot q = p^2 q$
$\{(s,f,s)\}$	$p \cdot q \cdot p = p^2 q$
$\{(s,f,f)\}$	$p \cdot q \cdot q = pq^2$
$\{(f,s,s)\}$	$q \cdot p \cdot p = p^2 q$
$\{(f,s,f)\}$	$q \cdot p \cdot q = pq^2$
$\{(f,f,s)\}$	$q \cdot q \cdot p = pq^2$
$\{(f,f,f)\}$	$q \cdot q \cdot q = q^3$

Again, the probabilities for other events are given by adding together the probabilities of the single-element events making it up. For example, then, the probability we get (exactly) one success in the three trials is

$$P(1 \text{ success}) = P(\{(s,f,f)\}) + P(\{(f,s,f)\}) + P(\{(f,f,s)\}) = 3pq^2$$

Similarly, the probability we get (exactly) two successes is $3p^2q$, the sum of the probabilities for the single-element events making it up.

If our experiment consists of n repeated independent Bernoulli trials, with probability of success equal to p for each trial, the same rule is used to compute the probabilities of occurrence of the single-element events: we multiply together the probabilities for the appropriate results on the given trials. For example, the probability we get all successes then is p^n; the probability that we get all failures is q^n. The probability that the *first* trial is a success and the remaining ones are all failures is pq^{n-1}, and so on. Once the values are known for p $(q = 1 - p)$ and n, we know the probabilities of occurrence of all the single-element events; these then can be used to compute the probability for any other event of interest.

Example A.6.1

Assume a fair die is rolled five times. By fair, we mean that the six faces are equally likely to occur on each roll. If we call the occurrence of a 6 a success (failure consists of the occurrence of any other number), our experiment then can be looked at as being $n=5$ repeated independent Bernoulli trials with probability of success $p = 1/6$, $q = 5/6$. The probability we get all 6's then is $(1/6)^5 = 1/7776$, the probability we get no 6's is $(5/6)^5 = 3125/7776$. The probability we get a 6 on the *first* roll and no more 6's is $1/6(5/6)^4 = 625/7776$. What is the probability we get exactly one 6 (on some roll)? The roll on which the single 6 occurs could be any of the five; the probability of any outcome including exactly one 6 is $1/6(5/6)^4$. Since there are 5 single-element events including exactly one 6 (the 6 could occur on the 1st, 2nd, 3rd, 4th, or 5th roll) we find the probability of exactly one 6 occurring is $5(1/6)(5/6)^4 = 3125/7776$.

Example A.6.2

Doug takes a 10-question multiple choice examination. Each question has 3 parts. Assume he is totally unprepared and must guess at the correct answer for every question. We can then look at his performance on the exam as an experiment consisting of $n = 10$ repeated independent Bernoulli trials with $p = 1/3$. The probability he gets all 10 questions correct is $(1/3)^{10} = 1/59,049$.

Suppose he will pass the exam if he gets 8 or more answers correct. What is the probability he will pass? This event is the union of many single-element events, all those elements in S for which he got exactly 8 correct or exactly 9 correct or all 10 answers correct. Any outcome with exactly 8 correct answers has probability $(1/3)^8(2/3)^2 = 4/59,049$. How many different elements are there in S, corresponding to his getting 8 questions correct? The 8 questions he answers correctly could be any 8 out of the 10.

Thus, the number of elements with 8 answers correct is $\binom{10}{8} = 45$. The probability that he answers exactly 8 questions correctly, then, is

$$\binom{8}{2} \left(\frac{1}{3}\right)^8 \left(\frac{2}{3}\right)^2 = \frac{180}{59,049}$$

In a similar way, any element for which he got 9 answers correct has probability $(1/3)^9 (2/3) = 2/59,049$; since there are $\binom{10}{9} = 10$ such elements in S, the probability he gets exactly 9 questions correct is

$$\binom{10}{9} \left(\frac{1}{3}\right)^9 \left(\frac{2}{3}\right) = \frac{20}{59,049}$$

The probability he passes the exam then is

$$\binom{8}{2}\left(\frac{1}{3}\right)^8\left(\frac{2}{3}\right)^2 + \binom{9}{1}\left(\frac{1}{3}\right)\left(\frac{2}{3}\right) + \left(\frac{1}{3}\right)^{10} = \frac{201}{59,049} = .0034$$

He has an extremely small chance of passing this exam by just guessing at all the answers.

Let us now examine the binomial random variable. It is defined for an experiment that consists of n repeated independent Bernoulli trials (see Definition 5.2). Formally, let X be the number of successes that will be observed on the n trials. We can think, then, of X as a function that associates a real number with every element of the sample space. (It is called a *real-valued function* of the elements of S.) The element of S that represents all successes has the number n associated with it. The element that represents all failures has the number 0 associated with it. Any element that represents one success (no matter which trial) and failures on all other trials has the number 1 associated with it, and so on. This binomial random variable is simply a rule associating numbers with outcomes.

The range of X is easily specified, no matter what the particular experiment may be. If we are going to perform n repeated independent Bernoulli trials, then certainly the smallest number of successes we might observe is 0 and the largest possible number is n; any integer between these extremes could occur as the observed value for X. Thus, the range of a binomial random variable is

$$R_X = \{x: x = 0, 1, 2, \ldots, n\}$$

The probability measure for the binomial random variable gives the probabilities that each of these elements of R_X is the observed value. That is, the probability measure gives

$$P(X=x) \qquad \text{for each } x \epsilon R_X$$

We shall now derive the binomial probability measure, which is discussed more fully in Section 5.1.

It is easy first to derive $P(X=0)$ and $P(X=n)$, the probabilities of the two extremes of R_X. We shall have $X=n$ if and only if we observe a success on every trial; we saw above that the probability of this occurring is p^n. Thus

$$P(X=n) = p^n = \binom{n}{n}p^{n-0}q^0$$

since $\binom{n}{n} = 1$ and $q^0 = 1$. This second expression is seen to fit into the general notation scheme useful for all $x \epsilon R_X$. We shall find $X = 0$ if and only if we get no successes in the n trials, that is, we must observe failures each time. The probability of this happening is q^n. Thus

$$P(X = 0) = q^n = \binom{n}{0} p^0 q^{n-0}$$

since $\binom{n}{0} = 1$, $p^0 = 1$.

What about the other elements of R_X? Let us first consider $P(X=1)$. We shall find $X=1$ if and only if exactly one success occurs in the n trials. Any outcome of the n trials that includes exactly one success has probability pq^{n-1}. The one success that is observed could occur on any of the n trials. Thus we have

$$P(X=1) = npq^{n-1} = \binom{n}{1} pq^{n-1}$$

since $\binom{n}{1} = n$. Consider now some general value $r \epsilon R_X$. Every outcome of the experiment that includes exactly r successes has probability $p^r q^{n-r}$; there are $\binom{n}{r}$ different elements of S, each containing r successes, since the successes could occur on any r of the n trials. Thus

$$P(X=r) = \binom{n}{r} p^r q^{n-r}$$

a formula that is valid for any $r \epsilon R_X$. This is the binomial probability measure; once we know the number of trials (n) and the probability of success for each trial (p), we can completely evaluate the binomial probability measure. We are assuming the experiment consists of n repeated independent Bernoulli trials when we use this measure.

Example A.6.3

Assume that a doctor has 15 patients, each with the same illness. He treats them all with the same medication. The probability that each of them will be cured with this medication is .7. What is the probability that they are all cured? What is the probability that at least 12 of them are cured?

We shall assume independence between the patients. Whether or not patient 1 happens to be cured has no effect on the cure of any other patient. Let X be the number of patients that are cured. Then X is binomial with $n = 15$, $p = .7$. The probability all 15 are cured, then, is

$$P(X = 15) = (.7)^{15} = .0047$$

The probability that at least 12 are cured is

$$P(X \geq 12) = P(X = 12) + P(X = 13) + P(X = 14) + P(X = 15)$$

$$= .1700 + .0916 + .0305 + .0047$$

$$= .2968$$

(The numerical values for these probabilities come from Table A.1; see Section 5.1 for a discussion of the use of the tables.) Section 5.1 gives many more examples of the binomial probability measure and its use.

We have seen one type of random variable and how it is defined, the binomial random variable. All of the other random variables discussed in this book can also be thought of as real-valued functions of the elements of a sample space S. The probability measure for the sample space, then, implies the appropriate probability measure for the random variable. The interested reader will find this type of discussion in *Introduction to Probability Theory and Statistical Inference*, Second Edition, by Harold J. Larson, John Wiley and Sons, New York, 1974.

Exercise A.6

1. Six fair dice are rolled one time. What is the probability each of the 6 different faces occur?

2. The game of odd-man-out is played as follows: Three people (Joe, Hugh, and Rachel) each flip a coin one time. Assume all three coins are fair. If two of the faces on the coins are alike, and the third is different, the person tossing the third coin is the odd man.
    (a) On one play of this game, what is the probability there will be an odd man?
    (b) What is the probability that Joe is the odd man on one play?
    (c) Given that there is an odd man on a given play, what is the probability that it is Hugh?

3. Assume that 12 people go deep-sea fishing on a sport fishing boat. Also assume that the probability is .1 that each of them will get seasick during the voyage, and that the cases of seasickness occur independently. (Can you see any reason to question this assumption?)
    (a) Let X be the number that get seasick during the voyage. What is the probability measure for X?
    (b) Compute the probability that two of them get seasick during the voyage.

4. A lady claims she can taste a cup of tea (with milk) and tell whether the tea or the milk was put into the cup first. She agrees to participate in an experiment to demonstrate her skill. We prepare 5 pairs of cups; in one

cup of each pair we put the tea in first and in the other cup we put the milk in first. She is then to examine each pair and identify which cup in the pair had the tea added first.

(a) Assume she has no special ability and guesses when she chooses the cup that had the tea added first. Let X be the number of cups she chooses correctly out of the 5 pairs. What is the probability measure for X?

(b) What is the probability she correctly identifies the tea-first cups in at least four of the five pairs?

5. Assume you are taking 4 courses this quarter and that you have probabilities .3, .4, .5, and .6 of getting an A in courses 1, 2, 3, 4, respectively. (Thus, the probability you get an A in course 2 is .4 and the probability you get an A in course 4 is .6.) If the grades you receive are independently determined, what is the probability you get

(a) Four A's?

(b) Three A's?

6. Assume the probability you are late to one or more classes, in any week, is .1. What is the probability that you are not late to any of your classes in an 11-week quarter?

7. Sixty percent of the people that enter an electronics store make one or more purchases while they are there. If 18 people enter the store during its first hour of business, what is the probability that at least 15 of them make a purchase?

8. Every item coming off a production line either is or is not defective. Assume that defectives occur independently and that the probability that any item is defective is .05. If 20 items from this line are inspected, what is the probability

(a) None are defective?

(b) One is defective?

(c) Two or more are defective?

Table A.1 Binomial Probability Measure;

Tabular Entry is $P(X = r) = \binom{n}{r} p^r q^{n-r}$

n	r	.05	.10	.15	.20	.25	.30	.35	.40	.45	.50
1	0	.9500	.9000	.8500	.8000	.7500	.7000	.6500	.6000	.5500	.5000
	1	.0500	.1000	.1500	.2000	.2500	.3000	.3500	.4000	.4500	.5000
2	0	.9025	.8100	.7225	.6400	.5625	.4900	.4225	.3600	.3025	.2500
	1	.0950	.1800	.2550	.3200	.3750	.4200	.4550	.4800	.4950	.5000
	2	.0025	.0100	.0225	.0400	.0625	.0900	.1225	.1600	.2025	.2500
3	0	.8574	.7290	.6141	.5120	.4219	.3430	.2746	.2160	.1664	.1250
	1	.1354	.2430	.3251	.3840	.4219	.4410	.4436	.4320	.4084	.3750
	2	.0071	.0270	.0574	.0960	.1406	.1890	.2389	.2880	.3341	.3750
	3	.0001	.0010	.0034	.0080	.0156	.0270	.0429	.0640	.0911	.1250
4	0	.8145	.6561	.5220	.4096	.3164	.2401	.1785	.1296	.0915	.0625
	1	.1715	.2916	.3685	.4096	.4219	.4116	.3845	.3456	.2995	.2500
	2	.0135	.0486	.0975	.1536	.2109	.2646	.3105	.3456	.3675	.3750
	3	.0005	.0036	.0115	.0256	.0469	.0756	.1115	.1536	.2005	.2500
	4	.0000	.0001	.0005	.0016	.0039	.0081	.0150	.0256	.0410	.0625
5	0	.7738	.5905	.4437	.3277	.2373	.1681	.1160	.0778	.0503	.0312
	1	.2036	.3280	.3915	.4096	.3955	.3602	.3124	.2592	.2059	.1562
	2	.0214	.0729	.1382	.2048	.2637	.3087	.3364	.3456	.3369	.3125
	3	.0011	.0081	.0244	.0512	.0879	.1323	.1811	.2304	.2757	.3125
	4	.0000	.0004	.0022	.0064	.0146	.0284	.0488	.0768	.1128	.1562
	5	.0000	.0000	.0001	.0003	.0010	.0024	.0053	.0102	.0185	.0312
6	0	.7351	.5314	.3771	.2621	.1780	.1176	.0754	.0467	.0277	.0156
	1	.2321	.3543	.3993	.3932	.3560	.3025	.2437	.1866	.1359	.0938
	2	.0305	.0984	.1762	.2458	.2966	.3241	.3280	.3110	.2780	.2344
	3	.0021	.0146	.0415	.0819	.1318	.1852	.2355	.2765	.3032	.3125
	4	.0001	.0012	.0055	.0154	.0330	.0595	.0951	.1382	.1861	.2344
	5	.0000	.0001	.0004	.0015	.0044	.0102	.0205	.0369	.0609	.0938
	6	.0000	.0000	.0000	.0001	.0002	.0007	.0018	.0041	.0083	.0156
7	0	.6983	.4783	.3206	.2097	.1335	.0824	.0490	.0280	.0152	.0078
	1	.2573	.3720	.3960	.3670	.3115	.2471	.1848	.1306	.0872	.0547
	2	.0406	.1240	.2097	.2753	.3115	.3177	.2985	.2613	.2140	.1641
	3	.0036	.0230	.0617	.1147	.1730	.2269	.2679	.2903	.2918	.2734
	4	.0002	.0026	.0109	.0287	.0577	.0972	.1442	.1935	.2388	.2734
	5	.0000	.0002	.0012	.0043	.0115	.0250	.0466	.0774	.1172	.1641
	6	.0000	.0000	.0001	.0004	.0013	.0036	.0084	.0172	.0320	.0547
	7	.0000	.0000	.0000	.0000	.0001	.0002	.0006	.0016	.0037	.0078

Table A.1 Binomial Probability Measure;

Tabular Entry is $P(X = r) = \binom{n}{r} p^r q^{n-r}$ (Continued)

n	r	.05	.10	.15	.20	.25	.30	.35	.40	.45	.50
8	0	.6634	.4305	.2725	.1678	.1001	.0576	.0319	.0168	.0084	.0039
	1	.2793	.3826	.3847	.3355	.2670	.1977	.1373	.0896	.0548	.0312
	2	.0515	.1488	.2376	.2936	.3115	.2965	.2587	.2090	.1569	.1094
	3	.0054	.0331	.0839	.1468	.2076	.2541	.2786	.2787	.2568	.2188
	4	.0004	.0046	.0185	.0459	.0865	.1361	.1875	.2322	.2627	.2734
	5	.0000	.0004	.0026	.0092	.0231	.0467	.0808	.1239	.1719	.2188
	6	.0000	.0000	.0002	.0011	.0038	.0100	.0217	.0413	.0703	.1094
	7	.0000	.0000	.0000	.0001	.0004	.0012	.0033	.0079	.0164	.0312
	8	.0000	.0000	.0000	.0000	.0000	.0001	.0002	.0007	.0017	.0039
9	0	.6302	.3874	.2316	.1342	.0751	.0404	.0207	.0101	.0046	.0020
	1	.2985	.3874	.3679	.3020	.2253	.1556	.1004	.0605	.0339	.0176
	2	.0629	.1722	.2597	.3020	.3003	.2668	.2162	.1612	.1110	.0703
	3	.0077	.0446	.1069	.1762	.2336	.2668	.2716	.2508	.2119	.1641
	4	.0006	.0074	.0283	.0661	.1168	.1715	.2194	.2508	.2600	.2461
	5	.0000	.0008	.0050	.0165	.0389	.0735	.1181	.1672	.2128	.2461
	6	.0000	.0001	.0006	.0028	.0087	.0210	.0424	.0743	.1160	.1641
	7	.0000	.0000	.0000	.0003	.0012	.0039	.0098	.0212	.0407	.0703
	8	.0000	.0000	.0000	.0000	.0001	.0004	.0013	.0035	.0083	.0176
	9	.0000	.0000	.0000	.0000	.0000	.0000	.0001	.0003	.0008	.0020
10	0	.5987	.3487	.1969	.1074	.0563	.0282	.0135	.0060	.0025	.0010
	1	.3151	.3874	.3474	.2684	.1877	.1211	.0725	.0403	.0207	.0098
	2	.0746	.1937	.2759	.3020	.2816	.2335	.1757	.1209	.0763	.0439
	3	.0105	.0574	.1298	.2013	.2503	.2668	.2522	.2150	.1665	.1172
	4	.0010	.0112	.0401	.0881	.1460	.2001	.2377	.2508	.2384	.2051
	5	.0001	.0015	.0085	.0264	.0584	.1029	.1536	.2007	.2340	.2461
	6	.0000	.0001	.0012	.0055	.0162	.0368	.0689	.1115	.1596	.2051
	7	.0000	.0000	.0001	.0008	.0031	.0090	.0212	.0425	.0746	.1172
	8	.0000	.0000	.0000	.0001	.0004	.0014	.0043	.0106	.0229	.0439
	9	.0000	.0000	.0000	.0000	.0000	.0001	.0005	.0016	.0042	.0098
	10	.0000	.0000	.0000	.0000	.0000	.0000	.0000	.0001	.0003	.0010
11	0	.5688	.3138	.1673	.0859	.0422	.0198	.0088	.0036	.0014	.0005
	1	.3293	.3835	.3248	.2362	.1549	.0932	.0518	.0266	.0125	.0054
	2	.0867	.2131	.2866	.2953	.2581	.1998	.1395	.0887	.0513	.0269
	3	.0137	.0710	.1517	.2215	.2581	.2568	.2254	.1774	.1259	.0806
	4	.0014	.0158	.0536	.1107	.1721	.2201	.2428	.2365	.2060	.1611
	5	.0001	.0025	.0132	.0388	.0803	.1321	.1830	.2207	.2360	.2256
	6	.0000	.0003	.0023	.0097	.0268	.0566	.0985	.1471	.1931	.2256
	7	.0000	.0000	.0003	.0017	.0064	.0173	.0379	.0701	.1128	.1611
	8	.0000	.0000	.0000	.0002	.0011	.0037	.0102	.0234	.0462	.0806
	9	.0000	.0000	.0000	.0000	.0001	.0005	.0018	.0052	.0126	.0269
	10	.0000	.0000	.0000	.0000	.0000	.0000	.0002	.0007	.0021	.0054
	11	.0000	.0000	.0000	.0000	.0000	.0000	.0000	.0000	.0002	.0005

Table A.1 Binomial Probability Measure;

$$\text{Tabular Entry is } P(X = r) = \binom{n}{r} p^r q^{n-r} \qquad \text{(Continued)}$$

n	r	.05	.10	.15	.20	.25	.30	.35	.40	.45	.50
12	0	.5404	.2824	.1422	.0687	.0317	.0138	.0057	.0022	.0008	.0002
	1	.3413	.3766	.3012	.2062	.1267	.0712	.0368	.0174	.0075	.0029
	2	.0988	.2301	.2924	.2835	.2323	.1678	.1088	.0639	.0339	.0161
	3	.0173	.0852	.1720	.2362	.2581	.2397	.1954	.1419	.0923	.0537
	4	.0021	.0213	.0683	.1329	.1936	.2311	.2367	.2128	.1700	.1208
	5	.0002	.0038	.0193	.0532	.1032	.1585	.2039	.2270	.2225	.1934
	6	.0000	.0005	.0040	.0155	.0401	.0792	.1281	.1766	.2124	.2256
	7	.0000	.0000	.0006	.0033	.0115	.0291	.0591	.1009	.1489	.1934
	8	.0000	.0000	.0001	.0005	.0024	.0078	.0199	.0420	.0762	.1208
	9	.0000	.0000	.0000	.0001	.0004	.0015	.0048	.0125	.0277	.0537
	10	.0000	.0000	.0000	.0000	.0000	.0002	.0008	.0025	.0068	.0161
	11	.0000	.0000	.0000	.0000	.0000	.0000	.0001	.0003	.0010	.0029
	12	.0000	.0000	.0000	.0000	.0000	.0000	.0000	.0000	.0001	.0002
13	0	.5133	.2542	.1209	.0550	.0238	.0097	.0037	.0013	.0004	.0001
	1	.3512	.3672	.2774	.1787	.1029	.0540	.0259	.0113	.0045	.0016
	2	.1109	.2448	.2937	.2680	.2059	.1388	.0836	.0453	.0220	.0095
	13	3.0214	.0997	.1900	.2457	.2517	.2181	.1651	.1107	.0660	.0349
	4	.0028	.0277	.0838	.1535	.2097	.2337	.2222	.1845	.1350	.0873
	5	.0003	.0055	.0266	.0691	.1258	.1803	.2154	.2214	.1989	.1571
	6	.0000	.0008	.0063	.0230	.0559	.1030	.1546	.1968	.2169	.2095
	7	.0000	.0001	.0011	.0058	.0186	.0442	.0833	.1312	.1775	.2095
	8	.0000	.0000	.0001	.0011	.0047	.0142	.0336	.0656	.1089	.1571
	9	.0000	.0000	.0000	.0001	.0009	.0034	.0101	.0243	.0495	.0873
	10	.0000	.0000	.0000	.0000	.0001	.0006	.0022	.0065	.0162	.0349
	11	.0000	.0000	.0000	.0000	.0000	.0001	.0003	.0012	.0036	.0095
	12	.0000	.0000	.0000	.0000	.0000	.0000	.0000	.0001	.0005	.0016
	13	.0000	.0000	.0000	.0000	.0000	.0000	.0000	.0000	.0000	.0001
14	0	.4877	.2288	.1028	.0440	.0178	.0068	.0024	.0008	.0002	.0001
	1	.3593	.3559	.2539	.1539	.0832	.0407	.0181	.0073	.0027	.0009
	2	.1229	.2570	.2912	.2501	.1802	.1134	.0634	.0317	.0141	.0056
	3	.0259	.1142	.2056	.2501	.2402	.1943	.1366	.0845	.0462	.0222
	4	.0037	.0349	.0998	.1720	.2202	.2290	.2022	.1549	.1040	.0611
	5	.0004	.0078	.0352	.0860	.1468	.1963	.2178	.2066	.1701	.1222
	6	.0000	.0013	.0093	.0322	.0734	.1262	.1759	.2066	.2088	.1833
	7	.0000	.0002	.0019	.0092	.0280	.0618	.1082	.1574	.1952	.2095
	8	.0000	.0000	.0003	.0020	.0082	.0232	.0510	.0918	.1398	.1833
	9	.0000	.0000	.0000	.0003	.0018	.0066	.0183	.0408	.0762	.1222
14	10	.0000	.0000	.0000	.0000	.0003	.0014	.0049	.0136	.0312	.0611
	11	.0000	.0000	.0000	.0000	.0000	.0002	.0010	.0033	.0093	.0222
	12	.0000	.0000	.0000	.0000	.0000	.0000	.0001	.0005	.0019	.0056
	13	.0000	.0000	.0000	.0000	.0000	.0000	.0000	.0001	.0002	.0009
	14	.0000	.0000	.0000	.0000	.0000	.0000	.0000	.0000	.0000	.0001

Table A.1 Binomial Probability Measure;

Tabular Entry is $P(X = r) = \binom{n}{r} p^r q^{n-r}$ **(Continued)**

						p					
n	r	.05	.10	.15	.20	.25	.30	.35	.40	.45	.50
15	0	.4633	.2059	.0874	.0352	.0134	.0047	.0016	.0005	.0001	.0000
	1	.3658	.3432	.2312	.1319	.0668	.0305	.0126	.0047	.0016	.0005
	2	.1348	.2669	.2856	.2309	.1559	.0916	.0476	.0219	.0090	.0032
	3	.0307	.1285	.2184	.2501	.2252	.1700	.1110	.0634	.0318	.0139
	4	.0049	.0428	.1156	.1876	.2252	.2186	.1792	.1268	.0780	.0417
	5	.0006	.0105	.0449	.1032	.1651	.2061	.2123	.1859	.1404	.0916
	6	.0000	.0019	.0132	.0430	.0917	.1472	.1906	.2066	.1914	.1527
	7	.0000	.0003	.0030	.0138	.0393	.0811	.1319	.1771	.2013	.1964
	8	.0000	.0000	.0005	.0035	.0131	.0348	.0710	.1181	.1647	.1964
	9	.0000	.0000	.0001	.0007	.0034	.0116	.0298	.0612	.1048	.1527
	10	.0000	.0000	.0000	.0001	.0007	.0030	.0096	.0245	.0515	.0916
	11	.0000	.0000	.0000	.0000	.0001	.0006	.0024	.0074	.0191	.0417
	12	.0000	.0000	.0000	.0000	.0000	.0001	.0004	.0016	.0052	.0139
	13	.0000	.0000	.0000	.0000	.0000	.0000	.0001	.0003	.0010	.0032
	14	.0000	.0000	.0000	.0000	.0000	.0000	.0000	.0000	.0001	.0005
	15	.0000	.0000	.0000	.0000	.0000	.0000	.0000	.0000	.0000	.0000
16	0	.4401	.1853	.0743	.0281	.0100	.0033	.0010	.0003	.0001	.0000
	1	.3706	.3294	.2097	.1126	.0535	.0228	.0087	.0030	.0009	.0002
	2	.1463	.2745	.2775	.2111	.1336	.0732	.0353	.0150	.0056	.0018
	3	.0359	.1423	.2285	.2463	.2079	.1465	.0888	.0468	.0215	.0085
	4	.0061	.0514	.1311	.2001	.2252	.2040	.1553	.1014	.0572	.0278
	5	.0008	.0137	.0555	.1201	.1802	.2099	.2008	.1623	.1123	.0667
	6	.0001	.0028	.0180	.0550	.1101	.1649	.1982	.1983	.1684	.1222
	7	.0000	.0004	.0045	.0197	.0524	.1010	.1524	.1889	.1969	.1746
	8	.0000	.0001	.0009	.0055	.0197	.0487	.0923	.1417	.1812	.1964
	9	.0000	.0000	.0001	.0012	.0058	.0185	.0442	.0840	.1318	.1746
	10	.0000	.0000	.0000	.0002	.0014	.0056	.0167	.0392	.0755	.1222
	11	.0000	.0000	.0000	.0000	.0002	.0013	.0049	.0142	.0337	.0667
	12	.0000	.0000	.0000	.0000	.0000	.0002	.0011	.0040	.0115	.0278
	13	.0000	.0000	.0000	.0000	.0000	.0000	.0002	.0008	.0029	.0085
	14	.0000	.0000	.0000	.0000	.0000	.0000	.0000	.0001	.0005	.0018
	15	.0000	.0000	.0000	.0000	.0000	.0000	.0000	.0000	.0001	.0002
	16	.0000	.0000	.0000	.0000	.0000	.0000	.0000	.0000	.0000	.0000
17	0	.4181	.1668	.0631	.0225	.0075	.0023	.0007	.0002	.0000	.0000
	1	.3741	.3150	.1893	.0957	.0426	.0169	.0060	.0019	.0005	.0001
	2	.1575	.2800	.2673	.1914	.1136	.0581	.0260	.0102	.0035	.0010
	3	.0415	.1556	.2359	.2393	.1893	.1245	.0701	.0341	.0144	.0052
	4	.0076	.0605	.1457	.2093	.2209	.1868	.1320	.0796	.0411	.0182
	5	.0010	.0175	.0668	.1361	.1914	.2081	.1849	.1379	.0875	.0472
	6	.0001	.0039	.0236	.0680	.1276	.1784	.1991	.1839	.1432	.0944
	7	.0000	.0007	.0065	.0267	.0668	.1201	.1685	.1927	.1841	.1484
	8	.0000	.0001	.0014	.0084	.0279	.0644	.1134	.1606	.1883	.1855
	9	.0000	.0000	.0003	.0021	.0093	.0276	.0611	.1070	.1540	.1855

Table A.1 Binomial Probability Measure;
Tabular Entry is $P(X = r) = \binom{n}{r} p^r q^{n-r}$ (Continued)

						p					
n	r	.05	.10	.15	.20	.25	.30	.35	.40	.45	.50
	10	.0000	.0000	.0000	.0004	.0025	.0095	.0263	.0571	.1008	.1484
	11	.0000	.0000	.0000	.0001	.0005	.0026	.0090	.0242	.0525	.0944
	12	.0000	.0000	.0000	.0000	.0001	.0006	.0024	.0081	.0215	.0472
	13	.0000	.0000	.0000	.0000	.0000	.0001	.0005	.0021	.0068	.0182
	14	.0000	.0000	.0000	.0000	.0000	.0000	.0001	.0004	.0016	.0052
	15	.0000	.0000	.0000	.0000	.0000	.0000	.0000	.0001	.0003	.0010
	16	.0000	.0000	.0000	.0000	.0000	.0000	.0000	.0000	.0000	.0001
	17	.0000	.0000	.0000	.0000	.0000	.0000	.0000	.0000	.0000	.0000
18	0	.3972	.1501	.0536	.0180	.0056	.0016	.0004	.0001	.0000	.0000
	1	.3763	.3002	.1704	.0811	.0338	.0126	.0042	.0012	.0003	.0001
	2	.1683	.2835	.2556	.1723	.0958	.0458	.0190	.0069	.0022	.0006
	3	.0473	.1680	.2406	.2297	.1704	.1046	.0547	.0246	.0095	.0031
	4	.0093	.0700	.1592	.2153	.2130	.1681	.1104	.0614	.0291	.0117
	5	.0014	.0218	.0787	.1507	.1988	.2017	.1664	.1146	.0666	.0327
	6	.0002	.0052	.0301	.0816	.1436	.1873	.1941	.1655	.1181	.0708
	7	.0000	.0010	.0091	.0350	.0820	.1376	.1792	.1892	.1657	.1214
	8	.0000	.0002	.0022	.0120	.0376	.0811	.1327	.1734	.1864	.1669
	9	.0000	.0000	.0004	.0033	.0139	.0386	.0794	.1284	.1694	.1855
	10	.0000	.0000	.0001	.0008	.0042	.0149	.0385	.0771	.1248	.1669
	11	.0000	.0000	.0000	.0001	.0010	.0046	.0151	.0374	.0742	.1214
	12	.0000	.0000	.0000	.0000	.0002	.0012	.0047	.0145	.0354	.0708
	13	.0000	.0000	.0000	.0000	.0000	.0002	.0012	.0045	.0134	.0327
	14	.0000	.0000	.0000	.0000	.0000	.0000	.0002	.0011	.0039	.0117
	15	.0000	.0000	.0000	.0000	.0000	.0000	.0000	.0002	.0009	.0031
	16	.0000	.0000	.0000	.0000	.0000	.0000	.0000	.0000	.0001	.0006
	17	.0000	.0000	.0000	.0000	.0000	.0000	.0000	.0000	.0000	.0001
	18	.0000	.0000	.0000	.0000	.0000	.0000	.0000	.0000	.0000	.0000
19	0	.3774	.1351	.0456	.0144	.0042	.0011	.0003	.0001	.0000	.0000
	1	.3774	.2852	.1529	.0685	.0268	.0093	.0029	.0008	.0002	.0000
	2	.1787	.2852	.2428	.1540	.0803	.0358	.0138	.0046	.0013	.0003
	3	.0533	.1796	.2428	.2182	.1517	.0869	.0422	.0175	.0062	.0018
	4	.0112	.0798	.1714	.2182	.2023	.1491	.0909	.0467	.0203	.0074
	5	.0018	.0266	.0907	.1636	.2023	.1916	.1468	.0933	.0497	.0222
	6	.0002	.0069	.0374	.0955	.1574	.1916	.1844	.1451	.0949	.0518
	7	.0000	.0014	.0122	.0443	.0974	.1525	.1844	.1797	.1443	.0961
	8	.0000	.0002	.0032	.0166	.0487	.0981	.1489	.1797	.1771	.1442
	9	.0000	.0000	.0007	.0051	.0198	.0514	.0980	.1464	.1771	.1762
	10	.0000	.0000	.0001	.0013	.0066	.0220	.0528	.0976	.1449	.1762
	11	.0000	.0000	.0000	.0003	.0018	.0077	.0233	.0532	.0970	.1442
	12	.0000	.0000	.0000	.0000	.0004	.0022	.0083	.0237	.0529	.0961
	13	.0000	.0000	.0000	.0000	.0001	.0005	.0024	.0085	.0233	.0518
	14	.0000	.0000	.0000	.0000	.0000	.0001	.0006	.0024	.0082	.0222

Table A.1 Binomial Probability Measure;

$$\text{Tabular Entry is } P(X = r) = \binom{n}{r} p^r q^{n-r} \qquad \text{(Continued)}$$

n	r	.05	.10	.15	.20	.25	.30	.35	.40	.45	.50
	15	.0000	.0000	.0000	.0000	.0000	.0000	.0001	.0005	.0022	.0074
	16	.0000	.0000	.0000	.0000	.0000	.0000	.0000	.0001	.0005	.0018
	17	.0000	.0000	.0000	.0000	.0000	.0000	.0000	.0000	.0001	.0003
	18	.0000	.0000	.0000	.0000	.0000	.0000	.0000	.0000	.0000	.0000
	19	.0000	.0000	.0000	.0000	.0000	.0000	.0000	.0000	.0000	.0000
20	0	.3585	.1216	.0388	.0115	.0032	.0008	.0002	.0000	.0000	.0000
	1	.3774	.2702	.1368	.0576	.0211	.0068	.0020	.0005	.0001	.0000
	2	.1887	.2852	.2293	.1369	.0669	.0278	.0100	.0031	.0008	.0002
	3	.0596	.1901	.2428	.2054	.1339	.0716	.0323	.0123	.0040	.0011
	4	.0133	.0898	.1821	.2182	.1897	.1304	.0738	.0350	.0139	.0046
	5	.0022	.0319	.1028	.1746	.2023	.1789	.1272	.0746	.0365	.0148
	6	.0003	.0089	.0454	.1091	.1686	.1916	.1712	.1244	.0746	.0370
	7	.0000	.0020	.0160	.0545	.1124	.1643	.1844	.1659	.1221	.0739
	8	.0000	.0004	.0046	.0222	.0609	.1144	.1614	.1797	.1623	.1201
	9	.0000	.0001	.0011	.0074	.0271	.0654	.1158	.1597	.1771	.1602
	10	.0000	.0000	.0002	.0020	.0099	.0308	.0686	.1171	.1593	.1762
	11	.0000	.0000	.0000	.0005	.0030	.0120	.0336	.0710	.1185	.1602
	12	.0000	.0000	.0000	.0001	.0008	.0039	.0136	.0355	.0727	.1201
	13	.0000	.0000	.0000	.0000	.0002	.0010	.0045	.0146	.0366	.0739
	14	.0000	.0000	.0000	.0000	.0000	.0002	.0012	.0049	.0150	.0370
	15	.0000	.0000	.0000	.0000	.0000	.0000	.0003	.0013	.0049	.0148
	16	.0000	.0000	.0000	.0000	.0000	.0000	.0000	.0003	.0013	.0046
	17	.0000	.0000	.0000	.0000	.0000	.0000	.0000	.0000	.0002	.0011
	18	.0000	.0000	.0000	.0000	.0000	.0000	.0000	.0000	.0000	.0002
	19	.0000	.0000	.0000	.0000	.0000	.0000	.0000	.0000	.0000	.0000
	20	.0000	.0000	.0000	.0000	.0000	.0000	.0000	.0000	.0000	.0000

Table A.2 Standard Normal Probability Measure $\mu = 0$, $\sigma = 1$; Tabular Entry Is $P(Z \leq z)$ for $0 \leq z \leq 3$

z	0	1	2	3	4	5	6	7	8	9
.0	.5000	.5040	.5080	.5120	.5160	.5199	.5239	.5279	.5319	.5359
.1	.5398	.5438	.5478	.5517	.5557	.5596	.5636	.5675	.5714	.5753
.2	.5793	.5832	.5871	.5910	.5948	.5987	.6026	.6064	.6103	.6141
.3	.6179	.6217	.6255	.6293	.6331	.6368	.6406	.6443	.6480	.6517
.4	.6554	.6591	.6628	.6664	.6700	.6736	.6772	.6808	.6844	.6879
.5	.6915	.6950	.6985	.7019	.7054	.7088	.7123	.7157	.7190	.7224
.6	.7257	.7291	.7324	.7357	.7389	.7422	.7454	.7486	.7517	.7549
.7	.7580	.7611	.7642	.7673	.7704	.7734	.7764	.7794	.7823	.7852
.8	.7881	.7910	.7939	.7967	.7995	.8023	.8051	.8078	.8106	.8133
.9	.8159	.8186	.8212	.8238	.8264	.8289	.8315	.8340	.8365	.8389
1.0	.8413	.8438	.8461	.8485	.8508	.8531	.8554	.8577	.8599	.8621
1.1	.8643	.8665	.8686	.8708	.8729	.8749	.8770	.8790	.8810	.8830
1.2	.8849	.8869	.8888	.8907	.8925	.8944	.8962	.8980	.8997	.9015
1.3	.9032	.9049	.9066	.9082	.9099	.9115	.9131	.9147	.9162	.9177
1.4	.9192	.9207	.9222	.9236	.9251	.9265	.9279	.9292	.9306	.9319
1.5	.9332	.9345	.9357	.9370	.9382	.9394	.9406	.9418	.9429	.9441
1.6	.9452	.9463	.9474	.9484	.9495	.9505	.9515	.9525	.9535	.9545
1.7	.9554	.9564	.9573	.9582	.9591	.9599	.9608	.9616	.9625	.9633
1.8	.9641	.9649	.9656	.9664	.9671	.9678	.9686	.9693	.9700	.9706
1.9	.9713	.9719	.9726	.9732	.9738	.9744	.9750	.9756	.9761	.9767
2.0	.9772	.9778	.9783	.9788	.9793	.9798	.9803	.9808	.9812	.9817
2.1	.9821	.9826	.9830	.9834	.9838	.9842	.9846	.9850	.9854	.9857
2.2	.9861	.9864	.9868	.9871	.9875	.9878	.9881	.9884	.9887	.9890
2.3	.9893	.9896	.9898	.9901	.9904	.9906	.9909	.9911	.9913	.9916
2.4	.9918	.9920	.9922	.9925	.9927	.9929	.9931	.9932	.9934	.9936
2.5	.9938	.9940	.9941	.9943	.9945	.9946	.9948	.9949	.9951	.9952
2.6	.9953	.9955	.9956	.9957	.9959	.9960	.9961	.9962	.9963	.9964
2.7	.9965	.9966	.9967	.9968	.9969	.9970	.9971	.9972	.9973	.9974
2.8	.9974	.9975	.9976	.9977	.9977	.9978	.9979	.9979	.9980	.9981
2.9	.9981	.9982	.9982	.9983	.9984	.9984	.9985	.9985	.9986	.9986
3.0	.9987									

Table A.3 Poisson Probability Measure; Tabular Entry is $P(X = r) = \dfrac{(\lambda t)^r}{r!} e^{-\lambda t}$

$$\lambda t =$$

$r =$	0.1	0.2	0.3	0.4	0.5	0.6	0.7	0.8	0.9	1.0	1.5	2.0	2.5	3.0
0	.905	.819	.741	.670	.606	.549	.497	.449	.407	.368	.223	.135	.082	.050
1	.091	.164	.222	.268	.303	.329	.348	.360	.366	.368	.335	.271	.205	.149
2	.004	.016	.033	.054	.076	.099	.122	.144	.165	.184	.251	.271	.256	.224
3		.001	.003	.007	.013	.020	.028	.038	.049	.061	.125	.180	.214	.224
4				.001	.002	.003	.005	.008	.011	.015	.047	.090	.134	.168
5							.001	.001	.002	.003	.014	.036	.067	.101
6										.001	.004	.012	.028	.050
7											.001	.003	.010	.022
8												.001	.003	.008
9													.001	.003
10														.001

Table A.3   Poisson Probability Measure; Tabular Entry is $P(X = r) = \dfrac{(\lambda t)^r}{r!}\, e^{-\lambda t}$

$\lambda t =$

(Continued)

$r =$	3.5	4.0	4.5	5.0	5.5	6.0	6.5	7.0	7.5	8.0	8.5	9.0	9.5	10.0
0	.030	.018	.011	.007	.004	.003	.002	.001	.001	.000	.000	.000	.000	.000
1	.106	.073	.050	.034	.023	.015	.010	.006	.004	.003	.002	.001	.001	.001
2	.185	.147	.113	.084	.062	.045	.032	.022	.016	.011	.007	.005	.003	.002
3	.216	.195	.169	.140	.113	.089	.069	.052	.039	.029	.021	.015	.011	.008
4	.189	.195	.190	.176	.156	.134	.112	.091	.073	.057	.044	.034	.025	.019
5	.132	.156	.171	.176	.171	.161	.145	.128	.109	.092	.075	.061	.048	.038
6	.077	.104	.128	.146	.157	.161	.158	.149	.137	.122	.107	.091	.076	.063
7	.038	.060	.082	.104	.123	.138	.146	.149	.147	.140	.129	.117	.104	.090
8	.017	.030	.046	.065	.085	.103	.119	.130	.137	.140	.138	.132	.123	.113
9	.007	.013	.023	.036	.052	.069	.086	.101	.114	.124	.130	.132	.130	.125
10	.002	.005	.010	.018	.029	.041	.056	.071	.086	.099	.110	.119	.124	.125
11	.001	.002	.004	.008	.014	.022	.033	.045	.058	.072	.085	.097	.107	.114
12		.001	.002	.003	.007	.011	.018	.026	.037	.048	.060	.073	.084	.095
13			.001	.001	.003	.005	.009	.014	.021	.030	.040	.050	.062	.073
14				.001	.001	.002	.004	.007	.011	.017	.024	.032	.042	.052
15						.001	.002	.003	.006	.009	.014	.019	.026	.035
16							.001	.001	.003	.004	.007	.011	.016	.022
17								.001	.001	.002	.004	.006	.009	.013
18										.001	.002	.003	.005	.007
19											.001	.001	.002	.004
20												.001	.001	.002
21													.001	.001

Table A.4   Student's *t*-Distribution; Tabular Entry is t_γ; Area to the left of t_γ is γ

ν	.80	.90	.95	.975	.99	.995
1	1.376	3.078	6.134	12.73	31.75	64.00
2	1.061	1.886	2.920	4.303	6.964	9.925
3	.978	1.638	2.353	3.182	4.541	5.841
4	.941	1.533	2.132	2.776	3.747	4.604
5	.920	1.476	2.015	2.571	3.365	4.032
6	.906	1.440	1.943	2.447	3.143	3.707
7	.896	1.415	1.895	2.365	2.998	3.499
8	.889	1.397	1.860	2.306	2.896	3.355
9	.883	1.383	1.833	2.262	2.821	3.250
10	.879	1.372	1.812	2.228	2.764	3.169
11	.876	1.363	1.796	2.201	2.718	3.106
12	.873	1.356	1.782	2.179	2.681	3.054
13	.870	1.350	1.771	2.160	2.650	3.012
14	.868	1.345	1.761	2.145	2.624	2.977
15	.866	1.341	1.753	2.132	2.602	2.947
16	.865	1.337	1.746	2.120	2.584	2.921
17	.863	1.333	1.740	2.110	2.567	2.898
18	.862	1.330	1.734	2.101	2.552	2.878
19	.861	1.328	1.729	2.093	2.540	2.861
20	.860	1.325	1.725	2.086	2.528	2.845
21	.859	1.323	1.721	2.080	2.518	2.831
22	.858	1.321	1.717	2.074	2.508	2.819
23	.858	1.320	1.714	2.069	2.500	2.807
24	.857	1.318	1.711	2.064	2.492	2.797
25	.856	1.316	1.708	2.060	2.485	2.788
26	.856	1.315	1.706	2.056	2.479	2.779
27	.855	1.314	1.703	2.052	2.473	2.771
28	.855	1.312	1.701	2.048	2.467	2.763
29	.854	1.311	1.699	2.045	2.462	2.756
30	.854	1.310	1.697	2.042	2.457	2.750
∞	.842	1.282	1.645	1.960	2.327	2.575

Table A.5 χ_2 **Distribution. Tabular Entry is** $\chi_{\gamma 2}$;
Area to the Left of χ_γ^2 **is** γ

γ

ν	.01	.05	.10	.25	.75	.90	.95	.99
1	.000	.004	.016	.102	1.324	2.706	3.843	6.637
2	.020	.103	.211	.575	2.773	4.605	5.992	9.210
3	.115	.352	.584	1.212	4.108	6.251	7.815	11.344
4	.297	.711	1.064	1.923	5.385	7.779	9.488	13.277
5	.554	1.145	1.610	2.674	6.626	9.236	11.070	15.085
6	.872	1.635	2.204	3.454	7.841	10.645	12.592	16.812
7	1.239	2.167	2.833	4.255	9.037	12.017	14.067	18.474
8	1.646	2.733	3.490	5.071	10.219	13.362	15.507	20.090
9	2.088	3.325	4.168	5.899	11.389	14.684	16.919	21.665
10	2.558	3.940	4.865	6.737	12.549	15.987	18.307	23.209
11	3.053	4.575	5.578	7.584	13.701	17.275	19.675	24.724
12	3.571	5.226	6.304	8.438	14.845	18.549	21.026	26.217
13	4.107	5.892	7.041	9.299	15.984	19.812	22.362	27.687
14	4.660	6.571	7.790	10.165	17.117	21.064	23.685	29.141
15	5.229	7.261	8.547	11.036	18.245	22.307	24.996	30.577
16	5.812	7.962	9.312	11.912	19.369	23.542	26.296	32.000
17	6.407	8.682	10.085	12.792	20.489	24.769	27.587	33.408
18	7.015	9.390	10.865	13.675	21.605	25.989	28.869	34.805
19	7.632	10.117	11.651	14.562	22.718	27.203	30.143	36.190
20	8.260	10.851	12.443	15.452	23.828	28.412	31.410	37.566
21	8.897	11.591	13.240	16.344	24.935	29.615	32.670	38.930
22	9.542	12.338	14.042	17.240	26.039	30.813	33.924	40.289
23	10.195	13.090	14.848	18.137	27.141	32.007	35.172	41.637
24	10.856	13.848	15.659	19.037	28.241	33.196	36.415	42.980
25	11.523	14.611	16.473	19.939	29.339	34.381	37.652	44.313
26	12.198	15.379	17.292	20.843	30.435	35.563	38.885	45.642
27	12.878	16.151	18.114	21.749	31.528	36.741	40.113	46.962
28	13.565	16.928	18.939	22.657	32.620	37.916	41.337	48.278
29	14.256	17.708	19.768	23.566	33.711	39.087	42.557	49.586
30	14.954	18.493	20.599	24.478	34.800	40.256	43.773	50.892

Table A.6 *F* Distribution. Tabular Entries are $F_{.95}$ (Regular Type) and $F_{.99}$ (Boldface); *g* is the Numerator Degrees of Freedom, *h* is the denominator Degrees of Freedom

h \\ $g\rightarrow$	1	2	3	4	5	6	7	8	9	10	15	20	30	50	100	200
1	161.4	199.5	215.7	224.6	230.2	234.0	236.8	238.9	240.5	241.9	245.9	248.0	250.1	251.8	253.0	253.7
	4052.2	**4999.5**	**5403.3**	**5624.5**	**5763.6**	**5859.0**	**5928.3**	**5981.2**	**6022.5**	**6055.9**	**6157.3**	**6208.6**	**6260.7**	**6302.5**	**6334.0**	**6349.9**
2	18.513	19.000	19.164	19.274	19.297	19.329	19.354	19.371	19.385	19.396	19.429	19.445	19.463	19.476	19.486	19.491
	98.502	**98.999**	**99.165**	**99.250**	**99.299**	**99.332**	**99.356**	**99.375**	**99.389**	**99.399**	**99.432**	**99.450**	**99.466**	**99.478**	**99.491**	**99.495**
3	10.128	9.5521	9.2766	9.1172	9.0135	8.9406	8.8867	8.8452	8.8123	8.7855	8.7029	8.6602	8.6166	8.5810	8.5539	8.5402
	34.116	**30.816**	**29.457**	**28.710**	**28.237**	**27.911**	**27.672**	**27.489**	**27.345**	**27.229**	**26.872**	**26.690**	**26.505**	**26.354**	**26.240**	**26.183**
4	7.7086	6.9443	6.5914	6.3882	6.2561	6.1631	6.0942	6.0410	5.9988	5.9644	5.8578	5.8026	5.7459	5.6995	5.6641	5.6461
	21.198	**18.000**	**16.694**	**15.977**	**15.522**	**15.207**	**14.976**	**14.799**	**14.659**	**14.546**	**14.198**	**14.020**	**13.838**	**13.690**	**13.577**	**13.520**
5	6.6079	5.7861	5.4094	5.1922	5.0503	4.9503	4.8759	4.8183	4.7725	4.7351	4.6188	4.5591	4.4957	4.4444	4.4051	4.3851
	16.258	**13.274**	**12.060**	**11.392**	**10.967**	**10.672**	**10.456**	**10.289**	**10.158**	**10.051**	**9.7222**	**9.5527**	**9.3794**	**9.2378**	**9.1299**	**9.0754**
6	5.9874	5.1433	4.7571	4.5337	4.3874	4.2839	4.2067	4.1468	4.0990	4.0600	3.9381	3.8742	3.8082	3.7537	3.7117	3.6904
	13.745	**10.925**	**9.7796**	**9.1483**	**8.7459**	**8.4661**	**8.2600**	**8.1017**	**7.9761**	**7.8741**	**7.5590**	**7.3958**	**7.2285**	**7.0915**	**6.9867**	**6.9336**
7	5.5915	4.7374	4.3468	4.1203	3.9715	3.8660	3.7870	3.7257	3.6767	3.6365	3.5107	3.4445	3.3758	3.3189	3.2749	3.2525
	12.246	**9.5466**	**8.4513**	**7.8466**	**7.4604**	**7.1914**	**6.9928**	**6.8401**	**6.7188**	**6.6200**	**6.3143**	**6.1554**	**5.9920**	**5.8577**	**5.7547**	**5.7024**
8	5.3177	4.4590	4.0062	3.8379	3.6875	3.5806	3.5005	3.4381	3.3881	3.3472	3.2184	3.1503	3.0794	3.0204	2.9747	2.9513
	11.259	**8.6491**	**7.5910**	**7.0061**	**6.6318**	**6.3707**	**6.1776**	**6.0289**	**5.9106**	**5.8143**	**5.5151**	**5.3591**	**5.1981**	**5.0654**	**4.9633**	**4.9114**
9	5.1174	4.2565	3.8625	3.6331	3.4817	3.3737	3.2927	3.2296	3.1789	3.1373	3.0061	2.9365	2.8637	2.8028	2.7556	2.7313
	10.561	**8.0215**	**6.9919**	**6.4221**	**6.0569**	**5.8018**	**5.6129**	**5.4671**	**5.3511**	**5.2565**	**4.9621**	**4.8080**	**4.6486**	**4.5167**	**4.4150**	**4.3631**
10	4.9646	4.1028	3.7083	3.4780	3.3258	3.2172	3.1355	3.0717	3.0204	2.9782	2.8450	2.7740	2.6996	2.6371	2.5884	2.5634
	10.044	**7.5594**	**6.5523**	**5.9943**	**5.6363**	**5.3858**	**5.2001**	**5.0567**	**4.9424**	**4.8492**	**4.5581**	**4.4054**	**4.2469**	**4.1155**	**4.0137**	**3.9617**

Table A.6 F Distribution. Tabular Entries are $F_{.95}$ (Regular Type) and $F_{.99}$ (Boldface); g is the Numerator Degrees of Freedom, h is the denominator Degrees of Freedom   (Continued)

h ↓ \ g→	1	2	3	4	5	6	7	8	9	10	15	20	30	50	100	200
15	4.5431	3.6823	3.2874	3.0556	2.9013	2.7905	2.7066	2.6408	2.5876	2.5437	2.4034	2.3275	2.2468	2.1780	2.1234	2.0950
	8.6831	**6.3589**	**5.4170**	**4.8932**	**4.5556**	**4.3183**	**4.1415**	**4.0045**	**3.8948**	**3.8049**	**3.5222**	**3.3719**	**3.2141**	**3.0814**	**2.9772**	**2.9235**
20	4.3512	3.4928	3.0984	2.8661	2.7109	2.5990	2.5140	2.4471	2.3928	2.3479	2.2033	2.1242	2.0391	1.9656	1.9066	1.8755
	8.0960	**5.8489**	**4.9382**	**4.4307**	**4.1027**	**3.8714**	**3.6987**	**3.5644**	**3.4567**	**3.3682**	**3.0880**	**2.9377**	**2.7785**	**2.6430**	**2.5353**	**2.4792**
30	4.1709	3.3158	2.9223	2.6896	2.5336	2.4205	2.3343	2.2662	2.2107	2.1646	2.0148	1.9317	1.8409	1.7609	1.6950	1.6597
	7.5625	**5.3903**	**4.5097**	**4.0179**	**3.6990**	**3.4735**	**3.3045**	**3.1726**	**3.0665**	**2.9791**	**2.7002**	**2.5487**	**2.3860**	**2.2450**	**2.1307**	**2.0700**
50	4.0343	3.1826	2.7900	2.5572	2.4004	2.2864	2.1992	2.1299	2.0733	2.0261	1.8714	1.7841	1.6872	1.5995	1.5249	1.4835
	7.1706	**5.0566**	**4.1993**	**3.7195**	**3.4077**	**3.1864**	**3.0202**	**2.8900**	**2.7850**	**2.6981**	**2.4190**	**2.2652**	**2.0976**	**1.9490**	**1.8248**	**1.7567**
100	3.9361	3.0873	2.6955	2.4626	2.3053	2.1906	2.1025	2.0323	1.9748	1.9267	1.7675	1.6764	1.5733	1.4772	1.3917	1.3416
	6.8953	**4.8239**	**3.9837**	**3.5127**	**3.2059**	**2.9877**	**2.8233**	**2.6943**	**2.5898**	**2.5033**	**2.2230**	**2.0666**	**1.8933**	**1.7353**	**1.5977**	**1.5184**
200	3.8884	3.0411	2.6497	2.4168	2.2592	2.1441	2.0556	1.9849	1.9269	1.8783	1.7166	1.6233	1.5164	1.4146	1.3206	1.2626
	6.7633	**4.7128**	**3.8810**	**3.4143**	**3.1100**	**2.8933**	**2.7298**	**2.6012**	**2.4971**	**2.4106**	**2.1294**	**1.9713**	**1.7941**	**1.6295**	**1.4811**	**1.3912**

ANSWERS TO EVEN-NUMBERED EXERCISES

CHAPTER 1, EXERCISE 1.2

2. (a) 140   (b) 784
4. 10
6. (a) 312.8   (b) 78.2
8. $x_1 = x_2 = \ldots = x_n$
12. (a) 1289   (b) 214.83
14. (a) $1090   (b) $181.67
16. (a) 20.93   (b) 3.70
18. 165 lb.

CHAPTER 2, EXERCISE 2.1

2. (c) 116
4. 3, 4, 5, 8, 3, 2
6. (a) 164.5, 193.5, 214.9   (b) 164.93, 194.73, 222.08
8. (b) 804.75, 865, 1199.5
10. 116, 130

EXERCISE 2.2

2. 31.57, 27.89
4. 382.22, 467.38, 352.02
8. 153.0, 116.05
10. 8.0, −22.40

EXERCISE 2.3

2. 122.74, 81.81
6. 16.12, 3.61

CHAPTER 3, EXERCISE 3.1

2. Discrete, $R_V = \{v: v = 0, 1, 2, \ldots, 7\}$
4. Continuous, $R_Y = \{y: 0 \le y \le \infty\}$
6. Discrete, $R_X = \{x: x = 0, 1, 2, \ldots, 9\}$
8. Continuous, $R_U = \{u: 0 < u < 6\}$
10. Continuous, $R_V = \{v: 6 \le v \le 12\}$

EXERCISE 3.2

2. (a) $1, 0, \frac{1}{2}$ (b) $0, 1, \frac{2}{3}$
4. (a) .30   (b) .95
6. (a) .095   (b) .905
8. (a) $R_X = \{x: x = 1, 2, 3\}$ (b) $P(X = 1) = \frac{1}{2}$;
   $P(X = 2) = \frac{1}{6}$; $P(X = 3) = \frac{1}{3}$

EXERCISE 3.3

2. .5
4. .25, .25, .75
6. (a) $\frac{1}{3}$ (b) $\frac{1}{3}$ (c) $\frac{1}{3}$
8. $\frac{2}{3}$

EXERCISE 3.4

2. $\dfrac{1}{12}$
4. Continuous, $R_W = \{w: w \ge 0\}$
6. (a) $\dfrac{1}{250}$ (b) 16/25   (c) 1/5
8. (a) 9   (b) $\frac{2}{3}$ (c) $\frac{2}{3}$
10. (a) $\frac{3}{4}$ (b) 0   (c) $\frac{1}{8}$

CHAPTER 4, EXERCISE 4.1

2. 3, 3
4. 1.98

6. 3.2
8. $-.053$
10. -1.499
12. $\frac{3}{4}$ ft

EXERCISE 4.2
2. 1.98, 1.19
4. (a) $2\,hp,\quad h(2p)^{1/2}$ (b) $\frac{1}{2}$
6. .999
8. $P(X = -1) = P(X = 1) = \frac{1}{2}$, No

EXERCISE 4.3

2. $(\frac{3}{4})^{1/2}, \frac{3}{4}$
4. 3.514
6. 4.5
8. 57071, 60000, 62929; 5858
10. (a) $.8(b - a)$ (b) $(.4)^{1/2}(b - a)$

EXERCISE 4.4
2. 1.359, 12.26
4. $.66, \dfrac{4}{9}$
6. (a) 133.3   (b) 18.3
8. 1.57

CHAPTER 5, EXERCISE 5.1
2. $\binom{5}{x}\left(\frac{1}{6}\right)^{x}\left(\frac{5}{6}\right)^{5-x}\ x = 0, 1, \ldots, 5;\quad \binom{5}{y}\left(\frac{1}{2}\right)^{y}\left(\frac{1}{2}\right)^{5-y}\ y = 0, 1, \ldots, 5$
4. $n \geq (\log(.01)/\log(.90)) = 44$
6. $\binom{20}{x}\left(\frac{1}{4}\right)^{x}\left(\frac{3}{4}\right)^{20-x}\ x = 0, 1, 2, \ldots, 20$
8. 9, .5
10. (a) .9887   (b) .893
12. .028
14. (a) 10   (b) .1762   (c) .5831

EXERCISE 5.2

2. (a) $-.385$ (b) .385   (c) .935   (d) .319
4. (a) B (b) D (c) A (d) F (e) B
6. .0668
10. .006

12. (a) .842   (b) −1.282   (c) .126   (d) 1.645
14. .0008, .7491, .2272

EXERCISE 5.3

 2. .135
 4. .865
 6. .1494
 8. .1353, .2707
12. .487

EXERCISE 5.4

 2. (a) .349   (b) 1
 4. (a) .8413   (b) .9987   (c) 0   (d) .008
 6. (a) 4   (b) .0115
 8. 13.5
10. .9707
12. 90
14. .9408

CHAPTER 6, EXERCISE 6.1

 2. (a) conceptual   (b) discrete
 4. (a) conceptual   (b) discrete
 8. Continuous
10. (a) .308   (b) .136

EXERCISE 6.2

 2. .405
 4. $(.5)^n$
 6. (a) 1   (b) .422   (c) .8655
 8. (a) .6826   (b) .1203

EXERCISE 6.3

 2. .596
 4. .9933
 6. 238, .894
 8. (a) 29.34, 34.24   (b) 23.67, 53.56

EXERCISE 6.4

2. .9876
4. .8413, .0228
6. .2877
8. 10, .0505, 0
10. .0869
12. .0191

EXERCISE 6.5

2. Normal; $\mu = 20$, $\sigma^2 = 18$
4. .8980
6. .3409
8. .9738
10. 0
12. .2389
14. $3.92\sigma_X$, $3.92\sigma_X/n^{1/2}$
16. .6985
18. -5.26

CHAPTER 7, EXERCISE 7.1

2. (a) 11.48, .212 (b) .206 (c) No
4. (a) .267 (b) .114
6. 532.9, 623.3, 485.3
8. .590, .328
10. .65
12. $p = \dfrac{1}{3}$ maximizes $3(p - 2p^2 + p^3) = \dfrac{4}{9} - 3\left(p - \dfrac{1}{3}\right)^2 \left(\dfrac{4}{3} - p\right)$
14. 525.60
16. (a) .25 (b) .3115 (c) .6329
18. .1446
20. (a) .615, .750, .400, .542 (b) .034, .034, .031, .045

EXERCISE 7.2

2. $2467.6 \le \mu \le 2496.4$
4. $298.3 \le \mu \le 309.7$
6. $0 \le p \le .006$
8. $.997 \le P(Y \ge 1) \le 1$
12. .0668, .1350

14. 2421, 6724
16. $146.73 \leq \mu \leq 154.77$
18. $121.20 \leq \mu \leq 129.12$
20. $.608 \leq p \leq .784$
22. 66,564
24. $.673 \leq p \leq 1$
26. $0 \leq p \leq .0754$
28. $.4934 \leq P(X \geq 1) \leq .6015$
30. $.054 \leq p \leq .074$
32. 3121
34. .858

EXERCISE 7.3

2. (a) 119.8   (b) 3.09
4. (a) $.150 \leq p \leq .177$ (b) $.264 < p < .348$
6. $.655 \leq p \leq 1$
8. .436

CHAPTER 8, EXERCISE 8.1

(When appropriate the value of a test statistic is underlined for identification)

2. .25
4. .4637
6. (a) $\bar{x}$ (b) reject when $\bar{x} \leq c = 700 - z_{1-\alpha} (50/30^{1/2})$
8. (a) .117   (b) .0197
10. .0455, two-sided test

EXERCISE 8.2

2. $24.42 \leq \bar{x} \leq 29.58$; $24.19 \leq \bar{x} \leq 29.81$
4. Reject; $\underline{29} \leq 29.42$
6. Accept; $66.56 \leq \underline{71} \leq 71.44$, .3810
8. Accept; $.378 \leq \underline{.5} \leq .522$
12. Reject; $\underline{9.25} \geq 9.15$
14. Reject; $\underline{150.75} \leq 156.66$
16. 9, 35
18. Accept; $\underline{10.83} \leq 11.5$
20. Accept; $\underline{5793} \leq 6400 \leq 15758$

24. Reject; $\overline{47.60} > \overline{40.96}$
26. Reject; $\overline{.8650} < .8651$
28. Reject; $\overline{.26} > .1824$
30. Accept; $\overline{.65} > .637$
32. Accept; $\overline{.212} \le \overline{.362} \le .455$
34. Accept; $.584 \le \overline{.696} \le .749$
36. Accept; $\overline{.050} \le \overline{.058}$
38. Reject; $\overline{100} > 93.49$
40. .0116
42. 0
44. 831

EXERCISE 8.3

2. Accept; $\overline{.63} < 12.0$
4. (b) Reject; $\overline{33.63} > 9.24$
6. Accept; $\overline{7.64} < 19.7$
8. Accept; $\overline{6.36} < 9.49$
10. Accept; $\overline{1.75} < 7.81$
12. Reject; $\overline{11.25} > 6.63$

EXERCISE 8.4

2. Reject; $\overline{2.63} > 2.62$
4. Accept; $\overline{14.3} < 16.9$

6. .4602
8. .2236

CHAPTER 9, EXERCISE 9.1

2. -2.5, $\quad -6.78 \le \mu_A - \mu_B \le 1.78$
4. $-1.08 \le p_M - p_H \le .050$
6. $2.14 \le \mu_I - \mu_A \le 3.46$
8. (a) .38 $\quad$ (b) $-.61 \le \mu_X - \mu_Y \le 1.37$
10. $-.40 \le D < .91$
12. $-.2073 < p_1 - p_2 < .0013$
14. $-7.92 < \mu_1 - \mu_2 < 3.92$

EXERCISE 9.2

2. Accept; $-.103 < \overline{-.005} < 1.03$

4. Reject; $\underline{.15} > .102$
6. (a) $|ts| \geq z_{1-\alpha/2}$ (b) $ts \leq -z_{1-\alpha}$ (c) $ts \geq z_{1-\alpha}$
8. Accept; $-2.262 < \underline{-2.05} < 2.262$
10. Accept; $-1.645 < \underline{-1.414} < 1.645$
12. Accept; $-1.833 < \underline{.26} < 1.833$
14. Reject; $\underline{-1.924} < -1.645$

EXERCISE 9.3

2. Reject; $\underline{36.28} \geq 6.93$
4. Reject; $\underline{127.68} \geq 3.55$
6. Accept; $-2.228 < \underline{-1.09} < 2.228$
8. Accept; $\underline{2.52} < 4.89$
10. Accept; $\underline{1.86} < 3.18$

EXERCISE 9.4

2. (a) $-.4$ (b) $.277$
4. (a) -6.33 (b) 1.164
6. (a) -2.8 (b) $-4.95 \leq \mu_1 - \mu_2 < -.65$
8. Accept; $-2.787 < \underline{-2.68} < 2.787$
10. Reject hypothesis of equal mileage at $\alpha = .0625$.

CHAPTER 10, EXERCISE 10.1

2. $.496, 40.77, 79.46$
4. (a) $145.03, -0.0633$ (b) $7.69, .0098$
6. (a) $-.027, 1.007$ (b) $.0618, .0317$
8. (a) $-4.4, 1.2$ (b) 1.67

EXERCISE 10.2

2. $2.924 \leq E(Y) \leq 3.016$
4. Accept; $-1.833 \leq .5 \leq 1.833$
6. Accept; $-2.262 < -.522 < 2.262$

EXCERCISE 10.3

2. $.765$
4. $-.956$
6. $.767$

EXERCISE 10.4

2. 36.23, 6.76, .509
4. (a) .306, .008   (b) .0001   (c) .024, .0004
6. $.736 \leq E(Y) \leq .756$
8. (a) 43.96, .1561   (b) 6.36, .0389
10. .9978
12. .9924

CHAPTER 11, EXERCISE 11.1

2. $-8363, 26534, 61432$
4. (a) 20.2   (b) 15.94, 26.06   (c) $t_{.2} \leq 18.1$
6. $t_{.952}$
8. (a) 192.9   (b) 116.23, 219.59

EXERCISE 11.2

2. Accept; 70% confidence interval $(12 \leq t_{.5} \leq 22)$ contains 20
4. Accept; $-2.576 \leq .035 \leq 2.576$
6. Reject; $\underline{-2.83} \leq -2.576$
8. Accept; $-1.96 \leq \underline{-1.04} \leq 1.96$
10. Accept; $-1.28 < \underline{1.46}$

EXERCISE 11.3

2. Accept; $-1.96 \leq \underline{-.73} \leq 1.96$
4. Accept; $-1.645 \leq \underline{-1.28} \leq 1.645$
6. $125 < t_{.50} < 203$

APPENDIX, EXERCISE A.1

2. (a) $\{5,6,7,8,9,10\}$ (b) $\{1,2,8,9,10\}$ (c) $\{8,9,10\}$
   (d) $\{8,9,10\}$ (e) $\{3,4,5,6,7,8,9,10\}$ (f) $\{1,2\}$
4. ϕ
6. No. Some citizens are not residents, and some residents are not citizens.

8.

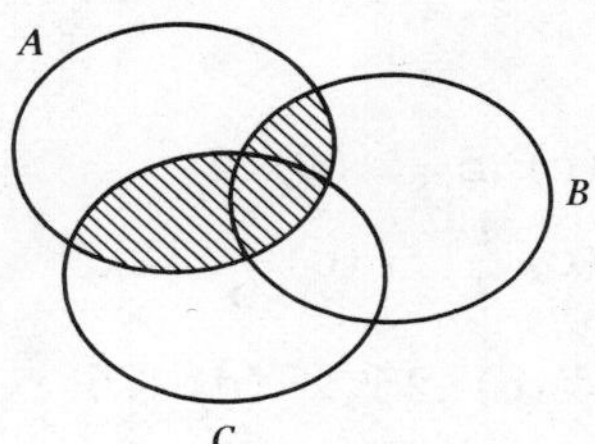

10.

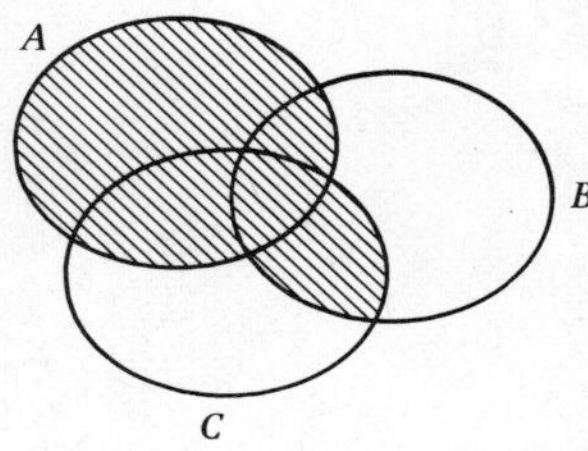

12.

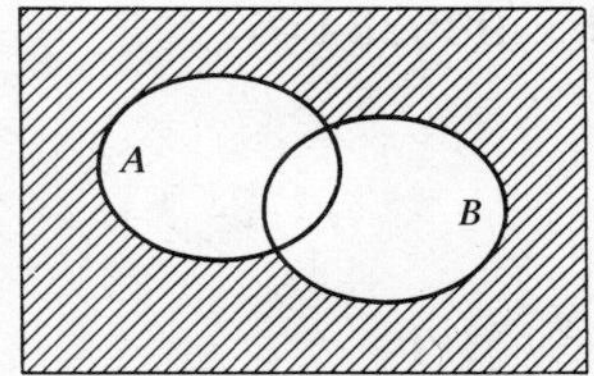

EXERCISE A.2

2. $\{HH, HT, TH, TT\}$
4. $\{0, 00, 1, 2, \ldots, 36\}$; {red, black, green}
6. $\{(x, y): x = A, 2, 3, \ldots, K; \; y = A, 2, 3, \ldots, K\}$
8. $\{0, 1, 2, \ldots, 10\}$

EXERCISE A.3

2. (a) .125   (b) .125   (c) .125   (d) .375
4. 648, 252
6. (a) .0222   (b) .20
8. $\binom{n}{0} = \binom{n}{n} = \dfrac{n!}{n!\,0!} = \dfrac{n!}{n!} = 1$
10. (a) $(8 \times 10^6)^{-1}$ (b) $(8 \times 10^5)^{-1}$ (c) $(4 \times 10^3)^{-1}$
16. 98280
18. .9
20. $1 - ((365!/340!)/(365)^{25}) = .5687$
22. $1 + 5 + 10 + 10 + 5 + 1 = 32$

EXERCISE A.4

2. (a) .3   (b) .7   (c) .5   (d) .2

4. (a) .5   (b) .5   (c) $\frac{1}{3}$ (d) $\frac{5}{6}$

6. (a) $\frac{7}{12}$ (b) .75   (c) .25   (d) .75

8. (a) .5   (b) $\frac{1}{3}$ (c) $\frac{1}{6}$

EXERCISE A.5

2. $\frac{1}{3}$

4. .25, .25, .25, .25

6. (a) .839   (b) .9995

8. (a) .00005476   (b) .98525476

10. .75

EXERCISE A.6

2. (a) .75   (b) .25   (c) $\frac{1}{3}$

4. (a) binomial $\left(n = 5, p = \frac{1}{2}\right)$ (b) $\frac{3}{16}$

6. .3138

8. (a) .3585   (b) .3774   (c) .2641

INDEX